Modernity, the Environment, and the Christian Just War Tradition

In this volume, Mark Douglas presents an environmental history of the Christian just war tradition. Focusing on the transition from its late medieval into its early modern form, he explores the role the tradition has played in conditioning modernity and generating modernity's blindness to interactions between "the natural" and "the political." Douglas criticizes problematic myths that have driven conventional narratives about the history of the tradition and suggests a revised approach that better accounts for the evolution of that tradition through time. Along the way, he provides new interpretations of works by Francisco de Vitoria and Hugo Grotius, and, provocatively, the Constitution of the United States of America. Sitting at the intersection of just war thinking, environmental history, and theological ethics, Douglas' book serves as a timely guide for responses to wars in a warming world as they increasingly revolve around the flashpoints of religion, resources, and refugees.

Mark Douglas is Professor of Christian Ethics at Columbia Theological Seminary. The author of *Christian Pacifism for an Environmental Age* (2019), his work has been supported by the Nohria Family Charitable Fund and the Center of Theological Inquiry, Princeton.

Modernity, the Environment, and the Christian Just War Tradition

MARK DOUGLAS
Columbia Theological Seminary

CAMBRIDGE
UNIVERSITY PRESS

CAMBRIDGE
UNIVERSITY PRESS

University Printing House, Cambridge CB2 8BS, United Kingdom

One Liberty Plaza, 20th Floor, New York, NY 10006, USA

477 Williamstown Road, Port Melbourne, VIC 3207, Australia

314–321, 3rd Floor, Plot 3, Splendor Forum, Jasola District Centre, New Delhi – 110025, India

103 Penang Road, #05–06/07, Visioncrest Commercial, Singapore 238467

Cambridge University Press is part of the University of Cambridge.

It furthers the University's mission by disseminating knowledge in the pursuit of education, learning, and research at the highest international levels of excellence.

www.cambridge.org
Information on this title: www.cambridge.org/9781009098939
DOI: 10.1017/9781009106610

First published 2022

A catalogue record for this publication is available from the British Library.

Library of Congress Cataloging-in-Publication Data
NAMES: Douglas, Mark, 1966– author.
TITLE: Modernity, the environment, and the Christian just war tradition / Mark Douglas, Columbia Theological Seminary.
DESCRIPTION: Cambridge, United Kingdom ; New York, NY : Cambridge University Press, 2022. | Includes bibliographical references and index.
IDENTIFIERS: LCCN 2021059614 | ISBN 9781009098939 (hardback) | ISBN 9781009106610 (ebook)
SUBJECTS: LCSH: Just war doctrine. | War – Religious aspects – Christianity. | War – Moral and ethical aspects. | BISAC: RELIGION / Ethics
CLASSIFICATION: LCC U22 .D68 2022 | DDC 241/.6242–dc23/eng/20220201
LC record available at https://lccn.loc.gov/2021059614

ISBN 978-1-009-09893-9 Hardback

Contents

Acknowledgments *page* vii

Introduction: Troubling a Tradition: The Failings of the Modern Narrative of the Christian Just War Tradition 1

1 Engaging the Other: Francisco de Vitoria and the Age of Conquest 27

2 Understanding the Self: Hugo Grotius and the Birth of the Secular 80

3 Shaping the State: The US Constitution as a Christian *Just War* Document 132

4 State Time/Secular Time: The US Constitution as a *Christian* Just War Document 174

5 Christian Just War Thinking and Modernity 223

6 Historical Roots and Roads Not Taken: An Environmental History of (Part of) the Christian Just War Tradition 246

7 Renarrating the Christian Just War Tradition 289

Afterword 321

Select Bibliography 327

Index 341

Acknowledgments

One of the major blessings involved in writing an interdisciplinary book such as this one – especially a book that is, at least in my mind's eye, part of a larger series – is the opportunity that researching and writing afford to meet, talk with, and learn from a wide variety of scholars, practitioners, and other interested parties who have shaped the ideas that have ended up in it. One of the minor burdens is trying to remember everyone and give them the credit they are due. In this latter obligation, I will certainly fail, but knowing that I will fail hardly absolves me of the responsibility to try to name and thank them. Here goes.

At various times over the last several years, I have presented sections of this work to a wide variety of scholarly audiences and conversation partners whose critical engagements with those sections have made them stronger. Among these, I begin with my colleagues at Columbia Theological Seminary who, during multiple Dean's Lunches, read the earliest draft of Chapter 7 and a late draft of Chapter 1. Their wisdom made both chapters stronger. The Reformed Theology Reading Group (James Calvin Davis, Roger Gench, Elizabeth Hinson-Hasty, and Doug Ottati) read and improved upon drafts of Chapter 2 and, if I remember correctly, Chapter 3. The New Haven Theology Discussion Group read a late version of Chapter 7 and offered both critical guidance and strong encouragement. And a small working group that was part of the "Agency in the Anthropocene" project at the University of Indiana read and gave feedback on Chapter 2 at a deeply enriching meeting at the University of Chicago.

Beyond the working groups, I presented parts of this book at three international conferences. I presented excerpts from Chapter 5 at a meeting of the International Society for the Study of Religion, Nature and Culture in Cork, Ireland; parts of Chapter 2 at the First Consultation on Religion, Gender, Violence, and Human Rights, sponsored by the Interdenominational Theological Center and the Faculdade Unida de Vitória, in Vitória, Brazil; and parts of Chapter 4 at the first Political Theology Network conference in Atlanta, Georgia. Interactions with colleagues at those three conferences not only strengthened those respective chapters, they offered occasions to network with scholars whose wisdom informs both this book and work beyond this book.

Colleagues at schools, churches, and civic groups around the United States invited me to present on the larger project of which this book is a part to diverse audiences. John Carlson at the Center for the Study of Religion and Conflict at Arizona State University gave me the chance to test some of my bigger ideas in front of a wider audience and Janna Hunter-Bowman invited me to join her class, virtually, at Anabaptist Mennonite Biblical Seminary. I also taught on those ideas at Trinity Presbyterian Church in Atlanta and Decatur Presbyterian Church. Sadly, the COVID-19 pandemic disrupted plans for presentations at several other schools and churches that were on the books until the world shut down, including Wheaton College and United Church of Chapel Hill. That those were among the worst expressions of disruption I experienced during the pandemic is a reminder of how privileged and fortunate I have been. That I was able to publish chapters related to work in this book in two other books that came from those engagements ("Law, Religion, Refugees, and the Modern Self," in *Religion, Gender, Violence, and Human Rights,* edited by Wanderley P. da Rosa and Reginaldo Braga [Vitoria, Brazil: Editora Unida, 2019] and "Global Migrations and Climate Refugees," *Global Migration and Christian Faith,* edited by M. Daniel Carroll R. and Vincent Bagote [Eugene, OR: Wipf and Stock, 2021]) are further signs of privilege and good fortune.

This book would have been immeasurably more difficult to write were it not for Columbia Theological Seminary's generous sabbatical policy, which allowed me to finish its first draft in the fall of 2019, and an even more generous grant from the Nohria Family Charitable Fund, which not only allowed me to buy out a semester of teaching and administrative responsibilities, but also funded a trip to Washington, DC, to talk with scholars in environmental security at a number of think tanks, including the Center for Climate and Security and The United States Institute of Peace.

Cambridge University Press continues to be a wonderful press to work with, having now published with them both this book and its predecessor, *Christian Pacifism for an Environmental Age*. My thanks to Beatrice Rehl at Cambridge University Press for her encouragement and, sometimes, patience with me. Thanks to the two reviewers for that press whose suggestions about additions and improvements to the book made its finished draft all the stronger. Thanks, also, to my wife, the Rev. Dr. Lindsay Armstrong, who not only put up with my grumbling about those reviewers' opinions but has been a constant encourager and conversation partner in the adventure that writing this book has been (as well as the adventure that our life together is). And, even more, special thanks to our daughter, Logan Armstrong Douglas, who spent part of the year between the end of her master's degree at the University of Glasgow and the beginning of her doctoral work at the University of St. Andrews preparing the bibliography and index for this book. Within the adventure that is Lindsay's and my life together, Logan is always the best part. I look forward to the soon-approaching time when Logan lights up the world with her own scholarship rather than toiling away on her dad's stuff.

Books like this one dwell in the past in order to make sense of the present and near future; as such, they are, I hope, viable responses to George Santayana's concern about repeating a forgotten past. They do not, though, do much toward materially addressing the types of contemporary concern that this book raises about war, refugees, environmental racism, political dysfunctions, and environmental destruction. During the research for and writing of this book, I have had the good fortune to learn from and work with a range of activists and organizers who are responding to these and related concerns. Azadeh Shahshahani at Project South in Atlanta; Codi Norred at Georgia Interfaith Power and Light in Decatur; Veronica Kyle at Faith in Place in Chicago; Doris Evangelista of Red Uniendo Manos in El Salvador; Alison Harrington at Southside Presbyterian Church in Tucson, Arizona; Nancy and Shelvis Smith-Mather of RECONCILE International in South Sudan; Jen Butler at Faith in Public Life in Washington, DC; Emma Lockridge of Michigan United; and Theresa Dardar of the Pointe-au-Chien Indian Tribe in Lafourche Parish, Louisiana are just a few of those activists and organizers I have been privileged to interact with, and they will have to stand in for a far wider range of people doing the vital work of repair during the first decades of an increasingly chaotic century while I have secreted away in a library office. It is to them – those named here and the many others whose names deserve to be included but are not – that this book is dedicated.

Introduction: Troubling a Tradition

The Failings of the Modern Narrative of the Christian Just War Tradition

> Augustine says in a sermon on the son of the centurion: If the Christian Religion forbade war altogether, those who sought salutary advice in the Gospel would rather have been counseled to cast aside their arms, and to give up soldiering altogether. On the contrary, they were told: "Do violence to no man; … and be content with your pay." If he commanded them to be content with their pay, he did not forbid soldiering.
>
> Thomas Aquinas, *Summa Theologica*

> Having seen what the sources of law are, let us come to the first and most general question, which is this: whether any war is lawful or whether it is ever permissible to war. This question, as also the others which follow, must first be taken up from the point of view of the law of nature.
>
> Hugo Grotius, *The Law of War and Peace*

> The triumph of just war theory is clear enough; it is amazing how readily military spokesmen during the Kosovo and Afghanistan wars used its categories, telling a causal story that justified the war and providing accounts of battles that emphasized the restraint with which they were being fought.
>
> Michael Walzer, *Arguing About War*

THE CONVENTIONAL NARRATIVE OF THE CHRISTIAN JUST WAR TRADITION

The conventional telling of the Christian just war tradition goes something like this:

People everywhere have always struggled to relate morality to warfare, but the incipient phase of the just war tradition really begins with the Romans (excepting a few unorganized antecedents in the Hebrew Scripture and from Aristotle, who coined the term "just war").[1] Yet while thinkers such as Cicero and Seneca pushed forward a particular way of integrating morality and warfare, attention to questions about when to fight and how to fight turned almost exclusively on prudential concerns revealed by natural law, which emphasize obligations toward self-defense and maintaining honor. The tradition picked up its greatest moral traction when St. Ambrose (340–397 CE) and, especially, St. Augustine (354–430) sought to reconcile the moral concerns of the Christian faith to Greco-Roman thought during the first few generations after the Edict of Milan made Christianity a legal religion. As Christianity became a religion with increasing political power, it had to think about how to wield that power while still maintaining some semblance of integrity to 300 years of church teaching that had mostly emphasized nonviolence. It was from Augustine that the tradition learned to highlight not only the just causes for going to war (self-defense, defense of an aggrieved party, the punishment of malefactors) but also the mandate of right intention as well as just cause. That is, we ought to be motivated by a just cause and not by some gain that we seek that can be masked by lip service to a just cause and our greater intention ought to be a just peace. It was also from Augustine that we learned of the obligations placed upon soldiers to act out of love, not malice, vengeance, or bloodlust.

A long, dark period followed Augustine in which questions about just war were mostly ignored in favor of intertribal warfare in Europe and the Crusades in the Middle East. The only advancement in thinking about war until the thirteenth century came through papal edicts such as "The Peace of God" (989) and "The Truce of God" (1027) in which a few specific *jus in bello* rules – notably, clarifications regarding the prohibition against taking the lives of noncombatants – were advanced, though more in the breach than in practice.[2] Yet by the time of Thomas Aquinas (1225–1274), lists of just war criteria were starting to be regularized. There were lists of *jus ad bellum* criteria such as just cause, right intention, and legitimate authority and *jus in bello* criteria such as discrimination (between combatants and noncombatants), proper motives, and a willingness to accept

[1] See Aristotle, *Politics*, trans. C. Lord (Chicago, IL: University of Chicago Press, 1984), Book 1, chs. 3–8.

[2] A few scholars (most notably James Turner Johnson) will sometimes include the work of Gratian (twelfth century) as a place where the tradition advances as Gratian joined Augustinian thought to canon law in his *Decretum Gratiani*. Most, though, skip Gratian.

offers of surrender without slaughtering those who surrender. Joining his Augustinian concerns about motives to his Aristotelian rational sensibilities, Thomas added proportionality to both these lists and clarified the meaning of legitimate authority while also shaping a Doctrine of Double Effect that could be applied to *jus in bello* criteria.

Thomas' disciples – including de las Casas (1474–1566), Vitoria (1483–1546), and Suárez (1548–1617) – continued to develop Thomistic approaches to just war during the age of exploration/conquest, particularly with regard to the way that indigenous peoples should be treated. By the end of Vitoria's life, the *jus ad bellum* and *jus in bello* just war criteria with which we are today familiar (i.e., the *jus ad bellum* criteria of legitimate authority, just cause, right intent, last resort, proportionality, and reasonable hope of success and *jus in bello* criteria of right intention, discrimination, and proportionality) were substantially in place and undergirded by appeals to tradition, Scripture, and natural law.

Hugo Grotius (1583–1645) helped the just war tradition make another great leap forward by situating it within the rules of international law in his classic text *On the Rights of War and Peace*. While Grotius was, like Vitoria, responding to colonization (and, less like Vitoria, to the wars of religion in Europe), the direction in which he took the tradition was not only new but would also shape its continued use to the present day. By connecting morality to law within the context of the nascent nation-state, Grotius both stabilized the criteria and also shaped them so as to give clear preference to the sovereign state as the primary actor in international affairs. So, for example, treaties as a form of moral agreements between states not only guaranteed sovereignty but also named specific consequences for actions that violate sovereignty.

Then just war thinking went on a kind of hiatus. Though Enlightenment and modern political philosophers from the seventeenth to the late nineteenth centuries (e.g., Locke, Kant, and Hegel) would mention rules of war from time to time, they did very little to advance the tradition. And while the nineteenth and early twentieth centuries saw the development of increasingly detailed codes of military behavior and laws for international relations, those working out such codes and laws – whether Francis Lieber (1798–1872), who designed the Field Army rules signed into law by Abraham Lincoln during the American Civil War; the Hague Conventions, which took Lieber's code and fit it into international treaty law in 1899 and 1907; or the Kellogg–Briand Pact of 1928, which fit military law into the League of Nations – did not so much advance just war thinking in new directions as apply established just war thinking to new contexts. As James Turner Johnson has argued, the turn toward international law initiated by Vitoria and Grotius led away from the normative foundations of the tradition.[3]

Finally, in the second half of the twentieth century, a number of scholars began to reclaim just war teachings and apply them to issues such as the existence of nuclear weapons and wars such as those in Vietnam and Iraq. Foremost among these scholars were Paul Ramsey (1913–1988) and Michael Walzer (1935), whose respective books *The Just War: Force and Political Responsibility* (1968) and *Just and Unjust Wars* (1977) revitalized the tradition and stimulated succeeding generations of research into just war thinking.

Now just war thinkers are everywhere. Indeed, the power of linking just war thinking to contemporary conflicts is so pervasive that Michael Walzer has declared victory: "The triumph of just war theory is clear enough; it is amazing how readily military spokesmen during the Kosovo and Afghanistan wars used its categories, telling a causal story that justified the war and provided accounts of the battles that emphasized the restraint with which they were being fought."[4]

Yet even as victory was being declared, just war thinking was revealing its own confusions and frustrations. New books and articles with titles such as *Rethinking the Just War Tradition*, *The Just War Revisited*, *Morality and War: Can War Be Just in the Twenty-First Century?*, and "On the Just War Tradition in the Twenty-First Century" proliferate, all of them attempting to fit just war theory into new sorts of conflict (terrorist activity by nonstate players, global economic forces that dwarf military ones, US military hegemony, weapons of mass destruction and cyberwarfare, etc.). This is where we are today: just war thinking is mostly victorious but often confused.

Or so goes the conventional Christian telling of the history of the just war tradition.[5]

[3] James Turner Johnson, *Morality and Contemporary Warfare* (New Haven, CT: Yale University Press, 1999), 51ff.

[4] Michael Walzer, *Arguing about War* (New Haven, CT: Yale University Press, 2004), 11. Tellingly, the chapter in which this quote appears is titled "The Triumph of Just War Theory (and the Dangers of Success)."

[5] Almost uniformly, books on Christian just war thinking follow this telling (in various ways and to varying degrees). Take, for example, these sentences that open James G. Murphy's *War's Ends: Human Rights, International Order, and the Ethics of Peace*: "The just war tradition is alive and well today. After emerging in ancient Rome and flourishing in medieval Europe, it experienced a partial decline in the early modern period, 1600–1900, but it has enjoyed a revival since World War I (1914–18)." James G. Murphy,

ASSESSING THE CONVENTIONAL NARRATIVE

There are understandable reasons for this telling of the history of the Christian just war tradition. For instance, the focus on great thinkers follows from the way scholarship is produced. Scholars spend much of their lives reading and at least the wiser ones are always acutely aware of how much they owe in their own thought to earlier scholars; as such, the focus on earlier writers reveals the unsurprising biases that come with offering intellectual history as the primary way to tell history. Moreover, this telling offers a conceptually neat and reasonably concise way to make sense of the jumbled thought that constitutes any tradition, let alone the just war one – which is doubly helpful since contemporary scholars tend not to be investigating the tradition so much as applying it, so their actual interests are usually developed elsewhere in their books. This telling feels familiar because it repeats the way modernity shapes history-telling: present builds on past in causally clear ways connected to identifiable and materially focused motives. And in all of this, it lets those telling the history of the just war tradition both situate themselves in the tradition and clarify the unique contribution they are making to the tradition.

Beyond its coherence and the way it connects scholars to the tradition, the conventional narrative also has considerable explanatory power. On

War's Ends: Human Rights, International Order, and the Ethics of Peace (Washington, DC: Georgetown University Press, 2014), p. 1. In just two sentences, Murphy captures the conventional narrative but he is far from alone in promoting that narrative. Among others who provide such a narrative – and this is simply a representative sampling; it is hardly exhaustive – see, e.g., Daniel M. Bell, Jr., *Just War as Christian Discipleship: Recentering the Tradition in the Church rather than the State* (Grand Rapids, MI: Brazos Press, 2009), chs. 1 and 2; David L. Clough and Brian Stiltner, *Faith and Force: A Christian Debate about War* (Washington, DC: Georgetown University Press, 2007), esp. 51ff.; David Fisher, *Morality and War: Can War Be Just in the Twenty-First Century?* (Oxford: Oxford University Press, 2011), ch. 4; and Brian Orend, *The Morality of War* (Toronto: Broadview Press, 2006), ch. 1. Among other anthologies of classic texts on morality and war that display the conventional telling of the history of just war thinking, see Arthur F. Holmes, ed., *War and Christian Ethics: Classic and Contemporary Reading on the Morality of War*, 2nd ed. (Grand Rapids, MI: Baker Academic, 2005); and Larry May, Eric Rovie, and Steve Viner, eds., *The Morality of War: Classical and Contemporary Readings* (Upper Saddle River, NJ: Pearson Prentice Hall, 2006). There are also articles that repeat the history, e.g., Nicholas Rengger, "On the Just War Tradition in the Twenty-First Century," *International Affairs* 78.2 (2002): 353–363. Indeed, the conventional telling is so dominant as to be nearly ubiquitous. Even the most subtle and prodigious historian of just war thinking, James Turner, regularly repeats the conventional telling (see, e.g., Table One in Johnson, *Morality and Contemporary Warfare*, 24). There are also ways in which Johnson escapes the conventional telling, however, of which more anon.

this telling, for instance, international law grows out of the just war tradition, initially through the work of Grotius but later through the Hague and Geneva conventions, the establishment of the United Nations and the International Criminal Court, and the creation of and reliance on the Universal Declaration of Human Rights to ground international law. That is, on this telling, the Christian just war tradition sires many of the most significant advancements in international engagement and human valuing that we've seen over the past century and so just warriors can look at their tradition with a feeling akin to parental pride.[6]

On this telling, the conventional narrative of the Christian just war tradition is an accretionary project that eventually and triumphantly leads to a theory capable of being used across time and in diverse contexts. Aristotle connects military thought to larger categories of political and prudential thinking; Augustine adds concerns about motives; Thomas contributes attention to proportionality; Vitoria systematizes the just war criteria and applies them to new populations and Grotius makes them function in a secular world; Ramsey and Walzer show how the criteria still matter in the twentieth century. Along the way – somewhere around the point at which Vitoria and Grotius were developing their work – an evolving tradition solidified into a theory with a clear set of generally applicable criteria. And criteria *function*. They become primary, powerful, and ready-at-hand tools for evaluating wars and the actions in them.

In all this, the conventional narrative reveals itself as an expression of a peculiarly modern process of telling history. There are clear links between causes and effects. Historians can be separated from their narratives such that everyone can tell a similar story. There need be no recourse to concepts such as providence, divine favor, or any other nonimmanent framing devices. And, above all, history is constituted by continuous and visible development: the story of accretion driven by the myth of progress animates the conventional narrative.[7]

[6] Albeit with the same kind of parental pride that also recognizes that their progeny haven't always pursued quite the path that their parents had planned or hoped for.

[7] Analyzing the origins, power, and shape of the modern myth of progress is its own complicated project and one I won't take up here. There are, after all, multiple expressions of such a myth: some are teleological, offering a quasi-providential narrative about advancement of ideas toward some great and final end; some are biological, offering a quasi-Darwinian narrative about a competition of ideas in which some outcompete others; some are historical, offering a quasi-Hegelian narrative about the synthesizing of old ideas into new ones; some are economic, offering a quasi-capitalist narrative about the

Yet telling the history of the just war tradition in this way also hides aspects of the tradition from us. So, for instance, the emphasis on the writings of earlier scholars reproduces both the "great man" theory of history,[8] in which particular individuals, through their own wisdom and power, are the primary actors that drive the movements of history,[9] and the Hegelian theory of history as progress over time through the clash of thesis and antithesis leading to synthesis. Both the "Great Man" and the Hegelian theories are modern approaches and, therein, hide the fits and starts, the ironies and disjunctions, the continuities and incompletions, the multiplicities of traditions and the tensions within and between them that mark the movements of traditions through time. Along the way, the conventional telling of the Christian just war tradition tacitly reinforces a theologically problematic way of understanding the work of God in history – though since the just war tradition (at least in the West) has mostly unmoored itself from its Christian connections, this concern carries little leverage for critiquing the tradition.

Moreover, telling the history of the Christian just war tradition in this way treats it principally as a history of ideas. But just war traditions are not simply about ideas; they are also about practices, technological changes, political structures, and the interactions of practices, technologies, structures, and ideas as they shape the values that undergird and find

growing fungibility of ideas such that more and more people benefit from them; and still others are scientific, offering a quasi-Kuhnian narrative in which the failures of old ideas and approaches are settled by dramatic paradigm shifts. All I am trying to get at when I name the conventional narrative's reinforcement of a myth of progress is that it reinforces the idea that change is gradual and functional: later expressions of the Christian just war tradition are improvements on earlier expressions of it because they are able to do work that those earlier expressions could not. The one distinctive component of a myth of progress that I see operative in the modern telling of the Christian just war tradition is the idea that as it moves forward through time, the tradition accumulates wisdom. It is an accretionary expression of the myth of progress, making it quasi-economic/Smithian in form. It doesn't know what to do with ideas that fall away. For more on the myth of progress and its problems, see, e.g., Larry Laudan, *Progress and Its Problems: Towards a Theory of Scientific Growth* (Berkeley, CA: University of California Press, 1977).

8 Until very recently, thinkers in the just war tradition have been almost entirely men. As such, I've mostly maintained masculine language. A prominent exception to this rule is Jean Bethke-Elshtain, who drew attention to the role women play in the context of war in her important book *Women and War* (Chicago, IL: University of Chicago Press, 1995).

9 The "great man" theory, developed in the mid-nineteenth century by Thomas Carlyle in *On Heroes, Hero-Worship, and the Heroic in History* (Whitefish, MT: Kessinger Publishing, 2006), argues that upon occasion, highly influential people (Plato, Jesus, Mohammed, Shakespeare, Napoleon, etc.) arise and, through force of charisma, intelligence, and imagination, turn history in new directions. Long discredited as a viable

expression in just war thinking. So, for instance, the development of the crossbow drove new just war concerns about acceptable and unacceptable weapons in the Middle Ages and the crossbow, itself, was the product of particular medieval wartime practices that favored attempts at taking enemy life at a distance, all of which was shaped by medieval political structures in which the church had enough political power to attempt to convert these concerns into rules.[10] The conventional narrative of the history of the Christian just war tradition gives short shrift to the range of material factors, political structures, and habits of action that shape it in favor of a problematically idealized and idea-heavy version of history. It is also blind to the way this complex array of forces is embedded in – and shaped by – interactions with the nonhuman natural world. It is that blindness that, ultimately, is the preoccupying concern of this book.

JUST WAR THINKING, JUST WAR TRADITIONS, AND JUST WAR THEORIES

At least partly as a result of the preoccupation with ideas and the exclusion of practices, technologies, and structures, this conventional approach to telling the history of the just war tradition places us within the context of a theory more often than a tradition.[11] Though the term "theory" has several meanings, the dominant one – and the one that those telling the history of the just war tradition advocate – is something like, "an integrated body of principles that are coherent and well

philosophy of history, the theory still pops up – usually unintentionally – when histories are told from the perspective of earlier important scholars. If pressed, almost all contemporary historians would admit that great leaders are as much a product of their contexts as they are drivers of change. Yet few are pressed, especially when recounting a history on the way to get to some other set of points.

10 And lest we treat "taking enemy life at a distance" as a natural evolutionary advancement rather than a socially constructed project, it's important to remember that one of the things that Shaka kaSenzangakhona (Shaka Zulu) such an important military figure in nineteenth-century southern Africa was his leadership in transforming the Zulu army from one that fought at a distance with *assengai*, or throwing spears, to a military that fought with *iklwa*, short stabbing spears.

11 Tellingly, when just warriors speak of its history they tend to refer to the "just war tradition" when referring to ideas and events prior to modernity and the "just war theory" when applying the tradition to contemporary concerns. I am not the only person to note the transition from just war tradition to just war theory. In their book *The Just War Tradition: An Introduction*, David D. Corey and J. Daryl Charles note that people have substituted the phrase "just war theory" for "just war tradition," which has led to the problem "that many just war writers today actually believe that the tradition can and should be abridged into a single (if complex) doctrine and, in extreme cases, that it

developed" and can be evaluated according to criteria such as clarity, coherence, comprehensiveness, simplicity, explanatory power, justificatory power, output power, and practicability.[12] Theories systematize, order, and unify diverse ideas in order to employ those ideas collectively to address issues that arise in the multiple and varied moral, political, scientific, and other spheres of human engagement where calls for judgments are made. For the modern mind, theory is an achievement. I will return to questions about just war theories momentarily. Here, though, I would suggest that before there was a modern just war theory, there was just war thinking and there were just war traditions.

With the term "just war thinking," I refer to any attempt – no matter how systematic, coherent, or protracted – to join strategic concerns about the use of force to moral concerns about the same as those concerns are expressed within the context of interactions between recognizable political bodies. Whatever else it is and no matter in what context or tradition it occurs, just war thinking links these two concerns as they are expressed by organized groups of people pursuing their collective interests. As such, just war thinking is grounded in two fundamental claims: that groups of people have interests they are willing to pursue even if such pursuit involves the use of force (or, at the very least, the threat of the use of force), and that these same groups want to identify themselves as behaving in morally defensible ways. Those interests can vary tremendously (control of resources, political power, self-defense, defense of allies, etc.). Likewise, their appeals to morality can vary tremendously, depending on the moral frameworks around which they orient their common lives.

This expansive understanding of just war thinking suggests that we can find evidence of it in any relatively stable political body that has

should be presented in the form of a 'checklist' of just war 'criteria' that captures once and for all what is just and unjust in war." Corey and Charles argue that the substitution ignores the foundations of just war thought, reduces the diversity of voices in the tradition, and diminishes the complexity of the tradition. While I am in broad agreement with their concerns, I do worry that they oversimplify the actual complexity of just war thinking by simplifying what theories do and ignoring the multiplication of theories. Closer attention to the genealogy of contemporary just war theories might have prevented such oversimplifications. Ironically, they have offered a functionalist criticism of the functionalist transition of just war thinking from tradition to theory. See David D. Corey and J. Daryl Charles, *The Just War Tradition: An Introduction* (Wilmington, DE: ISI Books, 2012), 7–10 (quote above on p. 7).

12 Tom L. Beauchamp and James F. Childress, *Principles of Biomedical Ethics*, 4th ed. (New York: Oxford University Press, 1994), 44 (quote); 45–47 (evaluative criteria).

endured over time. Such bodies, after all, have an interest not only in enduring – and so are willing to at least maintain resources necessary to defend themselves against external threats – but also in offering narratives about themselves that those persons who compose the body find sufficiently compelling that they are willing to fight on behalf of (and, perhaps just as importantly, not fight against) that body. Said differently, political bodies have strategic reasons both to use force to pursue their ends and to use arguments about pursuing the good.

The strength of just war thinking grows in proportion to the sophistication that political bodies use in joining strategic interests to moral arguments. One of the strongest ways of joining the two is through an appeal to justice. Whether that appeal takes the form of promoting those virtues that attend to the distribution of benefits and burdens that come with living in a community that interacts with other communities, of maintaining a set of basic rights for pertinently similar persons that should be defended when threatened, or of aligning the political structures of a society with some set of analogous structures identified with a transcendent/eternal society, such appeals provide both weight and shape to just war thinking.[13] Not all understandings of justice are equally compelling. Nor are all structures meant to promote and maintain justice equally enduring. But especially once a political body uses the language of justice as the means for joining strategic and moral considerations, it is de facto engaged in just war thinking.

Political bodies change over time. Established ideas are modified to address new technologies and insights; new leaders and adversaries arise; natural disasters befall them; economic fortunes change; and religious sensibilities wax, wane, and are replaced by new religious perspectives (and sometimes new religions). As political bodies change, the just war thinking that has helped to provide both political stability and moral coherence for them changes as well. Later ideas build upon earlier ones. New rules modify old ones. New practices based on established ideas emerge and new understandings of established practices gain traction. As they are compiled, these changes, expressed in terms of both continuity and discontinuity, constitute an emerging just war tradition.

A "tradition" has been famously defined by Alasdair MacIntyre as "an historically extended, socially embodied argument, and an argument precisely

[13] It is for this reason, incidentally, that I find debates about whether to associate the word "just" in the term "just war" with the word "justified" or the word "justice" to be misplaced. Just war thinking grows in significance as appeals to what can be justified (i.e., aligned with the good) are subsumed within appeals to justice and the particular maintenance and distribution of the good having to do with life together.

in part about the goods which constitute that tradition."[14] MacIntyre's definition gets at important qualities of traditions, including their perdurance, social embeddedness, attention to basic social goods, and manifestation in at least one kind of practice: argument among their members. It conceals, though, attention to the mechanisms that drive change in traditions even as it emphasizes the idea that the primary – if not exclusive – drivers of change are ideas and arguments about them. As a result, it not only favors a way of thinking about change over time as gradual, progressive, and consistent but also implies that ideas are the drivers of change. Given MacIntyre's own criticisms of modernity, it is hard to miss the irony of the fact that he is reproducing a part of the modern narrative about how to understand the movements of history when he offers his definition of tradition.

In responding to MacIntyre's criticisms of modern liberalism and its inattention to tradition, Jeffery Stout improves on MacIntyre's definition with his own: a political tradition

> inculcates certain habits of reasoning, certain attitudes toward deference and authority in political discussion, and love for certain goods and virtues, as well as a disposition to respond to certain types of actions, events, or persons with admiration, pity, or horror ... Its ethical substance, however, is more a matter of enduring attitudes, concerns, dispositions, and patterns of conduct than it is a matter of agreement on a conception of justice.[15]

Most basically, what Stout helps us get at here is the range of ideas, practices, attitudes, and structures that constitute traditions (whether political or otherwise). Beyond that, though, he also begins to highlight one of the complexities that comes with that range: finding coherence

[14] Alasdair MacIntyre, *After Virtue*, 2nd ed. (Notre Dame, IN: Notre Dame University Press, 1984), 222.

[15] Jeffery Stout, *Democracy and Tradition* (Princeton, NJ: Princeton University Press, 2004), 3. Stout links his ways of thinking about traditions to the work of Robert B. Brandom and, particularly, Brandom's work in *Making It Explicit: Reasoning, Representing, and Discursive Commitment* (Cambridge, MA: Harvard University Press, 1994) – though one suspects he might rely just as much on Brandom's *Tales of the Mighty Dead: Historical Essays in the Metaphysics of Intentionality* (Cambridge, MA: Harvard University Press, 2002) had their two books developed more sequentially than simultaneously. What Brandom is doing in both books – exploring a genealogy within analytic philosophy to explain his advocacy for inferentialism with regard to matters of linguistics and intentionality – is of interest to political philosophy because it provides a way of thinking about the movement of ideas through time (as exercises in "bebop history," to use Brandom's phrase from p. 90 of *Tales*) that gets at their nonrepresentational complexity and mutability without losing semantic coherence. Brandom and Stout both draw deeply from the work of the American philosophical tradition of pragmatism (a tradition with which I, too, align myself). There is much, I suppose, that I could do with their work as it pertains to just war thinking. This, though, is not that project.

among such a range is a far more complicated project than simply being able to name a framing argument. Traditions are discerned as much through the inferential processes of reasoning among those who participate in them as anything. The reason some group of persons willingly locates themselves within a tradition has to do with the particular language games they play with each other. Further, those of us who are not part of that tradition will recognize it as a tradition partly because we have reasons to trust that those within it are living lives about as coherent as our own.

This is not to say that the traditions we find ourselves in and the traditions others find themselves in will necessarily align: we may have significant disagreements about the wisdom of their ideas, the morality of their practices, the orientations of their attitudes, and the value of their structures. Nor is it to say that traditions do not change; indeed, it is to suggest that they are mutable precisely because they are always being constructed by those within them as they interact with each other in changing contexts, and they are stable because traditions, being complex things, do not change all at once. It is, though, to say that traditions, even as complex and changing things, help people align their lives with those of others through time.

Just war traditions, perhaps especially obviously because they explicitly address potentially violent conflicts, don't necessarily align with each other. Likewise, perhaps just as obviously, just war traditions change in part because nothing brings about change in society like the upheavals of violence and the threats of its use. They are complex – so complex, in fact, that there may be different ways of narrating them that can begin to make them sound like different, if overlapping, traditions (e.g., the existence of Roman Catholic and Protestant narratives about the Christian just war tradition that don't quite align). And they change over time in complicated ways.[16]

Just war traditions, then, are the congealing of just war thinking as that thinking is shaped by ideas, practices, attitudes, and structures (and,

[16] To see more of my arguments about how and why they change over time, see my *Christian Pacifism for an Environmental Age*, especially ch. 7. My arguments in this book follow from those there but partly to avoid redundancy and partly since that book explores a long time frame (from the first to the twenty-first century) and this one focuses in on one particular period (the transition from the late medieval to the early modern period in the North Atlantic), I will not repeat those arguments here. See Mark Douglas, *Christian Pacifism for an Environmental Age* (Cambridge, UK: Cambridge University Press, 2019).

I would add to Stout's understanding, technologies as they all interact with the nonhuman natural world). And because they have arisen in different places and times, they have proliferated. There are religiously based just war traditions (e.g., Jewish, Islamic, Hindu, Buddhist, Christian, etc.) as well as an array of secular just war traditions, including those that come from Stoic, cosmopolitan, and human rights perspectives.[17] Over the past few decades, research into these many just war traditions has, likewise, proliferated, perhaps signaling a recognition of the plurality of ways of ordering just war thinking; a sense of instability in contemporary international politics; and/or the proliferation of new causes, types, and exacerbators of larger-scale violence. Whether exploring religious traditions not conventionally linked to just war thought[18] or secular approaches previously underexplored,[19] the very proliferation of research into just war traditions highlights the importance of thinking again about just war traditions – including, in this instance, the Christian one – as the coherence of the idea of a single, encompassing, just war theory recedes.

MacIntyre's and Stout's definitions of tradition, after all, emphasize the very thing that theories lack: pluriformity within and between ways of thinking about just and unjust wars. Take, for example, MacIntyre's emphasis on the quality of argument. He insists that no one voice or idea can determine how a set of practices can be understood or applied. From within the context of theory – that is, from within "an integrated body of principles that are coherent and well developed" – differences of understanding and opinion need to be ironed out. As they are smoothed over, fundamental and competing claims about why this action or that idea matters are left behind. In the peculiarly Hegelian way that annihilates outlying ideas, theories synthesize and then obliterate divergent viewpoints and minority voices.

Add to that Stout's way of complicating the idea of a tradition to account for practices, attitudes, and structures. From within the framing orientation of theory, the contextual specificity of practices, attitudes, and structures mandates either the contextualization of a theory (i.e.,

[17] For a helpful annotated bibliography that explores the range of religious and nonreligious just war traditions, see Michael Farrell, *Modern Just War Theory: A Guide to Research* (Lanham, MD: Scarecrow Press, 2013).

[18] Tessa Bartholomeusz, *In Defense of Dharma: Just War Ideology in Buddhist Sri Lanka* (London: Routledge, 2002).

[19] Cécile Fabre, *Cosmopolitan War* (Oxford: Oxford University Press, 2012).

this theory works in this place but not that one) or the need to disregard such matters – which would otherwise certainly seem pertinent – as immaterial for structuring the rules about how people ought to behave when considering violence. Either contextualizing theories or disregarding inputs important for their applicability would seem to undermine the reasons to have a theory in the first place. As a result, the variety of ideas, practices, technologies, political structures, and their interactions that constitute just war thinking – once they are crammed into a theory – are all subsumed under a homogenized set of dehistoricized claims about *how* to behave regardless of who is doing the behaving, how they are connected to others in the past and present, what the context is within which they are making decisions about how to behave, which tools they have at their disposal, and how they make sense of those tools. Yet it is these practices, technologies, structures, and their interactions – and the arguments about them – that not only shape just war traditions but also fund the motives and resources through which those traditions move forward in time to face new contexts and problems. They shape the contexts in which questions about how to behave are linked to questions about why we should behave in such ways. When the contextually specific values and virtues that undergird and are manifest in just war thinking as it exists in particular times and places are buried under generic lists and generalizable instructions, valuable features of just war traditions are lost.

This is not to say that there is no place in just war thinking for a just war theory. The very existence of multiple just war traditions, each with their distinctive histories, guiding values, visions, virtues, and rules, creates a problem: what to do when two sides both make moral arguments about whether, when, and how to fight and those arguments don't align? One of the strengths of a just war theory, particularly of the sort that arose out of the Christian just war tradition as it entered the early modern period in the North Atlantic and shaped international law, was that it provided a generalizable set of rules about war that were publicly accessible, rationally coherent, and broadly applicable. Even when they were observed in the breach more than in practice, the rules of such a generalizable just war theory shaped the capacity of nations to make decisions that other nations could understand and that international bodies could use to hold violators of those rules to account.

The power of a just war theory such as that arising in early modern thought manifests in the degree to which the international community relies on a standard list of *jus ad bellum* and *jus in bello* criteria when individual nation-states make their arguments for and against war. Indeed, the utility of those criteria is so apparent that their analogs (and the reasoning that follows from them) have been developed in a variety of non-war-related arguments about, for example, the morality of lying,[20] the shape of policing,[21] and the treatment of subjects in medical research.[22] It is with good reason that Walzer describes the triumph of just war theory. Modern just war theory solves real-world problems that multiple just war traditions – and, perhaps especially, the Christian just war tradition – would leave achingly unresolved. Among them is the need to shape arguments that are not meaningful only for particular places or times.

THE WHISPERED HISTORICITY OF MODERN JUST WAR THEORY

It is worth noting that even though modern just war theory is generalizable and ahistorical (at least to the extent that it does not place moral weight on its origins when it is used to make decisions), it is still attentive to time. Time (or, more specifically, a particular modern and political understanding of time) courses through just war criteria: ideally, states should satisfy *jus ad bellum* concerns before entering war and addressing *jus in bello* ones; they should go to war only after other viable and less violent alternatives have been exhausted; they should ground *casus belli* arguments with clear claims about prior provocations; they should address new types of questions about *just post bellum* behaviors.[23] Understanding (or at least accepting) this specific political understanding of time is important in understanding and using modern just war theory.

Its temporal aporia has less to do with an inattention to time than with an inattention to its own history. Other than for historians of just war thinking, the theory's origins are of interest primarily to those who would either valorize the tradition from which it came as a way of signaling

20 Sissela Bok, *Lying: Moral Choice in Public and Private Life* (New York: Pantheon Books, 1978).

21 Benjamin N. Wisniewski, "Soldier, Officer, Citizen: Applying Just War Theory to Police Use of Force," Ph.D. Thesis, University of Missouri, 2017. Available at http://irl.umsl.edu/thesis/3[illegible].

22 Brett McCarty, "Medicine as Just War? The Legacy of James Childress in Christian Ethics," *Journal of the Society of Christian Ethics* 38.2 (Fall/Winter 2018): 57–74.

23 See more on the modern political conception of time and just war thinking in Chapter 4. For a sense of the modern political conception of time, see Elizabeth F. Cohen, *The Political Value of Time: Citizenship, Duration, and Democratic Justice* (Cambridge, UK: Cambridge University Press, 2018).

toward a defense of that tradition or criticize it for reinforcing particular Western conceptions of war, law, rationality, and justice. Understanding the history of modern just war theory should also matter as it is used – and even more so as questions about its utility are shaped by the difficulties of applying it to situations that it doesn't quite seem to fit.

Take, for example, a well-established *jus ad bellum* criterion such as legitimate authority. A focus of scholarly attention since at least the thirteenth century, legitimate authority has long functioned as among the more settled of criteria. And during an age of kings, this criterion was applicable in fairly straightforward ways. Thomas, for instance, argues that "[i]n order for a war to be just, three things are necessary. First, the authority of the sovereign by whose command the war is to be waged. For it is not the business of a private individual to declare war, because he can seek for redress of his rights from a tribunal of his superior ...."[24]

Yet since the rise of the modern nation-state, the legitimacy of a unified military/political authority that could claim sovereign power has not rested in an individual so much as in the state as a whole and the governing body of a state that seeks to maintain the integrity of its sovereignty through procedural means that diffuse the site of legitimacy. So, for example, in the United States, the civilian president is the commander-in-chief but Congress declares war and ultimately both the president and Congress are acting as representatives of the people as a whole, who vote them into or out of office based, in part, on how they make decisions about war. Where is legitimate authority actually housed in that setting and how is it legitimated? Indeed, debates about whether and how to declare war have grown so complex that the United States now usually enters conflicts without first declaring war. Is it therefore in violation of the criterion of legitimate authority? Certainly as conflicts since the Korean War (a conflict that still has not technically ended) have shown, the question "Who is the legitimate authority who can declare war and sue for peace?" becomes considerably more complicated to answer in a representative democracy than in a monarchy – a problem made even more complicated when disagreements about who the legitimate authority is drive the conflict, as is the case in civil wars and wars for independence, many of which have been fought in order to install or replace a representative democratic system.

[24] Thomas Aquinas, *Summa Theologica* 2-2, Q. 40, First Article, trans. Fathers of the English Dominican Province (Westminster, MD: Christian Classics, 1981), 1353.

These questions are further complicated by the existence of international bodies such as NATO, the United Nations, or even the European Union. On the one hand, such international bodies have very limited power to control the behavior of nations; indeed, their foundations rest on the recognition of the very state sovereignty that limits such international bodies. They do, though, have the power to offer support or disapproval – and therein moral legitimacy – to the actions of sovereign nations and to coordinate alliances among their member states. So: what constitutes legitimate authority in a world full of representative democracies that have chosen to bind themselves to each other through the creation of international bodies?[25]

Or, more briefly, take as another example the *jus in bello* criterion of discrimination. When armies faced each other across a front, soldiers needed to be within sight of the enemy, weapons could kill only a few people at a time, and battles were carried out almost exclusively on the ground, the process for discriminating between combatants and noncombatants was comparatively easy. Yet with the technological advent of weapons of mass destruction, three-dimensional warfare, drone surveillance and bombing, the dramatic urbanization of the earth's peoples, and the predominance of struggles for independence and civil wars among all forms of strife, the criterion of discrimination is significantly harder to apply or maintain.[26] Just war criteria can offer normative guidance but they still reveal their age.

As both examples suggest, attention to its history – which is to say, attention to just war traditions and to the Christian just war tradition in particular – lends coherence and depth to modern just war theory even, paradoxically, as modern just war theory was shaped, in part, as an attempt to get away from the authority of traditions in making difficult

[25] None of this is to say that questions of legitimate authority were easy to figure out before the modern nation-state – and just war theorists spent considerable time thinking about who could and could not be a legitimate authority (to which the short answer is that religious leaders [except the pope] and any political leaders who owed fealty to someone higher in the social hierarchy could NOT declare war, were not ultimately responsible for the welfare of their subjects, and could not, therefore, be legitimate authorities). For more on the complex relations between sovereignty and just war thinking, see James Turner Johnson, *Sovereignty: Moral and Historical Perspectives* (Washington, DC: Georgetown University Press, 2014); and Luis E. Lugo, ed., *Sovereignty at the Crossroads? Morality and International Politics in the Post-Cold War Era* (Lanham, MD: Rowman and Littlefield, 1996).

[26] I have addressed some of these issues regarding the applicability of criteria that arise at one point in history being used in another in "Changing the Rules: Just War Theory in the 21st Century," *Theology Today* 54.1 (January 2003).

decisions. This tension between tradition and theory is real and complicated. When we go looking for resources within the Christian just war tradition by which to address particular problems but do so from within the context of modern just war theory, we assume univocality where there is argument; clarity where there may be confusion; systematicity where there is ad hoc-ness; progress where there is discontinuity; set-apartness where there is interaction; and holism where there are only discrete, partial, and historically situated answers to vexing questions.

In ignoring the historically embedded quality of just war criteria when applying them, we lose access not only to the arguments that gave rise to them but also, therein, to resources for thinking about how to understand and apply those criteria in new historical contexts. What is it about authority that makes it legitimate (especially when elections can be rigged or ignored)? How important is it not only to not deliberately target noncombatants but also to pursue strategies that minimize the likelihood of noncombatant casualties? As contemporary wars have shown, these are large, important questions and we need to be able to do more in addressing them than simply attempting to apply static criteria. Yet to do so we need to recognize the degree to which we are looking back at just war traditions through lenses that are colored by a modern theory. We need, in short, to do a better job at reading history because the conventional narrative about the Christian just war tradition claims too much while explaining too little. Including attention to practices, technologies, political structures, and their interactions with the ideas of just war thinking will help us make better sense of the Christian just war tradition.

Correcting for these deficiencies that come with the modern narrative about the history of the Christian just war tradition is not overly complicated. History books abound with details about the wide ranges of practices, technologies, and political structures that manifest in particular times and places. The capacity to give attention to them exists in the modern social imaginary,[27] so filling in these gaps and lacunae

[27] A few points about the idea of a "modern social imaginary" and its role in this work are in order. In using the term, I am drawing from Canadian philosopher Charles Taylor, who first develops it in his book *Modern Social Imaginaries*. There, he describes social imaginaries as "the ways people imagine their social existence, how they fit together with others, how things go on between them and their fellows, the expectations that are normally met, and the deeper normative notions and images that underlie these expectations." Charles Taylor, *Modern Social Imaginaries* (Durham, NC: Duke University Press, 2004), 23. Social imaginaries are expressed by ordinary people through images, stories, and legends (rather, generally, than abstract ideas) and make possible "common practices and a widely shared sense of legitimacy." Taylor, *Modern Social Imaginaries*, 23. They are complex and multilayered, bringing into rough coherence the ideas,

may be pedestrian but it is not difficult. In part, this book takes up such a project, looking specifically at the momentous period during which the Christian just war tradition transitioned from its late medieval to its early modern form. Along the way, it argues for a narrative about the tradition that looks different from the conventional and modern one described at the beginning of this introduction.

THE CHRISTIAN JUST WAR TRADITION IN A NEW SOCIAL IMAGINARY

What happens when the modern social imaginary is, itself, part of the problem? What if it will not be enough to simply insert neglected components of the Christian just war tradition into a modern narrative about that tradition because that modern narrative is distorted by the failings of the modern social imaginary? And what if we are now exiting that modern social imaginary? How useful will a corrected but modern telling of the Christian just war tradition be in a different social imaginary – one where wars happen but now for reasons different from those of modern wars, fought in ways different than the ways of modern warfare, and made sense of through processes that are strange to the modern mind? Filling in gaps and holes won't be enough.

As a start, it will help to recognize that as the Christian just war tradition traveled through time, it changed to fit the various social imaginaries in which it operated. Imagine telling a history of the tradition in which social imaginaries changed from time to time as their visions of the way the world holds together no longer functioned, their answers to the pressing questions of the time no longer worked, and their central arguments no longer cohered. As social imaginaries changed, so would the tradition. During such periods – which I refer to as "axial periods" – even established ways of connecting ideas, practices, technologies,

practices, and background conditions (and the relations between them) through which people make such sense of and act in the world that those people seldom address at conscious levels. Social imaginaries change over time as new ideas and practices gradually penetrate and change existing social imaginaries. A social imaginary isn't simply the set of ideas that a particular society uses to order its common life (a "theory," though social imaginaries include theories), nor a lens through which a group of people see the world around them (a worldview, though they include worldviews), nor even a shared set of practices and the implicit agreements that have led to them (social habits, though social imaginaries include them as well). It is all of these plus their interactions plus the built environment – the stuff – that a given society creates and is shaped by. Given the way ideas, worldviews, habits, and technologies all participate in the shaping of just war thought, Taylor's description of the social imaginary is especially apropos to my topic.

structures, and their interactions change. Such times would be experienced as disruptions rather than as continuities; as apocalypses rather than progressions; as chaotic rather than consistent. The tradition would change in ways that didn't look so much like progression as revision.

Axial periods are, on my definition, periods of transition between social imaginaries when the arguments that constitute traditions rise toward consciousness because a tradition is being unsettled during such a transition. A cursory reading of Western history highlights such axial periods as the beginnings of the various ages into which that history is heuristically divided: the rise of Greek thought and its distribution around the Mediterranean through the conquests of Alexander, who wanted not only to control the lands around Greece but also to change them to resemble Greek culture; the rise of the Roman Empire that followed about 300 years later and established Roman law and culture throughout much of Europe, northern Africa, and parts of the Middle East; the collapse of that empire about 500 years later; the medieval period (which could, perhaps, be subdivided into periods before and after its engagement with Islam); the age of exploration/colonization/conquest beginning in the late fifteenth century and the nearly concomitant modern age as it began with the English, French, and American revolutions; and now something that continental philosophers, at least, refer to as "the postmodern age." As my subdivision of the medieval period begins to hint at, this timeline is entirely too simple. It misses the consistency of some ideas over long swaths of time, it overstates the impact of changes in governance and conceptions of the state in its telling, and it ignores the degree to which other histories – those that might focus on, for example, technological change, economic change, or religious engagement – could be told in ways that, while not entirely undermining this approach to dividing history, would certainly complicate it. And, obviously, it disregards the long and rich histories of non-Western cultures.

That said, this way of dividing up Western history is useful, in part because it does focus on who rules what and how they rule, in part because it orders history in ways that correspond to significant changes in social imaginaries in the West, and in part because it helps to clarify why the conventional telling of the history of the just war tradition has focused on great thinkers. These thinkers all thought and wrote during axial periods in Western history, which is to say they wrote during periods of transition in the social imaginary around them. These thinkers were not so much advancing just war thought through the singular power of their insights as they were manifesting the arguments that constitute just war traditions as those arguments were boiling up during times when social imaginaries were changing. What makes such scholars important isn't that they were wiser than everyone who had come before them; it's that they were struggling with how to make sense of the wisdom of those who had come before them when that wisdom had begun to seem less wise. And they were focused on ideas about war and peace because these axial periods are marked not only by conflict but also by changes in thought about whether, how, and why to fight.[28]

I believe we are living during an axial period.

IMAGING A NEW SOCIAL IMAGINARY

Many people would agree that we are living during an axial period, though there is less agreement about what is driving our movement through this period or how the next social imaginary will be shaped. Philosophers, grown weary of the limits of modern thought, have argued for a shift into the postmodern (which is really a kind of intensified late modernity). Technology specialists, marveling at the rate of change in, especially, computing power, have argued for a shift into an information age (in which the boundaries between the real and the virtual blur, as do those between the biological and the mechanical). Economists, alert to globalization and the explosive growth of finance as a global economic driver, have argued for a shift into a new kind of capitalism. Political scientists, concerned by increasing political instability and the resurgence of religious motives to drive public polities, have argued for a transition into a post-Westphalian world in which the nation-state is no longer the primary driver for making sense of political systems. All of these changes are, indeed, momentous and warrant attention, including by Christian just war thinkers since these changes are reshaping war. And it is worth noting that there is a peculiar kind of ego-stroking that comes with thinking that we live in especially momentous times.

Sitting behind all those other transitions, though, is one of such massive dimensions that we continue to struggle to make sense of it: the transition from the Holocene to the Anthropocene.[29] Climate change and its

28 For a more extensive exploration – including a theological perspective through which to make sense of the movements of history through social imaginaries – see my book *Christian Pacifism for an Environmental Age*, especially ch. 7.

29 The term "Anthropocene" – not to mention its attendant features and its discontinuity with the Holocene – is one about which there continue to be disagreements. This is not the place for defending whether such a term is appropriate, nor for exploring all the implications of the transition into the Anthropocene. Others are doing that work (see, e.g., Amanda H. Lynch and Siri Veland, *Urgency in the Anthropocene* [Cambridge,

destructive effects; the catastrophic loss of biodiversity; the proliferation of waste in our water, air, and land; and so forth: these are reshaping the world on a geological timescale and, in the process, will reshape philosophy, technology, economics, and politics in ways that are only starting to become visible. Other changes are those of a transition between eras; the shift into the Anthropocene is a change of epoch that is unmatched in written human history.

The shift into the Anthropocene will also shape a transition in the Christian just war tradition. Questions about whether, when, why, and how to wage war will be restructured as we face different types of, causes for, and understandings of war in the twenty-first century. That is, the tradition will change in ways that feel more discontinuous than continuous: established arguments will become less coherent, established answers less pertinent, and established connections to the past more tenuous. Simultaneously, experiences of discontinuity will drive searches for new ways of addressing the perennially difficult problems that come with violence and war.

How, though, to make sense of such transitions? In *Christian Pacifism for an Environmental Age*, I argued that as traditions move through time, they manifest both continuities and discontinuities with the past and that one helpful way to make sense of that movement is to think of the central questions that a tradition asks as having centripetal force, pulling it toward its center, even as the answers given to those questions exert centrifugal force, pushing it apart during axial periods. Questions conserve; answers reveal splintering over time. A moment's reflection helps make sense of this: answers necessarily have a kind of contextual specificity that is less true of questions and when contexts change, answers become less pertinent or useful even as questions resurface. In that book, I argued that the central questions of the Christian pacifist tradition are those having to do with matters of fidelity: Who is this God to whom we are faithful and what is this God asking of us in the context of that faith?

The central questions of the Christian just war tradition, meanwhile, are those of anthropology: What are people like? On what terms do we

MA: MIT Press, 2018] or Clive Hamilton, Christophe Bonneuil, and François Gemenne, *The Anthropocene and the Global Environmental Crisis: Rethinking Modernity in a New Epoch* [London: Routledge, 2015]). Suffice it to say here that when I use the term "Anthropocene," I am using it as shorthand to describe a time – an environmental age – in which human interactions with the nonhuman natural world will become the preoccupying concerns of everyone on earth because those interactions will shape their possibilities for flourishing and their experiences of suffering.

make sense of them? On what basis are they valued? Who counts as a person? How much do different people count? How do they live with each other? What do they do when they don't seem to be able to live with each other? How do they organize themselves? And how does the way people organize themselves help us make sense of what it means to be human? Or, asked differently: How do we make sense of self, state, and other in the Anthropocene?[30]

A MAP TO AID IN READING THIS BOOK

That last question is a complicated one, partly because it implies two preceding questions: How have we made sense of self, state, and other in the modern period? And how has the way we've made sense of self, state, and other in the modern period shaped us to begin to imagine life in a world of climate-shaped conflicts? Both of those questions need answering simply in order to get a better sense of where we are right now. This book takes up those two implied questions, splitting (broadly) into two parts: one part for each of the implied questions.

The first half of this book explores the way that those anthropological questions were answered during a particularly important time in the history of the Christian just war tradition: the transition from its late medieval to its early modern form. Because that transition was momentous, these were the very questions that the tradition, in its late medieval form, was failing to adequately answer and that Christian just war thinking was therefore focused on. The first chapter takes up the work of the Roman Catholic theologian Francisco de Vitoria, particularly as he worked to include a new "other" – the indigenous persons of the Americas and West Indies – into a framework that clarified whether Spanish explorers could kill these new others with impunity or whether these new others were to be treated the same as Europeans. The second chapter explores the work of the Protestant theologian Hugo Grotius as he participated in shaping a modern conception of the self through his work in developing international law: How does one have rights, why do they matter, and how do those rights stick to persons? The third and

30 Treat those terms – "self," "state," and "other" – loosely. I am not arguing that these terms are more fundamental to definitions than quasi-synonymous terms (and certainly not that the state is the most basic political structure). With these terms, I am trying to get at three big questions: How do we think about ourselves, how do we think about those unlike us, and how do we shape political structures?

fourth chapters then take up questions about how to understand a state by focusing not on a person but on a document: the Constitution of the United States of America. I focus on the Constitution partly in order to avoid repeating the errors of the "great man" myth of the conventional telling of the history of the Christian just war tradition and partly because the document has grown beyond the bounds of its framers such that it has, from its beginning, exerted a force beyond the capacities of those who wrote it. Chapter 3 explores the Constitution as a document that is best understood through the lenses of just war thinking. Not only its initial articles but also several of its first ten amendments (not to mention some of the amendments that would follow) are preoccupied with questions about when and through what structures the nascent constitutional republic can and should wage war and how it should behave when it does so. And Chapter 4 explores the Constitution as a *Christian* document, making the argument that it reveals a political theology marked by a peculiarly Christian posture of temporal deferral vis-à-vis its place in time and toward the rest of the world.

These first four chapters reveal two momentous if underexplored implications of the Christian just war tradition. First, in its transition from its late medieval to its early modern form, the Christian just war tradition did not simply fit into nascent modernity; it played a significant role in shaping modernity. Vitoria's thought not only shaped the patterns of conquest and colonization that are central to modern history but also promoted the very instrumental forms of reasoning that mark modern thought – including treating some persons instrumentally. Grotius' work not only initiated international law (and advanced modern law more generally) but also shaped the ideas of individual autonomy and state sovereignty that make such law coherent and that have defined modern political systems. And the US Constitution not only framed the modern state but also did so via attention to the universality of those rights within an immanent frame of reference that fetishized freedom and promoted proceduralism. Instrumentalism, autonomy, universality, and immanence: these are hallmarks of modern thought and each of them rises to prominence, at least in part, through the ministrations of Christian just war thinking. Chapter 5, which initiates the shift into the second half of this book, develops this implication.

The second momentous if underexplored implication of the Christian just war tradition as it transitioned from its late medieval to its early modern form is that in its transition, it shaped the conditions that have led to the environmental crises we face today. Instrumental thinking turned the nonhuman natural world into a collection of resources to be used (including turning human beings into resources, most egregiously via the slave trade). The emphases on individual autonomy and state sovereignty shaped persons for whom being free riders is an advantage and shaped politics that are ill-equipped to address environmental crises that do not respect borders. And the immanent frame associated with a posture of deferral, with its emphases on proceduralism and freedom, shaped a way of thinking about crises that is ill-equipped to understand, let alone address, the kind of slow-moving catastrophes that constitute contemporary environmental crises. Included in each of the first four chapters, then, are reflections exploring how the Christian just war tradition (inadvertently) shaped the environmental crises of the Anthropocene. Chapter 5 summarizes the implications of the way the Christian just war tradition shaped modernity and the environmental crises we face due to modernity.

One of the ironies implicit in the responsibility that Christian just war thinking has for shaping contemporary environmental crises is that the tradition did not need to go in the directions that it did as it transitioned from its late medieval to its early modern form. In Chapter 6, I offer an environmental history reading of that transition. There, I argue that the late medieval and early modern versions of the Christian just war tradition were shaped not only by ideas, practices, technologies, and political structures, but also by their interactions with the nonhuman natural world. Willis Jenkins has noted that "[a]ll human activity happens by multispecies collaboration, through more-than-human assemblages, from co-evolutionary affordances...."[31] To make sense of seemingly even the most human activities – including waging war – we need to attend to such collaborations, assemblages, and affordances. Not only did interactions with the nonhuman natural world shape the tradition, but also the way Christian just war thinkers understood the nonhuman natural world and their relationship(s) to it shaped it. And one of the implications of an environmental history reading of the Christian just war tradition's transition from its late medieval to its early modern form is that the range of ways that Christian just warriors understood the nonhuman natural world was diminished even as access to the resources of the nonhuman natural world proliferated.

[31] Willis Jenkins, "The Environment" (draft version) in Michael Kessler and Shaun Casey (eds.), *The Oxford Handbook of Political Theology* (New York: Oxford University Press, forthcoming).

Using my interpretation of the transition of the Christian just war tradition from its late medieval to its early modern form as a template, Chapter 7 concludes the book by renarrating the Christian just war tradition, correcting the narrative I laid out at the beginning of this introduction. In that chapter, I highlight some of the ways that attention to the practices, technologies, and political structures of just war thinking as it changed over time – and particularly as it changed during axial periods – helps make better sense of it. Along the way, I emphasize the tradition's interactions with the nonhuman natural world in order to better understand some of the forces that drove many of its most significant transitions as they occurred during those axial periods. If the conventional narrative has missed both the significance and the consequences of the Christian just war tradition, this last chapter starts to develop a process for both more fully appreciating and more critically engaging that tradition, especially as it continues to influence contemporary thinkers.

Questions about self, state, and other: At no time in modern memory have they been more pressing than they are right now. As we enter the Anthropocene and must address its wars, we are going to be better equipped if we have a richer understanding of the way those questions have been answered in the past. The practices, technologies, and political structures that have shaped Christian just war thinking: At no time in history have they been as complex and powerful as they are right now. As we enter the Anthropocene and face its wars and their weapons, we will make better judgments about them if we are clearer about how practices, technologies, and political structures are shaping these wars and weapons. Interactions between the human and the nonhuman natural world: At no time in human history has it been more important to become aware of our interconnectedness with the nonhuman natural world. As we enter the Anthropocene and respond to the violence that our entrance prompts, we will need to learn to recognize those interactions if we want to survive in it. The wisdom of just war thinking: At no time in the history of the tradition has it been more worthwhile to understand the sources and powers of that wisdom than right now. As we enter the Anthropocene and face new types of, causes for, and understandings of war, surfacing that wisdom will be as important as it has ever been. Renarrating the past in order to make sense of the present and begin to anticipate the future: This book pursues that project.

It begins with Francisco de Vitoria.

1

Engaging the Other

Francisco de Vitoria and the Age of Conquest

> Let our fourth proposition be: The Indian aborigines are not barred on this ground from the exercise of true dominion. This is proved from the fact that the true state of the case is that they are not of unsound mind, but have, according to their kind, the use of reason. This is clear, because there is a certain method in their affairs, for they have polities which are orderly arranged …
>
> Francisco de Vitoria, *De Indis et De Iure Belli Relectiones*

> [F]rom Victoria onwards a central feature of just war tradition is a claim to universality that takes itself seriously – unlike the just war theories of the Middle Ages, which in spite of a theoretical universal applicability were never in practice extended to regulate cross-cultural conflicts. The modern universalism has held, in whatever form it has appeared, that the just war tradition somehow expresses more than the cultural consensus of Western society …
>
> James Turner Johnson, *Just War Tradition and the Restraint of War*

> [L]and and body are connected at the intersection of European imagination and expansion, but what must be underscored is the point of connection – the Portuguese and Spanish, that is, the European. He is the point of connection. He stands now between bodies and land, and he adjudicates, identifies, determines.
>
> Willie James Jennings, *The Christian Imagination*

Francisco de Vitoria's two texts, *De Indis* and *De Iure Belli*,[1] mark the start of a vitally important transition in the Christian just war tradition as it exited a medieval social imaginary and entered a modern

[1] Francisci de Victoria, *De Indis et De Iure Belli Relectiones*, ed. Ernest Nys (Buffalo, NY: William S. Hein and Co., 1995).

one. Not only are there glimpses of early modernist just war thought in these texts, but they also find their starting point in one of the most acute questions in all of just war thinking: how to understand and engage an "other." Vitoria's surprisingly progressive answers to this question moved the tradition forward, powering its increasing political scope and moral significance. They also shaped failures and caused suffering for which the tradition is at least partly accountable and created lacunae that it must now overcome as it moves into the environmental age.

LOCATING VITORIA IN HIS TIME AND PLACE

Francisco (or Francisci) de Vitoria (or Victoria) was born in 1480 in Alava, Spain, the scion of two noble families, and raised in Burgos. Upon joining the Dominican order in the early 1500s, he was sent to Paris to continue his studies; while there, he studied the work of Thomas Aquinas with Peter Crockaert and was introduced to the works of Cajetan and Erasmus. Vitoria's work shows the influence of Cajetan in his continued engagement with the complex religious politics at the beginning of the Reformations and the influence of Erasmus in his humanistic treatment of indigenous persons in the Americas and West Indies. Having been promoted by the Dominicans, Vitoria returned to Spain to teach at the Dominican College of Saint Gregory at Valladolid. In 1526, the primary chair of theology at the University of Salamanca opened up and Vitoria successfully competed for it, taking the chair on September 7 of that year. He would hold the chair until he died in August 1546.

Separated from Vitoria by 500 years, it is easy to lose sight of how significant Vitoria was in his context. The University of Paris, for instance, was among the preeminent academic institutions of the time, but Vitoria did far more than repeat the wisdom of his professors when he returned to Spain. In the midst of dramatic social, political, and religious changes, Paris still pursued the same programs of study that it had done for decades, giving its education an anachronistic quality that Vitoria needed to move beyond. The University of Paris was, wrote Louis Delaruelle,

> always the formidable machine constructed in the Middle Ages for the manufacture of theologians. Everything there continues to be subordinated to this end. The study of literature consists entirely in that of grammar and is relegated to

the lowest grade of instruction. Logic is ever the science of sciences; disputation continues to be preferred to any deep study of authors.[2]

The conservativism of the University of Paris at the time is, for instance, visible in the institution's conflict with Erasmus. It defended the exclusive use of the Vulgate for interpreting scripture; Erasmus defended the novel (and startling, for its time) use of the Greek New Testament. In light of that conflict, Erasmus mocked the "theologasters" of Paris for their preoccupation with bygone ways of thinking.

Maintaining a medieval educational focus and pedagogical continuity was, for independent thinkers, undoubtedly incongruous during a period marked by the Renaissance, dawning humanism, explorations of Africa, Asia, and the Americas, and glimmerings of the Reformations. While Vitoria's work evinces the impact of his studies in Paris, especially through the way that Aristotle and Thomas Aquinas maintained a hold on his thought, that hold was not slavish and his *Relectiones*, or recorded lectures, reveal both his willingness to think beyond them and his attention to matters that his Paris training had hardly suited him for. As Ernest Nys would write, "it was obviously not among the masters of philosophy of theology in Paris that Franciscus de Victoria was enabled to acquire the precious possession wherein were united the spirit of research and of innovation, the tendency toward progress, the love of his neighbor, and the sentiment of solidarity."[3]

Neither was Salamanca an intellectual backwater. Politically, economically, intellectually, and militarily, Spain was the most powerful country in the Western world, Salamanca the most prominent university in Spain and the primary chair of theology among the most prestigious chairs at the university. The Holy Roman Emperor, King Charles V of Spain, consulted with Vitoria on questions having to do with King Henry VIII of England's marriage to and nullification of marriage from Catherine of Aragon (Charles's aunt), the German upstart Martin Luther, and the treatment of indigenous persons in the Americas and West Indies. Pope Paul III asked Vitoria to attend the first Council of Trent; failing health prevented him from doing so. Erasmus corresponded with Vitoria and, while they reached different conclusions about the sanity of war, that

[2] Louis Delaruelle, *Guillaume Budé: Les origines, les débuts, les idées maîtresses*, quoted in Ernest Nys, "Introduction" in *Francisci de Victoria: De Indis et Ivre Belli Relectiones*, trans. by John Pawley Bate (Buffalo, NY: William S. Hein & Co., 1995), 67.

[3] Nys, "Introduction," 68.

correspondence shaped Vitoria's thought.[4] Moreover, as James Brown Scott, Camilo Barcia Trulles, and Antonio Truyol Serra, among others, have argued, Vitoria's *Relectiones* initiated not only the Spanish School (or School of Salamanca) of international law (to be built upon by Francisco Suárez shortly later) but modern international law more broadly.[5]

THE SIGNIFICANCE OF VITORIA'S PLACE AND TIME

All of this happened at a time of dramatic global and conceptual change as Europe transitioned out of a medieval world and into what would become the modern one. The sixteenth century was not only the century of the Protestant Reformations and the Roman Catholic Counter-Reformation; it was a century marked by new forms of political philosophy (e.g., that of Machiavelli) and the birth of the modern nation-state, the exploration of the world and the colonization of the Americas and West Indies, the beginning of the scientific revolution (as shaped by luminaries such as Francis Bacon, Nicholas Copernicus, and Andreas Vesalius), modern forms of literature such as the essay (Montaigne) and chapbooks (Rabelais' *Gargantua and Pantagruel*), and the work

4 See Terence Martin, *Truth and Irony* (Washington, DC: Catholic University of America Press, 2015) for reflections on Erasmus' arguments about the insanity of war.

5 James Brown Scott argues for the primacy of Vitoria as the initiator of modern international law, though not without controversy: "Grotius is universally considered as the founder of International Law. This, like many general statements, is true enough but likely to mislead. He was not the founder nor was he the father of the science any more than Adam Smith was the founder or father of Political Economy as a Science.... We look beyond Grotius and see that the international law of today is rooted in a more remote past" to the Salamanca School (including, e.g., Suárez and Fernando Vazques de Menchaca). Of course, there are also precursors to the Salamanca School. The Law of Nations developed by Vitoria was preceded by writings by Joannes de Legnano, Christian de Pison, and, perhaps most prominently, Honoré Bonet. See James Brown Scott, quoted by Edmund A. Walsh, S. J., "Forward" in *The Spanish Origin of International Law: Lectures on Francisco de Vitoria (1480–1546) and Francisco Suarez (1548–1617)*, ed. James Brown Scott (Washington, DC: School of Foreign Service of Georgetown University, 1928), 9. There are also extensive debates about just how and how much Vitoria and the Salamanca School actually shaped international law (these debates moving far beyond the realms of this chapter). For this, see not only Scott but, e.g., Felix Cohen, "The Spanish Origin of Indian Rights in the Law of the United States," *Georgetown Law Journal* 31 (1942): 1–21; A. Nussbaum, *A Concise History of the Law of Nations* (New York, NY: Macmillan, 1954); and James Muldoon, "The Contributions of the Medieval Canon Lawyers to the Formation of International Law," *Traditio* 28 (1972): 483–493.

of inventors such as Da Vinci and Galileo. As James Mackintosh noted over 150 years ago, "The sixteenth [century] ... may be considered as the age of transition from the scholastic to the modern philosophy."[6] Vitoria wrote during – and about – an exciting time in European history.[7]

Vitoria is not known for writing a *Summa* although he did produce a range of writings on topics including a commentary on Thomas Aquinas and various works in ecclesial law.[8] His best-known writings, the *Relectiones,* are a collection of his public lectures compiled by his former students and, as such, are neither so orderly nor abstract as some other neo-Scholastic writing or later writings on international law. His writings on just war thinking have an occasional quality to them unlike, for example, the writings of Thomas Aquinas before him or Hugo Grotius after him. Some interpreters treat this as a weakness in his thought, arguing that Vitoria did not so much move the Christian just war tradition in a new direction as convene the tradition he received from Thomas Aquinas and prepare it for Grotius, who would move it, irrevocably, into the realm of modern (secular) international law.

I would argue, though, that this very occasionality moved the tradition in a new direction. Vitoria began his thought neither in a review and synthesis of just war thinking nor as a conventional application of just war thinking to problems like those around which it had grown during the Middle Ages. Instead, Vitoria began his exploration of just war thinking by attending to a new problem: whether Spain could wage war on the indigenous persons that populated the Americas and West Indies in order to establish colonies and gain the resources of this "new" world. This was not, in the first instance, a question of how to apply just war thinking to conflict with a group of people whose culture differed from that of Catholic Spain much as the cultures of the Moors or Turks had done. It was, instead, a question of whether just war thinking actually applied to a conflict with those whose very humanity was being called into question. As the tradition moved increasingly in the direction of the establishment of international law – with all the abstractness that such a

[6] James Mackintosh, "Dissertation on the Progress of Ethical Philosophy, Chiefly during the Seventeenth and Eighteenth Centuries," quoted in James Brown Scott, "Historical Introduction," *Spanish Origin of International Law*, 14.

[7] Antonio Truyol Serra has highlighted Vitoria in this context in his work *The Principles of Political and International Law in the Work of Francisco de Vitoria* (Madrid: Ediciones Cultura Hispanica, 1946).

[8] His two texts, *De potestate Ecclesiae* and *De potestate Papae et concilii*, preceded and anticipate the conciliar work of the Council of Trent.

move would entail – this attention to occasionality would persist in the establishment of legal precedents, alongside a growing emphasis on the significance of studying recent history to make sense of complex moral problems.

Spain in the fifteenth century was pro-slavery (its slaves included Moorish prisoners of war, Jews, natives of the Canary Islands, and Africans) and it fiercely defended the hierarchical orderings of human beings even prior to Columbus' trip across the Atlantic. As the new world opened up, it exported these values there, most notoriously via Pope Alexander VI's papal bull *Inter Caetera*, which gave Spain all lands west of a line of demarcation roughly 100 leagues west of the Azores.[9] In the first decades of the sixteenth century, questions about how to understand and treat the indigenous persons in the Americas and West Indies were settled neither in theory nor practice. Bartolomé de las Casas had not yet published his indicting *A Short Account of the Destruction of the Indies*, and the New Laws, which protected those indigenous persons from some of the more egregious forms of exploitation and mistreatment, would not be enacted until a decade after Vitoria gave the two lectures that would become *De Indis* and *De Iure Belli*.[10]

Not only was there no international law to refer to in adjudicating disagreements about how to treat indigenous persons, but nor were there clear precedents about how to understand, let alone treat those persons. For the Europeans, these were strange peoples speaking strange languages and displaying strange customs on strange lands full of new resources, and the strangeness of things in the new world was only

[9] This papal bull would serve as the basis for the "doctrine of discovery," used for hundreds of years afterwards – including in US Supreme Court Chief Justice John Marshall's opinion in *Johnson* v. *M'Intosh* in 1823, which gave the United States the legal standing to claim lands occupied by indigenous persons in North America.

[10] Vitoria's *Relectiones* consists of thirteen (of an original fifteen) lectures, collectively entitled *Relectiones Theologicae*, covering a range of matters, including homicide, marriage, and the relation between civil and ecclesial authority. The most important of these *relectiones* – certainly for this project – are the *relectios* "Of the Indians Lately Discovered" and "Of the Indians, or on the Law of War Made by Spaniards on the Barbarians." These two *relectiones* are available in *Francisci de Victoria: De Indis et Iure Belli Relectiones*, 101–187. I will refer to them as *De Indis* and *De Iure Belli*, respectively, which occur sequentially in that text. I would note, moreover, that relying on the (erroneous and problematic) convention of the times, Vitoria referred to the indigenous persons of the Americas and West Indies as "Indians" – an error repeated over time both through faithful translation and through lack of attention to these persons' own voices in describing themselves. There is, I should further note, a provocative separate literature on the translations of Vitoria's *relectiones* that is not pertinent to this work.

heightened through its contrasts to a Europe that was exploding with new technologies, worldviews, and self-assertiveness. Rumors of cannibalism, rampant human sacrifice, sexual laxity, and the absence of civilizational sophistication among indigenous persons only exacerbated the Spanish tendency to treat such persons in the new world as less than fully human. Spain carried its already established *encomienda* system of rewarding conquerors with land and the labor of those living on it into the new world, where it grew dramatically. And although the indigenous persons had defenders (including the Dominican Antony of Montesino in 1511 and de las Casas in 1519), Spanish desire for wealth and power tended to trump moral restraint in the new world.

DE INDIS AND *DE IURE BELLI*

This, then, was the issue that Vitoria confronted in *De Indis*: whether Spain could claim dominion over the indigenous persons of the Americas and West Indies and their land because they were not only savages but not fully human heathens (indeed, from one perspective, the issue was whether Spain had an obligation to claim dominion over them in order to compel them to become Christian). Vitoria began his lecture with three large questions:

> The whole of this controversy and discussion was started on account of the aborigines of the New World, commonly called Indians, who came forty years ago into the power of the Spaniards, not having been previously known to our world. This present disputation about them will fall into three parts. In the first part we shall inquire by what right these Indian natives came under Spanish sway. In the second part, what rights the Spanish sovereigns obtained over them in temporal and civil matters. In the third part, what rights these sovereigns or the Church obtained over them in matters spiritual and touching religion.[11]

It is already worth noting that the questions that drove *De Indis* both preceded and drove Vitoria's subsequent lecture, *De Iure Belli*. Implied in these questions – and quickly made explicit in the text that follows them – was the question of whether it would be permissible to wage war against the indigenous persons to achieve dominion. Answers to this question would turn on how to think about the indigenous people of the new world in the first place: Are they human and, if so, are they humans in the same way that the Spanish are? By what criteria will these questions be answered? And if they are humans but

[11] Vitoria, *De Indis*, 116.

not like the Spanish, what rules are there for engaging them, particularly in the context of exploration, evangelization, and conquest? Not only did his reflections in *De Iure Belli* grow out of the arguments that he made in *De Indis* – therein highlighting the degree to which Vitoria understands just war thinking as both applicable to and designed for addressing concrete and pressing issues of the day – but the concerns about the treatment of indigenous persons in the Americas and West Indies would then animate the whole of his just war deliberations. Said differently, questions about "the other" played the starring role in his just war work.

FOUR INSIGHTS FROM *DE INDIS* ON WAR AND THE OWNERSHIP OF POSSESSIONS

A walk through *De Indis* (which is not, in the first place, about war) reveals not only insights into Vitoria's understandings of the other but how those insights would lead him into surprisingly provocative claims in *De Iure Belli.* Four insights, in particular, are worth further attention in this traverse of the lecture he gave to answer the three questions he named in the quotation above.

After some generic throat-clearing on the significance of the questions, the legitimacy of multiple answers to the questions, the advisability of consulting with appropriate authorities, the responsibility of accepting the moral answers to these questions, and the wisdom of consulting with a theologian rather than a jurist on these issues, Vitoria dove into the three questions he names in the above quotation. In answer to the question of whether Spain can claim the lands and goods of the new world from the indigenous persons there, he pointed out that, "the people in question were in peaceable possession of their goods, both publicly and privately. Therefore, unless the contrary is shown, they must be treated as owners and not be disturbed in their possession unless cause be shown."[12] The weight of the issue, therefore, has to do with whether there is just cause to dispossess them of their lands and goods – and so a question about the economic and political rights of Spain slid toward becoming a question about whether there was just cause for war, beginning from a prima facie presumption in favor of the rights of the indigenous persons and their "ownership" of the things they possess.

[12] Ibid., 120.

That these indigenous persons were in mortal sin should not deprive them of their property (as arguments from scripture, the church, precedent, and logic made apparent to Vitoria). That they were not Christians should not deprive them of their property (as arguments from scripture, Thomas Aquinas, the church, precedent, and logic made apparent).[13] And that they lacked the capacity to reason and, therefore, should not own property, was fallacious: "[T]hey are not of unsound mind, but have, according to their kind, the use of reason. This is clear, because there is a certain method in their affairs, for they have polities which are orderly arranged and they have definite marriage and magistrates, overlords, laws, and workshops, and a system of exchange, all of which call for the use of reason; they also have a kind of religion."[14]

This answer is particularly instructive as it not only demonstrates Vitoria's commitment to a Thomistic understanding that being human means being able to reason but that the evidence of reason comes in human sociality. The capacity to shape a syllogism mattered far less than the ability to shape a society, and while there may have been other reasons for taking their property, arguments that they are subhuman or infantile could not be among them. This connection between reason and sociality would be important for Vitoria as he worked his way further into questions of war and it is the first insight that it will be important for us to keep in mind in interpreting *De Iure Belli*.[15]

Vitoria then proceeded to dismiss arguments for Spaniards to claim their property based on title. Neither the emperor nor the pope are lords of the earth even though they have been charged to govern its civil and spiritual affairs. The emperor and the pope may have jurisdiction in governing but not in owning – and the fact that aboriginal Americans do not recognize such jurisdiction does not give Spain the right to wage war. They could not lose their title if they refuse to accept the faith of Christ nor could war be waged against them because their ignorance was

[13] Vitoria's approach to biblical interpretation is, itself, significant as a sign about transitions into modern thought and modern "Christian" encounters with the other. See Yvonne Sherwood, "Francisco de Vitoria's More Excellent Way: How the Bible of Empire Discovered the Tricks of [the Argument from] Trade," *Biblical Interpretation* 21.2 (2013): 215–275.

[14] Vitoria, *De Indis*, 127.

[15] The reason/sociality connection here harks back to Aristotle – though as I read Aristotle, I understand the capacities to reason and to live in community as two separable components of what it means to be human, whereas for Vitoria, living in (complex) community is evidence of the capacity to reason. If anything, Vitoria's understanding of the human being is *more* oriented around politics than Aristotle's.

invincible (and, even if it were not, faith cannot be compelled by force). Moreover, short of demonstrations of the superiority of the Christian faith through miracles and clear signs that had not occurred and, just as significantly, because Spain had failed to treat them morally, the indigenous persons were justified in refusing to accept that faith. This answer, too, is particularly instructive because it sets up a problem that Vitoria would have to address when he delved further into the laws of war: What to do when persons of reason do not reach common conclusions? How is just cause coherent when there may seemingly be objectively valid but incommensurable claims about it (that is, there is the possibility of simultaneous ostensible justice)? This is the second insight that is important to attend to, since it shaped his work in *De Iure Belli.*

Could a violation of natural law justify taking property or waging war? While under conditions of invincible ignorance indigenous persons could not be condemned to lose life and property for failing to behave like Christians, and while their rejection of the Christian faith had, paradoxically, revealed their capacity for wisdom, any sins against nature (cannibalism, incest, etc.) were sins against universal law and indigenous persons could be condemned on such grounds. Vitoria's response to this question was complicated because natural law was, for him, the basis for the laws of nations that, themselves, would justify violence: *ius natural* grounds *ius gentium* and *ius gentium* grounds just cause.

To address this question, Vitoria developed a series of distinctions around it. Aboriginal Americans could be compelled to keep those laws that they, themselves, profess but not laws that others profess – including, seemingly, some natural laws: "Sins against nature" can mean "not only what is contrary to the law of nature, but also of what is against the natural order and is called uncleanliness."[16] And while coercive force could be used to address actions that are contrary to a law of nature (such as homicide), it could not be used to address sins against the natural order (fornication, idolatry, sodomy). Indeed, in a rather circular bit of reasoning, Vitoria argued that, since the laws of nature are related to the divine law revealed by the grace of God in Jesus, if coercive force could be used in response to sins against the natural order, it should be able to be used against those who do not accept the faith. And since faith cannot be compelled through force, neither can force be used against those who sin against the natural order. This answer is also instructive because it shaped how Vitoria would engage questions of the *ius*

[16] Vitoria, *De Indis*, 146.

gentium in the context of war, which is the third insight helpful for making sense of *De Iure Belli*.

By the end of this section of *De Indis*, the only reasons Vitoria could see for Spain to take the property of indigenous persons by title were (a) that they have voluntarily chosen to give it to Spain; and (b) because God has commanded it. Yet Vitoria had no evidence that either of these was the case.

DE INDIS AND JUSTIFICATIONS FOR THE USE OF FORCE IN TAKING PROPERTY

Having dispensed with illegitimate reasons to take property, Vitoria turned to other conditions that would support the cause of taking land and property, "the lawful and adequate titles whereby the Indians might come under the sway of the Spaniards."[17] Arguments here followed from claims he had already made and from widely accepted European expressions of property law (discussed in greater detail in Chapter 6). Because the indigenous peoples of the Americas and West Indies are fully human (i.e., capable of reason and shaped toward sociality) and therefore bound by the *ius gentium*, they had an obligation to allow the Spanish to travel through their lands without harming them (much as the French would be bound to allow the Spanish to travel through France by virtue of that same obligation).[18] They could not use force to expel the Spanish as that would not be just cause for war (assuming the Spanish are not harming them). Some goods (e.g., running water, the seas) are common to all and so the Spanish must be allowed access to them. They should allow trade with the Spanish and, so long as those trades are fairly pursued, should accept the Spanish making a profit from those trades. They must allow the Spanish to claim goods that are otherwise unclaimed.[19] They should allow persons born in the new world to become citizens there. Moreover,

[17] Ibid., 151.

[18] While the violation of a right to safe passage sounds like an unusual just cause for war, the matter is not without historical precedent in the history of Christian just war thinking. St. Augustine and, following him, Gratian argued for such a thing based on Numbers 21: 21–31, in which the Amorites refuse to let the Hebrew people pass through their land and so Israel conquers them instead.

[19] Vitoria is not making up these conditions; they spring out of precedents established within medieval property law (which is, itself, premised on arguments from natural law). Such conditions, then, are not arbitrary – though they are arguably cultural, depending on how one thinks about natural law. On medieval property law and natural resources, see Richard C. Hoffmann, *An Environmental History of Medieval Europe* (New York, NY: Cambridge University Press, 2014), ch. 7.

> if the Indian natives wish to prevent the Spaniards from enjoying any of their above-named rights under the law of nations ... the Spaniards ought in the first place to use reason and persuasion in order to remove scandal and ought to show in all possible methods that they do not come to the hurt of the natives, but wish to sojourn as peaceful guests and to travel without doing the natives any harm.[20]

Only after such processes have been exhausted could the Spanish use force to defend themselves – and even in this instance, the Spanish use of force should be limited to "permissible self-protection" since the indigenous persons may be driven by fear and ignorance.[21] Yet if conflict still arises, the laws of war pertain and the Spanish could pursue the rights of war associated therein, including taking the indigenous persons' goods, imprisoning them, deposing their leaders and setting up new ones (within the bounds of proportionality).

There were also religious reasons justifying the Spanish use of force against the indigenous persons of the Americas and West Indies. If the indigenous persons did not allow the gospel to be preached, the Spanish could first reason with them and then, if reason failed, make war against them. If native converts to Christianity were subjected to force or fear "to make them return to idolatry," and reason failed to convince, then war could be waged.[22] And if a large percentage of the indigenous population converted but their leaders did not, then "the Pope might for a reasonable cause, either with or without a request from them, give them a Christian sovereign and depose their other unbelieving rulers."[23]

For Vitoria, these religious reasons, while driven by the obligations to divine law for the Spanish (and, indeed, all Christians), are nevertheless grounded in the *ius gentium's* recognition of essential human sociality. So, for example, where converts are being forced to "return to idolatry," an ensuing war would be "based not only on religion, but on human friendship and alliance, inasmuch as the native converts to

[20] Vitoria, *De Indis*, 154.

[21] It is worth noting here that even in his defense of the indigenous persons, Vitoria displays a troubling, if unsurprising, level of cultural imperialism and hegemony: "It is, however, to be noted that the natives being timid by nature and in other respects dull and stupid ... may very excusably continue afraid at the sight of men strange in garb and armed and much more powerful than they ... unite their efforts to drive out the Spaniards or even to slay them." Vitoria turns this argument toward the possibility that in a conflict in which one side is just and the other is driven by invincible ignorance, it is possible that both sides could legitimately claim to have just cause for war: Vitoria, *De Indis*, 154.

[22] Vitoria, *De Indis*, 158.

[23] Ibid.

Christianity have become friends and allies of Christians and we are under an obligation to do 'good unto all men, especially unto such as are of the household of faith' (Galatians, ch. 6)."[24] And, while these religious reasons are unlikely to seem compelling to modern readers, it is also at least worth noting that Vitoria recognized that bringing force into any religious issue, and especially one shaped by an evangelical imperative, made it less likely that the imperative could lead to success: "For it may be that these wars and massacres and spoliations will hinder rather than procure and further the conversion of the Indians. Accordingly, the prime consideration is that no obstacle be placed in the way of the Gospel, and if any such be so placed, this method of evangelization must be abandoned and another one sought for."[25]

Having distinguished reasons that are clearly illegitimate and reasons that are clearly legitimate to make war on the indigenous persons of the Americas and West Indies, Vitoria concluded *De Indis* by raising two other *possible* reasons to go to war in the new world. The first would be if Spain became an ally of one of the indigenous tribes that was, itself, at war with another indigenous tribe, in which case Spain might join the conflict provided its ally had a right to wage war (e.g., was fighting in self-defense). That Vitoria spent so little time on this argument suggests that he found it a nonproblematic expression of long-accepted rules of just war.

There was, though, "another title which can indeed not be asserted, but brought up for discussion, and some think it a lawful one. I dare not affirm it at all, nor do I entirely condemn it."[26] The argument behind this is whether Spain might paternalistically wage war on the indigenous persons for their own benefit if/as they are found to lack the wisdom necessary to establish and maintain a lawful state for themselves and therefore need Spain to rule them (at least for some period of time) for their own welfare. Here, then, is a fourth insight that makes Vitoria's writing on war particularly interesting: Even in systematic assessments of reasons

[24] Ibid.

[25] Ibid. In an essay on religion and religious intervention in places of conflict and their relation to international law, Antje von Ungern-Sternberg argues that Vitoria "counted religious intervention (without using the term) among the arguments able to justify the use of force in America." As helpful as her essay otherwise is, this argument seems peculiarly tone-deaf to the actual *ius gentium* arguments that Vitoria was making. See Antje von Ungern-Sternberg, "Religion and Religious Intervention" in *The Oxford Handbook of the History of International Law*, ed. Bardo Fassbender and Anne Peters (New York, NY: Oxford University Press, 2012), 308.

[26] Vitoria, *De Indis*, 160.

to go to war or resist going to war, some uncertainty remains – and the most obvious uncertainty has to do with war fought against another party for the benefit of that party. I will return to this insight as well as the three I highlighted above momentarily. As this argument concludes the constructive portion of *De Indis*, though, I turn, briefly, to the basic arguments of *De Iure Belli* as they grew out of Vitoria's work in *De Indis*.

VITORIA'S *DE IURE BELLI* IN A NEW WORLD

Vitoria was clear that his lecture on the law of war was not intended to be comprehensive in structure or focus; instead, he proposed "to supplement the foregoing discussion of the titles, some just and some unjust, which the Spaniards may allege for their hold on the lands in question, by a short discussion of the law of war, so as to give more completeness to that *relectio*."[27] It is a paradoxical note, then, that though Vitoria is sometimes recognized within the Christian just war tradition as someone who systematized the tradition and offered an early, quasi-modern vision of it, he did not think of his work in those terms. Indeed, although all the modern just war criteria – both *jus ad bellum* and *jus in bello* – are either explicitly present in or can be inferred from this *relectio*, Vitoria himself suggested that he was writing to answer four basic questions that arose out of *De Indis:* "First, Whether Christians may make war at all; secondly, Where does the authority to declare or wage war repose; thirdly, What may and ought to furnish causes of just war; fourthly, What and how extensive measures may be taken in a just war against the enemy?"[28]

In Vitoria's answer to those questions, two inferences become apparent. First, his treatment of these questions – aside from those parts of his answers that pertain to the four insights named above – are conventional in form, rely on established authorities (scripture, Augustine, Thomas Aquinas, etc.), and are clearly the product of his age. That is, the most electric parts of Vitoria's reflections on just war thinking are those places where traditional just war thinking rubs against the pressing questions of what to do with those persons that Spain found in the new world. Second, for someone who spent a good deal of time in *De Indis* arguing against waging war on the indigenous persons of the Americas and West Indies, Vitoria was quite bellicose. He permitted private wars where they

[27] Vitoria, *De Iure Belli*, 165.
[28] Ibid.

were waged defensively (provided force was used in immediate response to an attack), wars to avenge earlier wrongs when the avenging party was a state ruled by a legitimate authority, the weighing of proportionality to include payment for damages incurred (including the victor of a just war collecting penalties for misconduct by the losing side), and the unintended but foreseeable deaths of innocent persons as collateral costs in war. Indeed, Vitoria allowed warriors "to despoil the innocent of goods and things which the enemy would use against us ... [including] to take the money of the innocent and to burn and destroy their grain and kill their horses, if this is requisite in order to sap the enemy's strength."[29] He even went so far as to allow the indiscriminate slaughter of those who would resist military actions during the heat of battle and the killing of "all the guilty"[30] when security and justice cannot be achieved by other means. Clearly, Vitoria's humanistic impulses are more apparent in *De Indis* than in *De Iure Belli*, and the relationship between his martial and humanistic impulses undermines the latter and continues to complicate how his work has been received.

Perhaps even more importantly, Vitoria's *De Iure Belli* demonstrates two qualities that reveal its prominence in shaping new ways of thinking about the rules of war and how these new ways of thinking would, in turn, reshape just war thinking. The first of these developing qualities is the link between war and the ownership of property. Wars have long been fought over control of land and other properties but, as I will argue below, Vitoria's linking of war to established European laws about property rights, once that linkage is carried into non-European lands, has the effect of reframing the laws of war. Lands throughout Europe were already owned; to wage war over them was, then, to engage in fights over ownership in which contending parties at least agreed on what it meant to control – and therein own – that land, and in which a sovereign's defense of their own land was a just cause. Europeans in the Americas and West Indies, though, saw much of the land in the "New World" as open for taking because nobody else owned it in a way that they recognized as ownership. Once Europeans claimed lands in the Americas and West Indies on behalf of their sovereigns, attacks on those lands by indigenous persons could be justifiably repelled, even though a probable actual cause for the attacks – indigenous anger at Europeans claiming what was not theirs and seeking to remove Europeans from that

[29] Ibid., 180.
[30] Ibid., 183.

land – went mostly unrecognized by Vitoria in *De Iure Belli*.[31] Questions of simultaneous ostensible justice aside, the very fierceness of fights over control of land in the new world suggests that old rules applied in new contexts can shape conflicts in need of new laws.

The second of these developing qualities as revealed in *De Iure Belli* is that Vitoria simply gave less attention to the impact of war on the nonhuman natural world than had many of his predecessors. Saving a deeper exploration of how his predecessors in the Christian just war tradition thought about the nonhuman natural world until Chapter 6, it is nonetheless important to signal here that Vitoria's willingness to "burn and destroy their grain and kill their horses, if this is requisite in order to sap the enemy's strength," while it conforms in principle to some of the language of his predecessors (i.e., the right to kill a horse because it is a war animal and to deprive the enemy of grain in order to halt their warring), takes on a different tone and shapes a different sensibility about war when resources for making a postwar life are widespread – as they were perceived to be in the new world – than that same language when used in a place with fewer resources. Resource abundance gives license to greater bellicosity. Again, late medieval Christian just war language, shaped in a European context, takes on new overtones and permits larger conflicts in a new one.

These two developing qualities with Christian just war thinking as it transitioned from its late medieval to its early modern forms – reframed connections between war and property ownership and reduced attention within just war thinking to the nonhuman natural world – would only become more prominent over time.

Let us return to the four insights about war that arose from *De Indis*, taking them in reverse order. The four were questions about whether one could wage war for paternalistic reasons; the significance of reframing the relationship between the *ius gentium* and the *ius natural*; the possibility that two sides within a war might both have a just cause because of simultaneous ostensible justice; and the idea that human nature and rationality could be viewed through the lens of human sociality. Each of these insights would have far-reaching ramifications not only in shaping the Christian just war tradition as it transitioned into its early modern form but in how just warriors understood their relationships with "the other" and the nonhuman natural world more generally.

[31] A similar set of problems, this time over control of the seas, would shape Hugo Grotius' work, to be discussed in Chapter 2.

PATERNALISTIC WARS, UNCERTAINTY, AND ENGAGEMENTS WITH THE OTHER

The last insight had to do with Vitoria's rather inconclusive conclusion about whether Spain could go to war for paternalistic reasons. Clearly, there are a range of deep and troubling theological, moral, logical, and practical problems with supporting such an argument for going to war. Little within just war thinking lends much support to such an argument – certainly not in light of criteria such as last resort or proportionality – especially once the tradition has criticized the self-deceptions involved in claims such as "we had to destroy the village in order to save it." The analogies that Vitoria relied on (the indigenous persons lack the capacity to order their own society; adults must care for infants and beasts; charity to neighbors involves looking out for their welfare regardless of their desires) are either wildly inappropriate or had already been undermined by his own earlier arguments. It is difficult to imagine a society capable of waging a defensive war and determined to do so that would, on the back end of defeat, be glad for its defeat and the loss of its way of life – though it is not impossible to imagine such a society, provided the war also shaped a capacity for repentance and change in that society (one thinks of post–World War II Germany). Vitoria had to turn to troubling arguments about natural order (e.g., "some help might be gotten from the consideration, referred to above, that some are by nature slaves, for all the barbarians are of that type and so they may in part be governed as slaves are")[32] that, were they true, would simultaneously undermine paternalistic claims about the capacities of indigenous persons to be governed until they grew into "adulthood."

Yet it is also worth noting that (a) Vitoria was writing prior to the rise of the modern nation-state and the fundamental claims of national sovereignty that are the centerpiece of such a state, and that paternalistic arguments about control of other states were not uncommon in the Holy Roman Empire (or the papacy, for that matter) – and Vitoria was a figure of his time; (b) the arguments for paternalistic reasons to go to war grow considerably more complex if the state against which one would wage war is riven internally or failed (one thinks of contemporary debates about whether there is a responsibility to protect or to pursue humanitarian intervention in violation of sovereign boundaries); and (c) Vitoria, who spoke with confidence throughout the *relectiones*, fills this

[32] Vitoria, *De Indis*, 161.

last section of his argument with hemming and hawing. None of this justifies his arguments, but it does at least suggest a tacit recognition on Vitoria's part that his just war thinking had not reached resolution on all matters. And if Vitoria, perhaps the preeminent Roman Catholic theologian of his time, was aware of the limits of just war thinking at that time, it might follow that maintaining some modesty in advancing the claims of just war thinking in our own time is appropriate.

Moreover, it is worth noting that, for Vitoria, this uncertainty expressed itself not in overstatements or vagueness but in honest questions and dissatisfied attempts to root around in the resources of the tradition and his own mind to answer them. We might be unhappy with his approach to the issue with which he ends *De Indis*, but it is still worth noting that a figure who is better known for synthesizing the just war tradition than for advancing it in novel ways and for (unintentionally) initiating its turn to international law did not think of the tradition as having an ending. Rather than seeing dramatic social, political, economic, and religious changes and retrenching against them, insisting that old answers to just war questions were adequate, Vitoria resisted the urge to cap the tradition and, instead, accepted its open-endedness. Just war thinking simply cannot have an ending or be a finished thing. Just as important in this context, this lack of an ending (and the questions shaped therein) has to do with how to think about and treat the other. Vitoria ends *De Indis* where he began, with questions about how to think of the other.[33] Whether modern expressions of just war thinking reflect such a willingness to accept the tradition's open-endedness in thinking about how it understands the other and encounters with the other is an important question as we leave modernity behind and enter

[33] I should note here that Vitoria's ways of thinking about "the other" at the beginning of the age of colonization are not only identifiably European; they are identifiably Spanish. So, for instance, English colonizing projects also engaged in deeply problematic ways with indigenous persons in the new world, but not in quite the same ways that Spanish colonizing projects did. These differences were shaped by economic, class, and religious differences and they were notably diverse (Roger Williams' engagements with native Americans in New England did not look like James Oglethorpe's in the south, nor with Arthur Philip's engagements with the aboriginal peoples of Australia). I am focused on Vitoria and his engagements with the other because of the wider impact of his thought on Christian just war thinking and on modernity itself. Toward the end of this book, I will also note another significant difference in the way southern Europeans (e.g., the Spanish and Portuguese) and northern Europeans (e.g., the English and Dutch) engaged the other; namely, that they experienced significantly different environmental conditions in the places they colonized. My thanks to James Davis for pointing out the possibility of regional differences in treatment of the other.

further into the social imaginary of the Anthropocene. Indeed, among the more hopeful manifestations of recent just war thinking has been the willingness on the part of many of its defenders to advance the tradition in new directions in light of new concerns (how to protect noncombatants in a world of weapons of mass destruction) and uncertainties (whether just war thinking can continue to treat the modern nation-state as the basic political body).

THE *IUS GENTIUM*, THE *IUS NATURAL*, AND THE COMPLEXITIES OF INTERNATIONAL LAW

The second to last insight from earlier in this chapter was that the *ius gentium*, which is the basis for international law in Vitoria's thought, was complexly related to the *ius natural*. Although it arises out of natural law and is a subset of natural law, the *ius gentium* maintained a kind of independence from the *ius natural* at political levels. One expression of this independence manifests itself in the way Vitoria thought of the law of the nations as pertaining to the political units that would later come to be understood as modern nation-states. For Vitoria, the *ius gentium* can be distinguished from the *ius natural* that applies to all persons regardless of their political location; it pertains, instead, to persons in their roles as participants in particular political systems, and those particular political systems look like states. This distinction is possible because of the divisions that Vitoria created within the natural law. By distinguishing the laws of nations from the laws of nature (and doing so on the basis of a distinction between actions that are contrary to the natural law and actions that are contrary to the natural order), Vitoria opened up a space in which some laws were universal and applied to all persons as individuals and others were context-dependent, the latter of which meant societies should be held to account for laws that they themselves recognized.

Its independence from natural law also allows the *ius gentium* to be complexly related to the vision of positive law that would come to dominate the development of modern international law. Even as the modern nation-state would become the basic subject around which international law would turn – and even as international law was, in its modern formulation, shaped by positive law – Vitoria's vision of the *ius gentium* assumed political differences between groups of people that would have to figure out how to get along with each other by appeals to consistency and community. These political differences mandated that when nations engaged each other, they would have to shape those engagements through appeals to laws that were mutually agreed upon rather than those that were universally applicable.

Vitoria's approach suggests that he wanted to have it both ways: to argue for a universal law that is context-specific and a natural law that has positive law components. In a subtly provocative set of claims worth quoting at length, Annabel Brett makes that very claim, arguing that,

> In his commentary on the *Secunda secundae*, dating from 1534–37, [Vitoria] engaged directly with Aquinas' positioning of the *ius gentium* as a kind of natural, as opposed to positive, law, consisting in natural reason. Here, Vitoria moved away from the Thomist position to make the *ius gentium* a kind of positive law, based on inter-human agreement. The universal character of that agreement meant that it retained a close proximity to natural law, both in being a result of natural reasoning processes and in functioning to protect the operation of natural law. Nevertheless, it had a human source in "the consensus of all peoples and nations" or "of the whole world," ... In his early *Relectio de potestate civili* of 1528, Vitoria spoke of this legislator of the *ius gentium* as "the whole world, which is in a sense [*quasi*] a commonwealth," thus apparently underlining the analogy between the *ius gentium* and civil law as types of positive rather than natural law. The *quasi* is important, however. Vitoria did not think of the whole world as one commonwealth like the particular commonwealth of France or Spain. If it were, a universal emperor, a *dominus mundi* would be a possibility. But in the later *Relectio de Indis* of 1539, Vitoria expressly denied this: There neither is nor can be any universal civil jurisdiction spanning all peoples, and consequently no universal civil law. In this work, the universal law of nations figures as a *ius inter gentes*, replacing the *ius inter homines* of the Roman law. By rewording the definition in this way, Vitoria was clear that the *ius gentium* was not a law that governs individuals in the first instance, but this does not mean that he saw it as a law between commonwealths or "States" either ... Rather, the *ius gentium* is part of an essentially temporal narrative of the separation of peoples "within" a natural pan-human community, a narrative that is different from that of the founding commonwealths. This is why Vitoria can both affirm that the *ius gentium* is the product of all the nations or of the whole world, and can move easily between its positive and natural character.[34]

Even more provocatively, Brett goes on to argue that Vitoria's attempt to have it both ways – to affirm natural law and positive law within his laws of the nations – led him to become an apologist for Spanish conquest.[35]

[34] Annabel Brett, "Francisco de Vitoria (1483–1546) and Francisco Suarez (1548–1617)" in *The Oxford Handbook of the History of International Law*, 1087–1088.

[35] There is, I suppose, an analogy here in the way Michael Walzer wrote of "supreme emergencies" that would justify the use of otherwise universally prohibited weapons such as nuclear bombs since what he meant by "supreme emergency" was a threat to a particular culture's continued existence and the culture that he was thinking of was that of Jewish

Although he rejected the use of those natural laws pertaining to the natural order as a basis for conquest, he did offer up as natural – and therefore universal – a set of laws he believed pertained to the essential qualities of human sociality (e.g., free communication, allowing others safe passage, participation in faith communities). In a move that foreshadows one that Immanuel Kant would make in much greater detail some 200 years later, Vitoria suggested something like *a priori* established moral obligations for all persons to honor universal laws that are categorical in scope and imperative in form but that account for the subjective differences between politically grounded persons living in diverse communities. The violation of such laws justified resort to coercive force, and "Vitoria's juridical universalism combined with the particularity of its application to the Indians undoubtedly operated to open their commonwealths to European invasion."[36]

Brett's claim about the "particularity" of Vitoria's application of his version of the *ius gentium* is not erroneous but it is complicated by the context in which Vitoria was applying his *ius gentium*. On the one hand, penalizing Vitoria's juridical/theological vision for its possible impact during a time when visions of God, human beings, and the larger world were kaleidoscopic (i.e., brilliant, changing, prismatic, and ambiguous) may be setting an unduly high bar. While all of us bear at least

Israel. Engaging the flashpoint of this tension between universal values and particular goods is, in Walzer as in Vitoria, the point over which the most critical ink has been spilled. See Michael Walzer, *Just and Unjust Laws: A Moral Argument with Historical Illustrations* (New York, NY: Basic Books, 1977), Ch. 16.

36 Brett, "Francisco de Vitoria (1483–1546) and Francisco Suarez (1548–1617)," 1089. Here, Brett partially echoes arguments made by Antony Anghie, *Imperialism, Sovereignty and the Making of International Law* (Cambridge, UK: Cambridge University Press, 2002). Anghie's more provocative larger thesis is that rather than thinking of the colonialist project as the product of the development of the Westphalian nation-state, we should think of it as the cause of that development – and Vitoria's work is the first piece of evidence in this regard: "Vitoria is an extremely complex figure. A brave champion of the rights of the Indians in his time, his work could also be read as a particularly insidious justification of their conquest precisely because it is presented in the language of liberality and even equality" (Anghie, *Imperialism*, 28). For further arguments about whether Vitoria should be read as an apologist for conquest, see China Miéville, *Between Equal Rights: A Marxist Theory of International Law* (Liden: Brill, 2005); Robert A. Williams, Jr., *The American Indian in Western Legal Thought: The Discourses of Conquest* (New York, NY: Oxford University Press, 1990); and, perhaps most helpfully, Georg Cavallar, "Vitoria, Grotius, Pufendorf, Wolff and Vattel: Accomplices of European Colonialism and Exploitation or True Cosmopolitans?" *Journal of the History of International Law* 10 (2008): 181–209 (and those they cite in their arguments). Brett takes something of a middle position between Vitoria's defenders (such as Scott and Nys) and Vitoria's critics.

some responsibility for the consequences of our ideas as they manifest themselves in actions and structures, such responsibilities adhere to the foreseeable consequences of such ideas within conceptual frameworks shaped by the social imaginaries within which we think. I doubt any of us would want to be damned for the unanticipatable consequences of our ideas and none of us can think from nowhere (or everywhere); reflexive consistency (not to mention the golden rule) suggests we should take care in damning others for theirs. We may conclude that Vitoria's convictions about, for example, political systems mirroring divine ones were absolutely wrong but we may also need to recognize that Vitoria was justified in holding those convictions from within the social imaginary within which he worked.[37]

Moreover, we ought to give some weight not only to the fact that Vitoria wrote *De Indis* to address a specific and fairly new problem (the treatment of indigenous persons in the Americas and West Indies) but also that his solution to that problem was more respectful of their identities and treatment than many other solutions being proffered at the time. So, for instance, had Vitoria explicitly made the argument that indigenous persons of the Americas and West Indies should be free to communicate with Europeans in Europe, to travel freely within Spain, France, Belgium, or England, and to proselytize in those places, some of the weight of Brett's condemnation would be lifted.

On the other hand, it was certainly the case that such freedoms were constrained in Europe (this was, after all, the age of the Spanish Inquisition and the beginning of horrific conflicts between European states), so Vitoria's arguments about the basic freedoms of Spaniards in the new world are peculiarly and noticeably oblivious to some fairly obvious failures of reflexive awareness. It is hard to dismiss the suspicion that even as he was arguing for the rights of indigenous persons in the new world to "own" their land and stuff, he still thought of the new world as mostly ungoverned, wide open, and available for the taking. *De Indis* may not have lined up well with Pope Alexander's *Inter Caetera*, but neither did it stand clearly against it; Vitoria should not be held responsible for the way that papal bull drove doctrines of discovery for the next several hundred years of conquest and colonization, but neither can he be held wholly innocent from the abuses of those doctrines.

[37] On the relation between truth and justification, see Jeffrey Stout, *Ethics after Babel: The Languages of Morals and their Discontents* (Princeton, NJ: Princeton University Press, 2001), esp. 24–28.

Part of what is revealed here is a problem endemic to appeals to any form of universal law, whether the natural law, a law of nations, or even universal declarations of human rights. While such appeals offer a moral basis from which to challenge the more egregious practices within particular cultures without simultaneously elevating a set of culturally specific norms from some other society, they do so by allowing a (nominally) neutral set of laws to trump any culturally specific norms. The standard upon which complicated political decisions are judged is blind to the norms, practices, and understandings that arise out of particular cultures and shape intercultural engagements. Even setting aside ideological criticisms of cultural bias inherent in claims about universal law, practical problems about the viability of appeals to universal laws that may feel like foreign impositions to indigenous persons. Political problems arising out of the power differentials between cultures that are necessary to impose international laws, not only reveal the way the universal trumping the particular can fund colonialist/imperialist projects, it will also shape one of the great tensions of modernity. Asserting the natural rights of indigenous persons on the basis of universal norms as a way of defending those persons undermines sensitivity to the very indigenous cultures in which such persons live and find meaning. Vitoria may not have intended for *De Indis* to serve as an apology for Spanish conquest in the new world but the effects of his approach did make such conquest more likely.[38]

Far more troublingly, Vitoria was silent on the treatment of indigenous persons of Africa who were already being enslaved and taken to the new world.[39] Spain was in the midst of deep debates about the enslavement of indigenous persons during the time of Vitoria's writings (and he certainly would have been aware of those debates) but he remained silent about these matters. Though arguments from silence may be weak – especially where those arguments are so contextually dependent – arguments from selective silence are damning. Vitoria's version of the *ius gentium* may

38 For a more extensive and nuanced assessment of this dynamic, see Williams, *American Indian in Western Legal Thought*, esp. 96–108.

39 The first ship to take Africans to the new world as slaves was a Portuguese vessel in 1526. The case of Spain is more complex if not more virtuous: Columbus brought persons from the West Indies back to Europe with him as slaves but the Spanish monarchs ordered them to be returned. The *encomienda* system was in effect in both Africa and the new world without restriction until the New Laws of 1542. And although the papal bull *Sublimus Dei* of 1537 made the enslavement of indigenous persons illegal, it was rescinded a year later.

only pertain to some aspects of human interactions but if it is genuinely a law of nations grounded in natural law, it has to apply everywhere. Perhaps the problem with "Vitoria's juridical universalism combined with the particularity of its application" is not simply the particularity of its application; it is that it was not actually juridically universal and so not actually a natural law-based *ius gentium*.

One generalizable lesson here is that just war traditions – and, given their theological starting points in human possibility, sin, and finitude, perhaps the Christian just war traditions in particular – should attend to the possible consequences of their ideas as they play out over time when they advance arguments about when to go to war and how to fight. Given that Vitoria offered *De Indis* as a restraint on Spanish claims to the new world and Spanish willingness to wage war to realize those claims only to have his work in *De Indis* used to justify those claims and undergird those wars, a corollary lesson is that any just war tradition that ignores the ironies, disjunctions, and paradoxes that arise as it moves through time can potentially undermine its own best visions. It follows that just warriors have a concomitant obligation to shape and describe their traditions in ways that necessarily attend to the traditions' potentials to lead to unanticipated consequences, that remain open to correction and revision in light of those consequences, and that make explicit their obligations toward modesty and honesty in describing their work and its potential. Just warriors may shape traditions that help to prevent, reduce, and mitigate suffering; they cannot shape traditions that are beyond reproach or regret.

Of course, failing to consistently apply an idea, damning as that failure may be, is not necessarily the same thing as having a bad idea. Vitoria may have failed to apply his defense of the other in *De Indis* in a way that was universal or internally consistent, but he was at least defending the status of indigenous persons in the Americas and West Indies against alternative dehumanizing descriptions of them and rapacious visions of the benefits that would come with claiming their land and its resources. He defended them as human beings with a status that was, in principle, equal to those of Europeans. And in this, it is important to remember that for Vitoria, the distinction between universally valid laws and context-dependent laws revolved around the way he understood human beings: as rational and social. That is, where laws contrary to the Roman Catholic vision of natural order are recognized by another society, then as long as that society adheres to its own laws (and is therefore rational) and coheres over time (and is therefore social), its failure to conform itself to Roman Catholic understandings of the natural order should not count against it. Only those actions that undermine its own capacities for rationality and sociality would violate the universally valid natural law and justify taking up arms against that country.

Yet once such criteria for assessing the legitimacy of another society are in place, the distance between a natural law-based and a positive law-based version of international law is even greater in theory than in practice. As Antonio Cassese notes, during the early stages of international law, "[t]here only existed a core of rules: those on the conclusion of treaties; the exchange of, and the rights and privileges accruing to diplomats; the free use of the high seas; the capture of pirates; and the resort to force (admissible both to protect one's own interests and to vindicate one's rights); as well as some rudimentary rules of warfare."[40] These rules can fit within either Vitoria's vision of international law (at least when it is consistently applied) or what would become Hugo Grotius' vision of international law. It would follow, then, that at least as long as Vitoria's understanding of the *ius gentium* was accepted, anyone who wished to argue for one or the other theory as a basis for international law would need to explain how that theory matters in practice.[41] It also suggests that the conceptual distance between Vitoria and the Spanish origins of international law and Grotius and the Westphalian origins of international law are not so great as has been suggested – nor is the distance between at least one explicitly religious vision of these origins and the supposedly more "secular" vision offered by Grotius.

The limitations of Vitoria's *ius gentium* approach to thinking about just wars through attention to the other bring into relief at least one other problem with early international law that would be shared by both natural law-based and positive law-based approaches: their inattentiveness

40 Antonio Cassese, "States: Rise and Decline of the Primary Subjects of the International Community" in *The Oxford Handbook of the History of International Law*, 54.

41 Mary Ellen O'Connell, for instance, argues that natural law provides a constructive basis for international law as applied to the context of war because it situates her understanding of war within a larger vision of peace and pacifism that is so fundamental to her perspective as to all but eliminate not only positive law as a sustainable basis for establishing rules for war but war as capable of being a rule-governed activity in the first place. Whether she is successful in her arguments turns, in my opinion, on the degree to which one accepts the viability and singularity of that larger vision. See, among other places, Mary Ellen O'Connell, "Peace and War" in *The Oxford Handbook of the History of International Law*, or Mary Ellen O'Connell, *The Power and Purpose of International Law: Insights from the Theory and Practice of Enforcement* (New York, NY: Oxford University Press, 2011).

to the interplay of language and power. At least since Michel Foucault's work linking knowledge and power,[42] scholars throughout the humanities and social sciences have attended to the complex ways that processes of description, definition, and delineation are also means through which power and differentials in power are expressed and reinforced. As described above, Vitoria's focus on the status and location of indigenous persons in the Americas and West Indies as a starting point for making sense of just war thinking and, concomitantly, international law not only begins with attention to them as "others" that need to be made sense of in order to establish treatment of them, but also makes sense of them according to an understanding of human beings that he accepts as universal (i.e., rational and social) while arising out of a Thomistic theological framework. That is, he treats rationality and sociality as neutral categories without recognizing that how indigenous persons understand rationality, sociality, and the relationships between rationality and sociality, and even whether they would use such constructs as basic to their own self-descriptions of what it means to be human may reveal differences between those indigenous persons and a sixteenth-century Spaniard. Vitoria, after all, had not been to the new world or observed the cultures there prior to writing *De Indis* about how to treat *de indis*.

Truism though it may be, one of the dangers of starting explorations of just war thinking with attention to the "other" is that it risks overcoming the failings of "othering" them by "saming" them: insisting that they are like us. "They are fundamentally like us" is, paradoxically, as objectifying a statement as "they are fundamentally different from us." Neither statement gets at the complex way that "they" are both like and unlike "us" – and neither statement begins with the project of letting them speak for themselves. While explorations of what it means to be human may be fairly innocuous at abstract levels, those explorations can have real, lasting, and pernicious consequences when they are expressed in the contexts of war and international law. It is not simply that Vitoria's "juridical universalism combined with the particularity of its application" could be used to justify the European invasion of the Americas and West Indies; it is that Vitoria's juridical particularity combined with the universality of its application in international law and the just war tradition helped to initiate and justify the hegemonic power of Western thought on international law and the just war tradition from the

[42] See, e.g., Michel Foucault, *Power/Knowledge: Selected Interviews and Other Writings, 1972–1977*, ed. Colin Gordon (New York, NY: Vintage Books, 1980).

age of conquest to the present.[43] This is a perpetual problem for natural law thinkers to address – even those like Vitoria that prove themselves nimble in its revisions and uses.

SIMULTANEOUS OSTENSIBLE JUSTICE, HUMANIST THOUGHT, AND THE LIMITS OF REFLEXIVITY

The second insight offered above is that reasonable persons can reach irreconcilable conclusions and that these irreconcilable conclusions will have far-reaching consequences for just war thinking; the door was now open to what James Turner Johnson refers to as "simultaneous ostensible justice."[44] Vitoria's argument that the indigenous persons of the new world would, by virtue of their invincible ignorance, be justified in thinking they could be in the right should they go to war – even if the Spanish explorers and missionaries actually were right – does more than highlight the limitations of epistemologies traversing cultural differences. Setting aside the insulting quality of claiming that others are ignorant, the "invincible" quality of that phrase implies that Vitoria can imagine a war in which both sides can legitimately claim to have a just cause. As Johnson notes, after Vitoria, "it is possible to conceive of wars in which both belligerents are deceived by such ignorance into thinking

43 See, e.g., Anthony F. Lang, Jr., Cian O'Driscoll, and John Williams, eds., *Just War: Authority, Tradition, and Practice* (Washington, DC: Georgetown University Press, 2013), 117–121, for the connection between just war thinking and colonialism.

44 Johnson connects Vitoria's notion of invincible ignorance to his (Johnson's) idea of simultaneous ostensible justice in several places throughout his corpus. So, e.g.,

> Considering the case of the Indians of the New World and whether they had the right to defend themselves against the Spanish explorers and missionaries, Victoria invoked the concept of invincible ignorance. It is possible in war, he argued, that one side may *be* in the right, while the other, by invincible ignorance, may *believe* itself in the right. Indeed, it is possible to conceive of wars in which *both* belligerents are deceived by invincible ignorance into thinking they have a just cause, when in fact neither does. In such cases it is not a question of "objective" versus "subjective" just cause, because even an objective third party [*sic*] observer may be unable to disentangle conflicting claims so as to adjudicate the matter. So, Victoria argues, both sides should be treated as if they had the just cause, and both sides must consider themselves scrupulously bound by the limits of the *jus in bello*, the law of war.
>
> The *jus ad bellum*, here focused closely upon the element of just cause, resolves itself into a doctrine of simultaneous ostensible justice.

James Turner Johnson, "Just War Theory: What's the Use?" *Worldview* (July/August, 1976): 43. See also James T. Johnson, *Can Modern War Be Just?* (New Haven, CT: Yale University Press, 1984), 21.

that they each have a just cause, when in fact neither does. In such cases the question is not of objective versus subjective just causes, because even an objective third-party observer may be unable to disentangle conflicting claims so as to adjudicate the matter."[45] Among *jus ad bellum* criteria, just cause is bedrock both historically and conceptually. Complicating just cause in the way Vitoria does not only destabilizes that bedrock; it forces the other criteria to carry additional weight in adjudicating questions about when and why to go to war and how to fight once at war.

The question of how to adjudicate just cause when belligerents come from different cultures did not begin with Vitoria. Just warriors had struggled to address matters of just cause not only when faced with conflicts between substantially different cultures (e.g., medieval conflicts between Christian and Muslim forces, where those two religions carry distinct and irreconcilable understandings of war and the justification for going to war) but even where there is substantial overlap between cultures (e.g., the conflict between France and England during the Hundred Years' War that was complicated by different understandings about the rights of accession). Yet such instances were addressed by the parochial qualities of particular just war traditions, by ignoring just war thinking when applied to conflicts between substantially different cultures, or by leaving such questions unresolved. By beginning his exploration of just war thinking through his attention to the other, Vitoria pushed the problem of multiple incommensurable just causes to the fore and contributed to initiating one of the greatest Reformations of just war thinking in its history: the shaping of international law through the secularizing of the tradition. After Vitoria, questions that had been ignored or deferred demanded answers.

Toward such answers, begin by noting that Vitoria treated the problem through the lenses of his particular religious perspective: Multiple incommensurable just causes are brought on by invincible ignorance, a category that makes sense within his Thomist framework. This religious perspective is visible from the very beginning of his argument: Answering a question about whether to go to war with indigenous persons in the Americas and West Indies was inextricably linked to Vitoria's arguments that religious differences cannot justify war because faith cannot be compelled. The logic of Roman Catholic natural law thinking resolves tensions associated with cultural differences. On their face, Vitoria's answers to questions about the treatment of indigenous persons

[45] Johnson, *Can Modern War Be Just?*, 20–21.

are the consequence of his traditional training and so his attention to the problem of multiple incommensurable just causes is simply a logical consequence of his Thomism and its form of natural law. While he may have been among the first to work out these implications of Thomas Aquinas' thought, the implications of this working out of Thomism are hardly revolutionary. If anything, they were all but inevitable given not only his own background but also the degree to which religious perspectives so pervaded Western cultures that a nonreligious (and, for Vitoria, non-Roman Catholic) argument would be nonsensical. Vitoria, after all, was writing at a time that preceded the dawn of the "secular age" described by Charles Taylor. We should be shocked if his treatment of the issue did not come across as explicitly religious, and not only because he was a trained theologian but also because at that time even most jurists and political philosophers generally looked at the world through religiously tinted lenses.[46]

What makes Vitoria's focus on multiple incommensurable just causes more interesting is the degree to which it also reveals the impact of humanist thought on his work.[47] Not only does the question of contending and opposed just causes begin with his arguments in *De Indis* about the full humanity of the indigenous persons of the Americas and West Indies based on their capacities for rationality and sociality, but, from his perspective, evidence of their rationality is not contingent on their ability to accept the wisdom of the gospel. Indeed, he treats any belligerent responses to the Spanish mistreatment of them – undermining as it does the coherence of the gospel witness – as further evidence of their rationality. Implicit in his argument are the assertions that a moral sense is evidence of rationality and that such a moral sense need not be based on a particular religious vision (let alone a Christian one). That is, Vitoria argues not that the indigenous persons' rights need to be respected because they practice other religions, but that their religious practices are immaterial to their rights. Bear in mind that at the time

[46] Tangentially, this is part of what makes Vitoria's Italian contemporary, Machiavelli, so much more revolutionary: His political framework treats religious perspectives as epiphenomenal to political interests. It is, by the way, worth wondering who – Machiavelli or Vitoria – offered a description of human nature and politics that has actually proven to be more accurate.

[47] "Thanks to being brought into contact with [Josse van Assche and Erasmus, among other humanists], Franciscus de Victoria undoubtedly found his natural leanings strengthened and received help from this beneficent influence for taking in hand the defense of the just cause of the Indians." Nys, "Introduction," 69.

of his writing, the Spanish Inquisition was actively forcing Jews and Muslims to convert to Christianity (and, by the middle of the sixteenth century, forcing Spanish Protestants to convert to Roman Catholicism). The humanist turn from cosmic-sized theological metaphysics toward human-sized moral concerns is on full display here. While religious and humanist impulses blur into each other in Vitoria's arguments (as well as the arguments of those who would come after him), Vitoria's work gives a clear example of how those two impulses are being reweighted during the lead-up to modernity.

Vitoria raises the problem of invincible ignorance in the context of his arguments about legitimate titles by which Spain may wage war against indigenous persons, the fifth proposition of which is that "[i]f the Indian natives wish to prevent the Spaniards from enjoying any of their above-named rights under the law of nations, ... the Spaniards ought in the first place to use reason and persuasion in order to remove scandal and ought to show in all possible methods that they do not come to the hurt of the natives, but wish to sojourn as peaceful guests and to travel without doing the natives any harm."[48] His arguments about invincible ignorance and "holding the war to be a just war on both sides"[49] are premised, then, on the existence of a *ius gentium*, not a particular religious perspective. This is further evidenced by the fact that his analogy ("just as the French hold the province of Burgundy with demonstrable ignorance, in the belief that it belongs to them, while our Emperor's right to it is certain, and he may make war to regain it, just as the French may defend it,")[50] is an intra-European, intra-Catholic conflict: Neither sharing a religious perspective nor differing in religious perspectives is material to the conditions by which multiple sides may legitimately claim just cause. By the time Vitoria reaches his arguments about whether a war can be just on both sides in *De Iure Bello*, his conclusions are already determined by his concerns in *De Indis*: "Assuming a demonstrable ignorance either of fact or of law, it may be that on the side where true justice is the war is just of itself, while on the other side the war is just in the sense of being excused from sin by reason of good faith, because invincible ignorance is a complete excuse."[51]

48 Vitoria, *De Indis*, 154.
49 Ibid., 155.
50 Ibid.
51 Vitoria, *De Iure Bello*, 177.

While holding the line that only one contending party will actually have just cause, Vitoria's humanism allows for the possibility that both parties may legitimately argue that their causes are just. This follows from the fact that these legitimate contending just causes for war are a condition of the *ius gentium* which, as noted above, offers the primary grounds for determining just cause, and the *ius gentium* is not only universal in scope but deaf to both explicitly religious (e.g., divine command) and "order of nature" (e.g., revealed law) arguments through which one might legitimate war. Not only do concerns about human welfare drive decisions about going to war, but also capacities for human agency make such decisions possible. Indeed, awareness of the limits of just cause in the face of his humanist impulses seems to drive Vitoria's repeated arguments against going to war in *De Indis,* his hope that other forms of conflict resolution will be attempted prior to going to war, and his insistence that, in the face of multiple contending just causes, wars should be defensive and acts of war should be proportionate to the rights of self-defense. Even his more bellicose writings in *De Iure Bello* arise out of a humanist resistance to treating war as a natural condition of human existence and, therein, a legitimate means for furthering political power.

Twin paradoxes follow from Vitoria's arguments about multiple contending just causes. First, the incorporation of humanist thought into just war thinking in the sixteenth and seventeenth centuries does not lead to greater concern for human beings in a way that also leads to reductions in violence. Where just causes can be multiple and the constrained morality of the *ius gentium* displaces the authority of an emperor, a pope, a single dominant religious tradition, or a universally agreed upon *ius natural*, war could be waged with greater levels of righteous furor and fewer checks on the legitimate reasons to wage such wars – which is precisely what would occur during the wars of religion that would follow Vitoria at the dawn of an international order framed by the sovereignty of modern nation-states. And, second, the appeal to a natural law-based *ius gentium*, though oriented by the universalizing impulse that marks natural law, had the effect of casting into sharper relief differences between cultures rather than highlighting their commonality. Obviously, this was, in part, due to the increased levels of contact with other cultures during a time of exploration, evangelization, and conquest. But it was also due to the humanist impulses that motivated Vitoria's writings and led him to advance the possibility that both sides in a war could legitimately claim just cause.

The emphasis on human agency that marks the humanist turn in Western cultures is further reinforced by attending to what Vitoria does with the criterion of just cause once he has argued that just causes can be plural and opposed: Rather than dismissing just cause as a category worth attention or use, he operationalizes it by reinforcing the roles that other *jus ad bellum* criteria play in order to retain it. The obligations of the prince now include seeking the wise counsel of others (including from those beyond his immediate circle in order to hear other perspectives) in order to determine just cause: The criterion of legitimate authority takes on just cause inflections since determining just cause will now involve pursuing levels of reflection that take seriously the possibility of being dramatically wrong. Pursuing processes short of violence in the pursuit of conflict resolution grow in importance as both sides can legitimately claim just cause: The criterion of last resort takes on additional weight. Establishing metrics by which to assess the consequences of a war in which both sides claim to have just cause mandates closer attention to the consequences of a war that can be exacerbated by the pursuit of righteous grievances on both sides: The criterion of proportionality becomes more complicated.

One strong piece of evidence for this conclusion is that Vitoria answers the just cause question that drives *De Indis* ("Is Spain justified in waging war against the indigenous persons of the Americas and West Indies?") through giving extended attention to titles (legitimate authority), rejecting coerced faith (right intention), and so forth. Indeed, notable in its absence is any sustained argument in his lectures about traditional just causes. Self-defense and defense of persecuted groups are assumed to be legitimate just causes rather than causes that mandate extensive arguments.

This suggests that, at least for Vitoria, just cause is not simply one *jus ad bellum* criterion among others; it is part of the substructure of just war thinking that makes the tradition and the other *jus ad bellum* criteria coherent. Maintaining the language of just cause in the face of invincible ignorance – which Vitoria certainly did – complicated it. Rather than surrender the criterion of just cause in the face of such complexity, though, Vitoria implicitly reframed it and, in so doing, suggested a different way of relating the criteria to each other. That is, Vitoria's attempt to answer questions about what to do with just cause in the face of invincible ignorance also suggests that we need not think of *jus ad bellum* criteria as independent and/or lexically ordered.[52] The criteria are not a set of discrete boxes to check off. They are, instead, interdependent and arise collectively out of the way that a particular expression of just war thinking reveals itself as it addresses fundamental questions, such as "who are these others and what are our obligations to them?"

RATIONALITY, SOCIALITY, AND THE (MODERN) SHIFT FROM METAPHYSICS TO POLITICS

Finally, the first insight offered above had to do with the way Vitoria connected rationality and sociality in his description of human beings. At the outset of exploring some implications of this insight, it is important to admit that Vitoria's treatment of the question of what it means to be human in his *Relectiones* is deeply frustrating. On the one hand, questions about human nature are essential to *De Indis* and *De Iure Bello*. One simply cannot make sense of his attempts to defend the rights

[52] James Turner Johnson has argued that one of the effects of Vitoria's foregrounding the possibility of multiple incommensurable just causes was "to undercut the *jus ad bellum* for nearly three centuries but at the same time to give more weight to the moral and legal efforts aimed at managing force in war, the *jus in bello*." Johnson's argument makes sense if we see the *jus ad bellum* criteria as independent but necessary components of the process by which to address questions of whether to go to war – as if they are each load-bearing columns upon which to build the legal/moral superstructure of just war thinking. If one falls, they all fall. On Johnson's view, the shift that Vitoria initiates (and Grotius expands) from *jus ad bellum* to *jus in bello* criteria helps to explain how international law not only grows out of but comes to replace religious just war thinking:

> In the time of Gratian and his successors ... one of the most fundamental questions in the restraint of violence was the establishment of responsible and just government; accordingly, right authority, just cause, and the other *jus ad bellum* concepts were at the center of concern. Only after the issue of sovereignty and authority to use force had been settled did the focus shift to incorporate fully the *jus in bello* concerns ... Still, in the late-medieval consensus on just warfare, these two kinds of interests were held in balance – a state of affairs that obtained right through the great Neo-Scholastic just war theorists of the sixteenth and early seventeenth centuries, Franciscus de Victoria and Francisco Suarez. With Grotius, though, the balance shifted decisively toward the *jus in bello*, a condition characteristic of the next three centuries of development in international law.

James Turner Johnson, *The Quest for Peace: Three Moral Traditions in Western Cultural History* (Princeton, NJ: Princeton University Press, 1987), 202. Johnson and I do not disagree on the impact of Vitoria's thought in this regard on the development of international law: *jus ad bellum* concepts are not as central to international law as *jus in bello* concepts are. However, as I will be arguing later, the birth of international law does not lead to the end (even for a time) of the religious development of just war thinking and *jus ad bellum* concepts do express themselves in realms beyond international law in ways that have been mostly ignored.

of indigenous persons in the Americas and West Indies without attending to his recognition that their rights are implied in their full humanity: They are neither animals, nor infants, nor subhumans. Moreover, the very project of treating war as an activity that has the capacity to be rule-governed strongly suggests that human beings are capable not only of extreme violence toward each other but also of constraining and ordering that violence toward appropriate and inappropriate ends. On the other hand, his explorations of human nature in these *Relectiones* are unsystematic, brief, and more often implied than explicit.

Vitoria certainly thought of himself as standing within the tradition of Thomism that he learned from his studies in Paris, so beginning with Thomas Aquinas makes sense. Famously, Thomas Aquinas describes human nature in the *Prima Pars* of his *Summa Theologica,* QQ 75–89, as a union of body and soul; part physical and part spiritual, human beings stand between animals (physical but not spiritual) and angels (spiritual but not physical): they are "metaphysical amphibians."[53] As embodied souls, human beings have intellectual and appetitive powers (including the will) that shape habits of rational thought and action that are properly oriented toward the pursuit of union with God. And, via their intellective natures, human beings carry the image but not the likeness of God, which distinguishes them from other animals.[54] It is this capacity for abstract and rational thought that so preoccupies Thomistic interpretations of what it means to be human.

As Vitoria turns to the treatment of indigenous persons, it is worth noting that he simply bypasses questions about whether to treat them as fully human. He assumes their humanity in order to explore questions about the kind of human beings they are. Are they like slaves? Simpletons? Children? Heretics? Are they intrinsically irrational? Are they intrinsically wicked? Are they, in their sinfulness, categorically different from Christians? Should there be affirmative answers to any of these questions, there would be clearer grounds for waging war against them. Yet Vitoria answers all these questions – at least tentatively – in the negative and, as he answers them, he relies on classic Thomistic language about the powers of reason and the image of God, the retention of spiritual power even after mortal sin, and so forth. Yet much as he

53 The phrase is Eleonore Stump's. See her "Non-Cartesian Substance Dualism and Materialism without Reductionism," *Faith and Philosophy* 12 (1995): 514.

54 For a much deeper exploration of Thomas' understanding of human nature, see Robert Pasnau, *Thomas Aquinas on Human Nature* (Cambridge, UK: Cambridge University Press, 2002).

retains the Thomistic emphasis on rationality for assessing the humanity of indigenous persons, Vitoria modifies it. For Vitoria, reason is not measured primarily by capacities for purely abstract thought so much as for practically abstract thought: Rationality reveals itself in the social consequences of what human beings can do:

> [T]hey are not of unsound mind, but have, according to their kind, the use of reason. This is clear, because there is a certain method in their affairs, for they have polities which are orderly arranged and they have definite marriage and magistrates, overlords, laws, and workshops, and a kind of religion. Further, they make no errors in matters which are self-evident to others; this is witness to their reason. Also, God and nature are not wanting in the supply of what is necessary in great measure for the race. Now, the most conspicuous feature of man is reason, and power is useless which is not reducible to action. Also, it is through no fault of theirs that these aborigines have for many centuries been outside the pale of salvation, in that they have been born in sin and void of baptism and the use of reason whereby to seek out the things needful of salvation. Accordingly, I for the most part attribute their seeming so unintelligent and stupid to a bad and barbarous upbringing, for even among ourselves we find many peasants who differ little from brutes.[55]

At least from the perspective of his Thomistic thought, Vitoria's claims about indigenous persons' capacity for the use of reason and, therein, their right to own and keep their own property is peculiar in at least two ways. First, his qualifier in that first sentence ("according to their own kind") suggests the pluralization of rationalities: Different fully human persons can be reasonable in ways that do not align with each other. On the one hand, this conclusion might follow from Thomas Aquinas' conclusion that there is not only one active intellect in all persons; that "there are as many active intellects as there are souls, which are multiplied according to the number of men."[56] On the other hand, Thomas Aquinas' argument in that article is against treating the intellect as a single thing that is shared by all people since such sharing would undermine his emphasis on the independence of embodied persons from each other. Vitoria's answer, though, suggests he is not concerned with separating individuals from each other in order to maintain the integrity of embodied souls so much as he is concerned with separating collections of people from each other in order to maintain the integrity of distinct cultures.

This is not to say that Vitoria relativizes all forms of reason; he is elsewhere quite clear that Spanish reasoning is superior to that of indigenous

[55] Vitoria, *De Indis*, 127–8.
[56] Thomas Aquinas, *Summa Theologica* I.79.5.

persons, if only because Spanish reasoning has greater capacity to see and understand the implications of the *ius gentium* for cross-cultural interactions. Yet even if Spanish reasoning is superior to that of indigenous persons of the Americas and West Indies, such superiority is one of degree, not kind. Vitoria does not explore the origins of these distinct rationalities; his pluralization of rationalities does, though, complicate not only a political question such as "on what grounds can we wage war against them?" but also an epistemological question such as "how can we understand what they are thinking if we don't reason the same way they do?" Frustratingly, Vitoria gives this question a pass by focusing on the other peculiarity of the claim above: that indigenous persons are clearly capable of reason because they shape and live in societies.

It is this emphasis on sociality as a mark of human rationality that is the most provocative aspect of Vitoria's theological anthropology. The contrast with Thomas Aquinas is striking. On the one hand, the evidence of rationality in human beings is, for Thomas Aquinas, adduced by attending to the way individuals are made: The unique combination of body and soul gives rise to capacities for freely understanding and pursuing the good. The evidence of rationality in human beings for Vitoria, on the other hand, is adduced by attending to the way societies are made: Families are created and maintained via marriage, polities are created and maintained via magistrates and laws, economies are created and maintained via systems of exchange, and religion is maintained via corporate practices of faith.

How does this matter? To begin with, social structures provide much clearer and apparent evidence of participation in the human race than does the capacity for contemplation. Moreover, it avoids the conceptual cul-de-sacs associated with questions such as whether someone who is an infant, mad, or mentally incapacitated still counts as fully human: Participation in social structures is evidence enough of one's humanity because it is evidence of one's participation in communities that can reason. More provocatively, though, it shapes a new kind of abstraction in which social structures (which can be dissociated from individuals) come to stand in for any independent and individual capacity for reason. The capacity for reason becomes something that inheres in groups as much as it does in the individuals that constitute groups.[57]

[57] To be clear, evidence of sociality may be a more apparent mark of human rationality than evidence of some form of abstract reasoning; it is not, though, a less ambiguous good than abstract reasoning. Social systems can be more or less inclusive, more or less violent, more or less oppressive, etc.

Implicit in this approach is an answer to the discomforting question with which Vitoria concluded *De Indis*: whether Spain might be justified in waging war against the indigenous persons of the Americas and West Indies for their own benefit because one might propose that "[a]lthough the aborigines in question are … not wholly unintelligent, yet they are little short of that condition, and so are unfit to found or administer a lawful State up to the standard requirement by human and civil claims."[58] Were such to be the case, Vitoria argues they would "have no proper laws nor magistrates, and are not even capable of controlling their family affairs; they are without any literature or arts, not only the liberal arts, but the mechanical arts also; they have no careful agriculture and no artisans; and they lack many other conveniences, yea necessaries, of human life."[59] In effect, Vitoria sets such a low bar for participation in human society that wars fought for paternalistic reasons are ruled out in practice even if not in theory.

More importantly, this connection between rationality and sociality builds upon and reinforces something implicit in Vitoria's emphasis on the *ius gentium*: Wars are waged between states and the capacity for those wars to be just turns on the degree to which a state, collectively (or at least a collection of representatives of the state), can discern the law of nations. These wars find their justification in the establishment of peace, order, and security – all social goods. For this reason, alone, rulers must turn to their counselors in order to reveal that their authority is legitimate. While it is highly doubtful that Vitoria had in mind a modern vision of the state (of the sort that would arise after the revolutions of France, England, and the United States, in which polities are structured in terms of representative democracies), Vitoria's turn toward sociality as a basic mark of human nature does initiate a trajectory that can lead in that direction. Said differently, Vitoria's approach to addressing the question of how to understand the other vis-à-vis oneself and one's own group begins to shape a vision of the state that will grow in prominence over the next few centuries.

None of this absolves Vitoria for his blindness to the treatment of indigenous persons from Africa,[60] or the way his thought was used to

58 Vitoria, *De Indis*, 161.

59 Ibid.

60 See Willie Jennings, *The Christian Imagination: Theology and the Origins of Race* (New Haven, CT: Yale University Press, 2011) and J. Kameron Carter, *Race: A Theological Account* (New York, NY: Oxford University Press, 2008) for provocative reflections on the creation of "race" as a concept that happens concurrent to the age of exploration and conquest and shapes modernity.

justify colonialist expansions of the Spanish Empire in the new world. Nor, it is worth noting, does all this make war rarer or even harder to justify on other traditional grounds. It does, though, mandate that those who would go to war against another nation must recognize that they go to war against other human beings. Practices of dehumanizing the enemy in order to make killing them easier – practices that are so endemic to war – are ruled out.

All that said, at least three other comments about Vitoria's treatment of questions about the other are worth noting here. First, given the power of Spain at the time, it is surprising how quickly and clearly Vitoria comes to the defense of the indigenous persons of the Americas and West Indies. His march through the long range of unjustifiable reasons for attacking them gains its authority in a rhetorical strategy that all but explicitly claims that his conclusions are so obvious as to be beyond argument. The strength of his argument is based on the full humanity of those indigenous persons where that humanity is revealed through processes that are (or at least should be) apparent to even casual observers of their cultures. Moreover, in basing his conclusions on their full humanity, Vitoria also trusts the power of a moral argument to resolve a political debate (the contrast with his contemporary, Machiavelli, is apparent here). Moral arguments carry political weight when they come from persons with public influence living in cultures where morality and politics are interrelated. And while any system that binds morality and politics to each other is justly open to criticism for using its moral claims to illicitly justify, rationalize, and/or otherwise give cover to immoral political acts, any system that denies such binds opens itself to crises of legitimacy that are both felt by its members and visible to outside parties. Vitoria's work may have provided an apologetic for conquest, but it is at least worth questioning what rhetorical alternatives were available in his day that also would have achieved so vigorous a defense of the full humanity of the indigenous persons of the Americas and West Indies as a way of addressing a just war question.

Second, orienting his just war ideas around the practical question of how to treat indigenous persons not only added complexity to those ideas, it also shaped the way he thought about the other. At one level, the non-hypothetical quality of the questions he faced helped him avoid treating them simply as abstractions: Their full humanity is demonstrated through their readily observable social configurations and, as full humans, they participate in the *ius gentium* that frames relationships between societies. There is, for instance, very little romanticizing of the indigenous persons about whom he wrote; to the degree that romanticizing other cultures is a means through which Western cultures force other cultures to fit within a Western cosmology – and often comes with the simultaneous degradation of those cultures – the absence of such romanticization is, somewhat paradoxically, to Vitoria's credit (and distinguishes him from, e.g., de las Casas). Indeed, if anything, Vitoria romanticized Spain and Spanish/Roman Catholic cultures: The absence of reflexivity in his writings suggests he never asked himself questions about whether Spain would agree to the same *ius gentium* derived international laws that he insisted societies of indigenous persons in the Americas and West Indies must accept.

At another level, though, the means through which Vitoria recognized the full humanity of those indigenous persons in order to answer just war questions helped to initiate a modern form of abstracting the other. The *ius gentium* into which Vitoria welcomed indigenous persons of the Americas and West Indies may have honored most intracultural distinctions (i.e., by generally allowing cultures to maintain their own political, economic, familial, and religious systems) but it collapsed cultural differences once multiple cultures engaged each other. Vitoria's law of nations reinforced the cultural imperialism that comes with thinking that a particular Western, Catholic social imaginary is universally normative (so, e.g., there must be the right to proselytize individuals and it must be possible to alienate property from persons through processes of trade and ownership). Participation in the human race and participation in Vitoria's vision of the *ius gentium* were synonymous. This kind of universalizing impulse that so marks the shift to modernity replaced earlier us–them/Greek–barbarian/Christian–heathen models of objectifying others but it hardly did away with the objectification of others. Vitoria's work helped change the direction of the conversations about how to think about and treat others, but it did not make those conversations any less problematic.[61]

Third, it is worth reemphasizing in a condensed way something that should be clear from, though perhaps diffused through, this chapter: Vitoria's treatment of these questions was both novel and imperfect.

[61] The next chapter will take up the place of refugees in modern thought. It is, though, worth hinting here that the understanding and status of refugees-as-others is complexly related to the very origins of modernity. For an especially thought-provoking exploration of this connection, see Nicholas Terpstra, *Religious Refugees in Early Modern Europe: An Alternative History of the Reformation* (New York, NY: Cambridge University Press, 2015).

It was novel – that is, Vitoria's questions took the tradition in a new direction – because Vitoria took the resources of his tradition (notably, Aristotle and Thomas Aquinas) and did not so much apply them as reshape them to fit the questions and concerns of his time. As the very questions Vitoria was asked to address reveal, the indigenous persons of the Americas and West Indies were seen as somehow different not only from Spaniards and other Europeans, but also from Moors, from persons of sub-Saharan lands, from Turks, and from those in the Orient. And it was imperfect not only because those resources only partially helped him address such questions and not only because he was an imperfect user of those resources, but also because traditions always change in imperfect ways. Not only is Vitoria's work full of blind spots, assumptions, and errors – not to mention a distastefully patronizing tone – but also the form of his work – public lectures on particular questions – actually brings those things closer to the surface than would a more systematic and abstract genre or approach.

Stepping back from these three final points (and the four particular insights and explorations of their implications that constitute the bulk of this chapter), Vitoria's writings help us see how the Christian just war tradition moves through time. Vitoria's writings on war are occasioned by a set of dramatic environmental, social, and technological changes: the beginning of the age of exploration that is, itself, the consequence of the need for new sources of natural resources; the advent of new technologies (in, e.g., cartography, navigation, shipbuilding); growing economic resources to fund exploration and the desire for resources that such exploration promises; changing ideas about the state and its relationship to the church. These changes pulled a particular and very practical question to the fore: How should Europeans understand and treat indigenous persons in the Americas and West Indies? Vitoria advanced the Christian just war tradition not primarily because of some unique individual capacity for insight or other greatness on his part but because he came along at the right time, was fortunate to have the right background, and was in the right location to speak to that question. And in so doing, he not only took the tradition in a new direction, but he also helped to initiate its move into the nascent area of international law in the process.

More generally, Vitoria serves as an example of how attention to the other shapes the movement of the Christian just war tradition through time. Others within the tradition (e.g., Augustine, Paul Ramsey) would provide further interesting examples of such attention to the other within the tradition as well. Moreover, just as Vitoria's work models the ambiguities, limitations, possibilities, and failures of the tradition as it moves through time, so exploration into the way other Christian just war thinkers engaged questions about the other would reveal such things as well. Such is always the case when traditions are as marked by discontinuity as by continuity.

So: How has the other come to be understood in modern forms of the Christian just war tradition?

THE DAMAGING TRIUMPHS OF POLITICS AND OBJECTIFICATION: OTHERING THE OTHER

When Clausewitz argued that "war is a mere continuation of policy by other means,"[62] he gave voice to one of the central ideas of modern warfare: that wars are fundamentally political events oriented around political goods (e.g., the acquisition and maintenance of power). In a political world dominated by the modern nation-state, this suggests that the most basic form of war – the Platonic ideal of war, as it were – is one between two or more states as those states seek to gain political advantage vis-à-vis each other in pursuit of their own power. Each state uses the resources that best situate it to maintain and pursue its political power. Large populations can mean large armies. Sophisticated technology can mean advanced weapons. Great wealth can mean buying armies and weapons. And since the maintenance of political power involves creating systems through which to order society and distribute responsibilities, modern wars between states presume the fungibility of resources, sophisticated systems of social control (e.g., bureaucracies, including in the military), and the tendency for resources to pool around particular nuclei of power. As Azar Gat has noted, during the early modern period,

> [a]s power and resource accumulation reinforced each other in a positive loop mechanism, massive social power structures emerged. Control over resources meant that a host of dependants were tied to the rich and powerful in an asymmetrical relationship. This, in turn, meant that for a highly differential share of

[62] Carl von Clausewitz, *On War*, trans. by Michael Eliot Howard and Peter Paret (Princeton, NJ: Princeton University Press, 1989), 88. The difference between "policy" and the more commonly used word "politics" is negligible in this instance; Clausewitz meant to connect war to the political realm. More interestingly, the phrase might be better translated as follows: "War is a mere continuation of policy with (*mit*) other means." This translation suggests not that the politics of war are different than other forms of politics but that war brings additional resources/means into politics.

> the benefits or for fear of sanction, or both, they supported the rich and powerful as the latter further increased their wealth and power by taking advantage of economic opportunity and through the exploitation and extortion of others ... State emergence was the culmination of this process, when a single power nucleus won against all others in the often-violent intrasocial competition, or else power nuclei joined to regulate the competition among them. Either way, one power structure established command over a population, institutionalizing power, driving other social power nuclei into subordination, and introducing hitherto unprecedented levels of hierarchical organization, coercion, systematic resource extraction, and force mobilization, while competing with neighboring state structures.[63]

The types of war that had dominated premodern Europe – those driven by tribal differences, ideological/religious loyalties, and, most notably, the control of land for agriculture – receded conceptually if not in practice. Tribal identity became national identity; ideological and religious loyalty were employed toward political ends; and land became just one more among a possible range of resources through which to accrue power (and not necessarily a good one, since having more land meant defending more land). Political structures grew in scale to pursue the acquisition of power in order to administer an increasing range of human activities that could be put in the service of political ends.

Where political power is the currency that serves and sustains the state, the means for acquiring and maintaining power multiply. Hereditary titles, ideological fervor, control of natural resources or of geographically strategic locations, control over the means of production or distribution of goods, the ability to lend significant sums of money, access to technological innovations, gossip, networks of informants, the ability to do favors, a talent at controlling otherwise restive populations, defensible borders, and even the capacity to provide strategic insights into the movements of political power in complex systems: All of these can be used for the pursuit of political power and in service to the control of the state.

Prioritizing political power makes the sources of power fungible. As the nation-state became the prototypical political structure in the modern period, its institutional complexity did not so much lead to the creation of new sources of power (all political structures rely on a range of sources of power) as make the fungibility of those sources more apparent and the convertibility of those sources more significant.

[63] Azar Gat, *War in Human Civilization* (Oxford: Oxford University Press, 2006), 401–402.

The fungibility of the sources of power in political systems that favored the multiplication of such sources mirrored other dramatic multiplications occurring at the end of the fifteenth and beginning of the sixteenth centuries. Without necessarily intending to, Luther and Calvin multiplied religious traditions in Europe, Vesalius and Copernicus multiplied scientific understandings there, mercantilism and the writings of Jean Bodin (not to mention the impending expansion of fiat money) multiplied the number of things that could be measurably valued and traded there, and Columbus and Cabral multiplied the number of continents known to Europeans. All these multiplications, in their own ways, destabilized the teleological clarity through which medieval Europeans made sense of their place in the universe. There is no need for a theory of the divine right of kings unless that right can be questioned, no need for the concept of the secular unless religious faith can involve choice, no need to defend a Ptolemaic conception of the universe until someone says the sun does not revolve around the earth, and no Machiavellian political theory unless political power can flow. The loss of a hegemonic teleological vision of the world increases not only the range but the significance of fungible sources of power. When it becomes less obvious that thought must conform to a preexistent order, the possibilities for multiple expressions of strategic thinking grow.

Christian just war thinking changed in ways consistent with these multiplications and the fungibility of power that came with this new vision of war as politics. It found itself dealing with new types of others, a self that could be coherently understood as separated from the world around it, and the expansionist impulses of new states. It turned toward the abstractions of a unifying theory about whether, when, and how to fight when traditions of thought on such matters splintered. It offered the guidance of law as a means by which to locate questions of war within the larger political purposes and projects of the state. It took on a secular veneer as the theologically framed teleology of the Middle Ages lost explanatory power. It sought to provide general moral guidance that both honored important values (e.g., the values of human life and just order) and increasingly treated those values as measurable against each other.

Along the way – and partly through the ministrations of Francisco de Vitoria and the Salamanca School – it reshaped the way it thought not only about particular others it encountered but about what constituted otherness in the first place. In the process, it contributed to shaping a project of alienation that continues to bear consequences to the present.

Three hundred years after Vitoria, Karl Marx would make much of the way that modern capitalism alienated workers from their labor, but the precondition of alienating workers from their labor is the possibility that people can be alienated from the world – and the possibility that people can be alienated from the world is a characteristic of modernity that was shaped, in part, by the transformations of Christian just war thinking in the sixteenth and seventeenth centuries. The multiplications of early modernity helped make such alienation possible: Human beings could exist as objects of reflection without necessarily fitting them into the larger cosmology that human beings merely inhabit. This change shaped the fertile ground in which notions of autonomy would grow. Though the Renaissance may have returned the focus of philosophical reflection to the human world and the Reformations may have shaped the focus of theological engagements with the divine around the priesthood of all believers, neither the Renaissance nor the Reformations, in themselves, mandated such an alienation. The multiplications of early modernity, though, push toward such an alienation.

This helps make sense of the peculiar and damaging place in which Vitoria conceptually located the indigenous persons of the Americas and West Indies: Although he may have recognized them as fully human, he never ceased to treat them as "other." The essence of alienation is the capacity to shape a perspective in which even those recognized as fully human still are not given a place in the same conceptual world as the one that recognizes them. Whether by his own intent or not, Vitoria could be used as an apologist for war against these indigenous persons because the only way they could avoid being "other" was to become European (i.e., maintaining the values, practices, and ways of understanding that shaped the conceptual world of the Spanish Empire) – and this was impossible.

The project of alienation also helps make sense of the peculiar and damaging absence of a place in Vitoria's reflections for persons indigenous to sub-Saharan Africa. Where people can be alienated from the world, they can also be removed from it by reducing them to commodities, becoming one more collection of fungible goods to be traded in the interests of accruing and maintaining power. Slavery had existed for millennia and throughout much of the world in a variety of forms (debt slaves, indentured servitude, war slaves, etc.). The early modern state, though, beginning with Spain and Portugal, prioritized race-based chattel slavery, transporting Africans to the new world when the indigenous persons of the Americas and West Indies proved more susceptible to the diseases that the Europeans brought with them than did the indigenous persons of sub-Saharan Africa. Societies have always created others (one thinks of the Greeks referring to all non-Greeks as barbarians because when they spoke, they said "Bar-bar-bar-bar"); modern European societies linked otherness to commodification through this project of alienation.

And this project of alienation helps make sense of the impulses not only to explore the new world but to colonize it. For the powerful European states, the new world was composed not of new territory to be explored so much as new resources to be collected and controlled. Even as people can be alienated from the world, the nonhuman parts of the world can be thought of as inert collections of goods to be acquired and traded because they need not be included among the list of qualities that make humans human. Early modern humans were not understood to be composed of the dust to which they would return but rather were seen as the kind of beings that could, uniquely, separate themselves conceptually from the rest of the natural world around them. The otherest other is anything that can be viewed as disposed of without making any necessary impact on the one disposing of it.[64] Little wonder that race-based chattel slavery grew in the new world in ways that it never did in Europe: Seeing the nonhuman world as composed only of natural resources and seeing others only as labor resources are of a piece.[65]

64 There is an extensive set of arguments sitting behind this paragraph having to do with the modern transformation of "nature" from the space in which humans exist into objects to subdue/control/monetize. On these arguments, one might look to Sigurd Bergmann, *Creation Set Free: The Spirit as Liberator of Nature* (Grand Rapids, MI: Eerdmans, 2005), esp. Part One and the various scholars that Bergmann addresses there. In his famous essay, "The Historical Roots of Our Ecological Crisis" (*Science* 155 [1967]: 1203–1207), Lynn White argued that Christianity was largely to blame for the conditions that led to ecological destruction in the twentieth century. I would argue that White is right only to the extent that one expression of Christianity in early modern thought played a role in shaping the conditions for modernity and modern impulses bear far more responsibility for environmental catastrophe. Francis Bacon bears far more responsibility than St. Paul for the messes in which we have found ourselves.

65 While it is beyond the scope of this project to explore, these arguments about the transformation of the "other" and the way others become resources/commodities in early modernity suggests further correlative (if not causal) links between some of the most deleterious consequences of modern thought including racism, misogyny, and environmental degradation. Modern expressions of racism and misogyny cast persons of color and women not only as less valued than white men, but also as more connected to the very natural/earthy world that has been disenchanted and commodified. For more on the relation between racism and environmental degradation, see, e.g., James H. Cone, "Whose Earth Is It Anyway?" *Cross Currents* 50.2 (Spring/Summer, 2000): 36–46; Emilie Townes, *In a Blaze of Glory: Womanist Spirituality as Social Witness* (Nashville: Abingdon, 1995); and Leonardo Boff, *Cry of the Earth, Cry of the Poor*

The transformation of war into a political project and the recognition of political resources as fungible helps to fund a view of the world in which the other must stay other and, as other, can be exchanged with anything else that is also other. Just war thinking always addresses questions about the other; early modern just war thinking answers those questions by connecting otherness to projects of commodification in a world driven by political strategies and fungible power.

ALIENATION AND THE SHAPING OF PROPERTY WITHIN THE EARLY MODERN JUST WAR TRADITION

Why highlight early modern answers to questions of the other at a time when these questions are, again, coming to the fore? Two answers are readily apparent. First, we continue to be shaped by modern perspectives. The strengths of modern expressions of the Christian just war tradition as it engages questions of the other continue to manifest themselves today. Modern just war thinking as theory is portable in its capacity to engage any other in any context. Modern just war thinking as international law has shaped processes through which to recognize and promote the inalienable rights of all persons. Modern just war thinking as engagement with the other has addressed the significance of cultural differences without losing sight of the possibility of constructing grander narratives through which to find commonalities across differences.

Likewise, the weaknesses of modern expressions of the Christian just war tradition as it engages questions of the other continue to manifest themselves today. Modern just war thinking as theory has shaped just war criteria into a set of rules, the application of which is inattentive to differentials in power and the histories of particular people. Modern just war thinking as international law not only assumes a legal sameness among all peoples but also privileges a particular way of thinking about people and systems favored by Western cultures. Modern just war thinking as engagement with the other makes it possible to treat at least some

(Maryknoll, NY: Orbis, 1997). Scholars of the Environmental Justice Movement, begun with the Memphis Garbage Collectors' Strike led by Martin Luther King, Jr., in the late 1960s and developed through the work of Hazel M. Johnson in Chicago in the 1970s, have been especially attentive to the connections between race and environment. For a much more extensive exploration of environmental racism and an extensive bibliography on the subject, see Laura Pulido, "Rethinking Environmental Racism: White Privilege and Urban Development in Southern California," *Annals of the Association of American Geographers* 90.1 (2000): 12–40.

others (whether human or nonhuman) as resources among a larger pool of resources. As Christian just war thinking changes to address wars in the Anthropocene, it will be better served if it understands where it has most recently come from than if it ignores its modern expressions or swings back, pendulum-like, against them.

Much has been made of the impact of Cartesian dualism on modern thought. A species of mind/body dualism that has haunted Western philosophy since Plato, Descartes' dualism was driven by his arguments about the dualism of substances: Mind and body are both substances (the former composed of an immaterial substance and the latter composed of a material substance) and the important questions have to do with whether and how these substances interact. Since things that are material but do not have mind (which, for Descartes, meant everything but human beings) can only interact in mechanistic ways with other things that have only material substance, Descartes effectively doubled his dualisms: mind/body and human/everything else. Any number of environmentalists trace environmental calamities to Cartesian dualism and the failings of human beings to adequately register their intrinsic connections with the world or to treat the world as, itself, somehow animated by a mind/spirit that is its own.

As significant as Cartesian dualism may have been in shaping Western thought, there were other forces at work in early modernity that manifest the same human/everything else dualism as the one in Cartesian thought. Among these has been early modern just war thinking and its treatment of others, which, as I have argued above, largely begins in the work of Francisco de Vitoria. Indeed, early modern just war thinking may have played a far larger role than it is given credit for. Early modern just war thinking about the other, after all, immediately manifested itself in political forms as European colonizers used its arguments repeatedly both against those being colonized and with regard to the resources that drove their colonizing projects in the first place.[66]

Among the most significant manifestations of these political forms was the delineation of private rights into reconstructed ideas about private property that would be carried from Europe to the Americas and West Indies, therein not only literally but metaphorically funding dominant patterns of colonization and conquest through which this human/everything else dualism triumphed. In his provocative and wide-ranging

[66] By extension, Descartes' substance dualism may have played a far less significant role than it is given credit for; I have consistently found that Cartesian thought plays a better role as exemplar of a particular modern idea than as a driver of modern thought.

essay "Empire and International Law: The Real Spanish Contribution," Martti Koskenniemi argues that

> ... the principle Spanish contribution [to international law] is not in those express arguments [having to do with Spanish treatment of indigenous persons] but in the development of a whole vocabulary that has since come to delineate the imperial dimensions of international law. At the heart of this vocabulary stand three notions already familiar from Roman law and medieval Christianity – *dominium, ius gentium,* and the *bellum iustum.* I would like to suggest that the most important Spanish contribution to the practice of empire lies in the recovery of those three notions and in giving them a meaning through which it was possible to react not only to the *duda Indiana* – the concerns of conscience raised by the Spanish activities in the Indies – but to three further transformations that took place in the period: the formation of centralized political communities – states – that demanded absolute loyalty from their citizens; the emergence of a global economic system based on private ownership and the search for profit; and continuous warfare, not only against the infidel, but among Christian rulers themselves. The notions of *dominium, ius gentium,* and the *bellum iustum* were now deployed to give a legal and moral articulation to the transformation of early modernity...[67]

Setting aside, for the moment, Koskenniemi's larger arguments about *dominium* (questions about which I will return to in Chapter 4) and *bellum iustum* (questions about which pervade this book), I here focus on the way that the Salamanca School and, most notably, Francisco de Vitoria shape a *ius gentium* that funds "the emergence of a global economic system based on private ownership and the search for profit."

Thomas Aquinas had argued that the natural law revealed that all persons were born free and all property should be held in common.[68] At the same time, though, Thomas Aquinas thought that the existence of private property not only stimulates industry (which can contribute to the common good) but also inhibits conflict (which also contributes to the common good). He therefore suggested that the basis for private property arose out of human reason as an addition (*adinventio*) to natural law that could be framed within the *ius gentium.* Since Vitoria makes the *ius gentium* do even more work than Thomas Aquinas had, it is an unsurprising, if evitable, move that Vitoria would emphasize private property. Conditions in Europe and the new world favored that very move.

[67] Martti Koskenniemi, "Empire and International Law: The Real Spanish Contribution," *University of Toronto Law Journal* 61.1 (2011): 11–12.

[68] "The possession of all things in common is ascribed to the natural law; not in the sense that the natural law dictates that all things should be possessed in common, and that nothing should be possessed as one's own; but in the sense that no division of possessions is made by the natural law." Thomas Aquinas, *Summa Theologica* 2.2, Question 66, Article 2.1.

So Vitoria's approach to questions about the ownership of goods in the Americas and the West Indies followed from Thomas Aquinas' justification within the *ius gentium* for things to be owned by individuals when the *ius natural* favored all things being held in common. It also followed from expanding notions of property law and systems of finance as they were being expressed in Europe and from the political reality of economic forces that both drove and funded exploration, conquest, and colonization. The "facts on the ground" with which Vitoria was dealing (e.g., that European financial markets turned on the possibility of people owning and trading things; that there had to be a difference between investing money for profit via the collection of interest and usury; that gold and silver from the Americas and West Indies drove the expansion of increasingly complicated financial systems in Europe; that defending the full humanity of indigenous persons on the basis of their sociality meant treating trade and commerce as fundamental human endeavors) had the effect of advancing a theory of *dominium* as a sphere in which economic freedom belongs, universally, to human beings through the *ius gentium* and, therefore, the defense of private property and the right to own things were tantamount to a defense of one's own humanity. Although Vitoria does not advance an argument about the right to go to war in defense of private property in the way that Hugo Grotius will a century later, the preconditions for Grotius' approach had already been framed by Vitoria's attempt to answer questions about the treatment of the indigenous persons in the Americas and West Indies. The nonhuman natural world had to be composed of stuff that could be bought, owned, sold, and traded because there was no other way, in Vitoria's thought, to make sense of how to be human. Vitoria and the Salamanca School reacted "to the expansion of trade and commerce by producing a legal and political vocabulary that takes that expansion as largely natural and articulates the social world created through it as a universal network of inter-individual relations."[69]

Vitoria's defense of the humanity of indigenous persons in the Americas and West Indies on the basis of a universal *ius gentium* grounded in human sociality had the peculiar, if unsurprising, effect, then, of reinforcing the very particular European visions of private goods, financial systems, and the right to take up arms to defend such things that therein exacerbated not only conflicts in the Americas and West Indies but also the reduction of all things in the nonhuman natural

[69] Koskenniemi, "Empire and International Law," 35.

world that Europeans found there to "resources."[70] Just as importantly, viewing the Americas and West Indies as a collection of resources during a time of increasingly globalized trade and the concomitant escalation of the power of markets through which to make sense of the world has meant that everything in the nonhuman natural world – and not just those things in the new world – could be viewed as "resources."

Vitoria rarely referenced the nonhuman natural world in *De Indis* (and even more rarely in *De Iure Belli*) and so it is hard to make strong claims about the way he viewed that world. However, on those rare occasions that he did, his references mesh with this strong division between human beings and everything else. So, for instance, he cites Genesis 1 to suggest that human dominion over animals is due to the fact that human beings are, uniquely, made in the image of God.[71] Then, later in *De Indis*, he writes that "wild beasts do not have dominion over themselves. Therefore much less over other things. The proof of this assumption is that they may be killed with impunity, even for pleasure; and so Aristotle (*Politics, I*) says that the chase of wild beasts is just and natural."[72] This position is a departure from Thomas Aquinas, who maintained the position that animals are beneath human beings and may be used but nevertheless abjured cruelty toward animals lest such cruelty lead to cruelty toward other human beings.[73] That Vitoria held animals in less regard

70 Koskenniemi notes that "Vitoria portrays trade and commerce as part of the 'natural partnership and communication' among humans. From the beginning of time, everyone was allowed to visit and travel through any land he wished [and t]his right was clearly not taken away by the division of property (*divisio rerum*). It was a practical consequence of this that all nations were to show hospitality to strangers and everybody had the right to 'all things that were not prohibited or to the harm or detriment of others.'" Koskenniemi, "Empire and International Law," 26. There is an irony here in that such rights were hardly visible in Europe even if they were being enforced through appeal to arms in the new world. Tzvetan Todorov has argued that the lack of indigenous persons' voices in describing the narratives of exploration, conquest, and colonialization reveals that Columbus "discovered America, but not the Americans." Tzvetan Todorov, *The Conquest of America: The Question of the Other* (New York, NY: Harper Collins, 1982), 49. Part of what Koskenniemi reveals in the way Vitoria thought about property and law in the new world is that Vitoria exported the property laws of Europe as if there were no Europeans.

71 Vitoria, *De Indis*, 121.

72 Ibid., 126.

73 "And if any passages of Holy Writ seem to forbid us to be cruel to dumb animals, for instance to kill a bird with its young; this is either to remove man's thoughts from being cruel to other men, and lest through being cruel to animals one becomes cruel to human beings; or because injury to an animal leads to the temporal hurt of man, either of the doer of the deed, or of another; or on account of some signification: Thus, the Apostle expounds the prohibition against *muzzling the ox that treadeth the corn.*" Thomas Aquinas, *The Summa Contra Gentiles* III.ii, 112. Trans. by the English Dominican Fathers from the latest Leonine edition (London: Burns Oates and Washbourne Ltd., 1923), vol. 4, p. 92.

than Thomas Aquinas did signals how little concern he had for them; if anything, he sounds more like Descartes than Thomas Aquinas – which may, itself, be a sign worth reading as further evidence of the way Vitoria shaped modern thought through alienation of the other.[74]

THE CHANGING IMPACT OF ALIENATION WITHIN THE JUST WAR TRADITION

Another thing we can learn from this history of the Christian just war tradition's transition into its early modern form as we now transition into the Anthropocene involves paying attention to how the tradition has changed in the past. Some insights in this regard are generic and applicable regardless of the changes it undergoes. As the tradition encounters the other, it will be shaped by larger forces (environmental changes, technological advances, political arrangements, unexpected events, philosophical and theological perspectives, social trends) that condition how the other is encountered. There will be unanticipated consequences to the arguments it makes and the answers it gives to questions about the other, and those consequences will reveal imperfections and a distinct location in history. Its longer pattern of including more people among those who must be valued and thinking about why they should be valued may continue, perhaps even reaching into the nonhuman natural world.

Other insights into how Christian just war thinking might change as we move into the Anthropocene are specific to its movement during this time. As the conception of war changes during this time, it may move away from Clausewitzian understandings of war, with their focus on politics and their

74 Manuel Jiménez Fonseca argues that Vitoria's understanding of nature, property, and colonization are all of a piece: that Vitoria's approach to dealing with the nonhuman natural world is oriented around the very types of European property law that were exported to the Americas and West Indies for the purposes of taking over resources there and conquering any indigenous persons who would attempt to thwart that project. Though I am less convinced that the theological arguments advanced by Vitoria and de las Casas were primarily intended as justifications for colonization and plunder than he is, my arguments in this chapter otherwise largely align with his. See Manuel Jiménez Fonseca, "The Colonization of American Nature and the Early Development of International Law," *Journal of the History of International Law* 11 ([illegible]): 109–115. On the relation between nature, property, and international law in Vitoria, see also Ileana Porras, "Appropriating Nature: Commerce, Property, and the Commodification of Nature in the Law of Nations," *Leiden Journal of International Law* 27 (2014): 641–660, who argues that "nature became visible to the early modern Europeans who contributed to a new tradition of *jus gentium* (the law of nations) only in the moment they articulated it as a material thing subject to appropriation, reducible to property, and capable of entering the stream of commerce. In other words, conceptually transforming nature into what we today would call a natural resource" (642).

processes shaped by the fungibility of power.[75] Wars in the Anthropocene look different. So, for example, as control of basic goods such as potable water and basic foodstuffs becomes more tenuous and capricious and access to these basic goods becomes more complicated, we might see more resource wars that could defy the fungibility of power by making some sources of power politically untradeable. As environmental crises defy political boundaries, we might see fewer interstate or even regional wars and more alignments among "most affected peoples" (e.g., those living at sea level; those who face the influx of climate refugees; those with limited access to the technologies of renewable energy) willing to use violence to pursue their goals.

And, more conceptually, as environmental metaphors displace economic ones as the means through which human beings make sense of the world around them, the tendency to treat people as alienated from the world and to treat both people and things as commodities might recede. In their place, we might find a vision of people as interdependent and the nonhuman natural world as so intrinsically valued that any costs it would bear as the result of war would need to be included in calculations about whether, when, and how to fight. For instance, where modern Christian just war thinking focused on *jus ad bellum* and *jus in bello* criteria, Christian just war thinking in the environmental age may need to further develop and even emphasize *jus ante bellum* and *jus post bellum* criteria. Why, for instance, wouldn't the same motivations that shaped earlier just war thinking (e.g., the pursuit of justice, the stability of political systems, the goal of peace) mandate global action – perhaps even extreme action – to care for the natural world in order to avoid, constrain, or mitigate violence? Might it be a military obligation to preserve glaciers, reduce greenhouse gasses, prevent environmentally costly drilling and mining operations, expand humanitarian operations after severe storms, or create sanctuaries to protect ecosystems during the sixth extinction? Such questions await.[76]

[75] John Mueller argues that we have already seen the end of conventional (Clausewitzian) war – though for quite different reasons – in *The Remnants of War* (Ithaca, NY: Cornell University Press, 2004).

[76] Indeed, such questions are already being asked within the burgeoning field of environmental security studies – a field of study to which I hope to return in the next volume of this series. See, e.g., Simon Dalby, *Security and Environmental Change* (Malden, MA: Polity, 2009); Harald Welzer, *Climate Wars: Why People Will Be Killed in the 21st Century* (Malden, MA: Polity, 2017); Christian Parenti, *Tropic of Chaos: Climate Change and the New Geography of Violence* (New York, NY: Nation Books, 2011); Michael T. Klare, *Resource Wars: The New Landscape of Global Conflict* (New York, NY: Owl Books, 2001); and Jay E. Austin and Carl E. Bruch, eds., *The Environmental Consequences of War: Legal, Economic, and Scientific Perspectives* (Cambridge, UK: Cambridge University Press, 2000).

At very least, the history of Christian just war thinking reveals that questions about the other – including who and what is valued and how much – will be reasked as we move further into the Anthropocene. How those questions are answered will reveal as much about the forces that are shaping how we move into the new age as the answers that were given to such questions reveal about how we moved through past ones. Having Vitoria available as example will help.

Later in this book, I will return to Vitoria and the insights of this chapter as well as the way the Christian just war tradition addresses questions of the other more generally with an eye to the way such questions intersect with – and will increasingly interact with – the natural world. Before making the turn to the natural world, though, I need to do more with the relation between theological anthropology and the Christian just war tradition. Questions about how to think of the other are not the only questions of theological anthropology, after all. Such questions about the nature of the other always carry implications for how we think about ourselves (and vice versa). A century after Vitoria, another of the great figures of the Christian just war tradition arose to reshape that tradition into modern terms; in his writings, he revealed a preoccupation with that very question. His name was Hugo Grotius.

2

Understanding the Self

Hugo Grotius and the Birth of the Secular

> The second book, having for its object to set forth all the causes from which war can arise, undertakes to explain fully what things are held in common, what may be owned in severalty; what rights persons have over persons, what obligation arises from ownership; what is the rule governing royal successions; what right is established by a pact or a contract; what is the force of treaties; what of an oath private or public, and how it is necessary to interpret these ...
>
> Hugo Grotius, "Natural Laws and the Laws of War"

> As against earlier contract theories, the one we find with Grotius and Locke starts from the individual. Being in a political order to which one owes allegiance presupposes, on this view, that one has given it one's consent.
>
> Charles Taylor, *Sources of the Self*

> There is a way of reading the past that allows it to answer back, to criticize the present, to enhance our freedom. By doing justice to the otherness of the past, we discover that familiar habits of thought are not universal, and if they are not universal, they are not necessary either. They are subject to change.
>
> Jeffrey Stout, "Justice and Resort to War"

There is no greater exemplification of the impact that the Christian just war tradition had on the world as it moved from its late medieval to its early modern forms than its birthing of modern international law. The splintering of political authorities, reconceptualization of political identities, and expansion of political geographies in the West mandated new ways of thinking about relations between states. Early modern jurists, trying to get their footing in this new world, turned to the Christian just war tradition – including, notably, Francisco de Vitoria – for resources to shape that thinking. Leading their way was Hugo Grotius.

Grotius, the author of the first great modern texts on international law, was, understandably, concerned about conflicts between states during times of upheaval. His work *De Jure Belli ac Pacis* (*On the Rights of War and Peace*) not only helped found modern international law, but it also funded trajectories of thought for matters having to do with law and conflict that carry influence to this day. That work also reveals blind spots in Grotius' thought – including those having to do with the movement of peoples – that grow increasingly onerous in the twenty-first century.

Born not quite a century after and fewer than 1,000 miles from the birthplace of Francisco de Vitoria in Western Europe, Hugo Grotius would seem, nevertheless, to be the product of a different world than the one that shaped Vitoria. While Vitoria grew up in a world in which Roman Catholicism exercised a near monopoly within Christianity, Grotius grew up in a world where various flavors of Protestantism were not only live options for Christians, but participation in intra-Protestant debates carried significant ecclesial and secular implications (including, for Grotius, imprisonment for being on the losing side of one of those debates). While Vitoria was trained in medieval theology through medieval/scholastic systems of education in Paris, Grotius was trained as a Protestant humanist at the University of Leiden – a school that was not even founded until after Vitoria's death and that was created by William, Prince of Orange, as part of his Protestant revolt against the Spanish Empire. While Vitoria's work presumes European stability and so focuses on conflicts between Europeans and indigenous persons in the Americas and West Indies, Grotius focuses on intra-European ecclesial, national, and pan-European strife.[1] In the eventful hundred-year gap between Vitoria and Grotius one finds, among other events, the growth of Protestantism and the European wars of religion, the birth of modern nation-states, and the dramatic expansion of colonialism.

A STARTING POINT FOR LOCATING GROTIUS: THE OVERLAPS OF VITORIA AND GROTIUS

As different as their ages seem to have been, the overlaps between the two are noteworthy and suggest that they both participated in the

[1] Indeed, where Vitoria wrote his *De Indis* in defense of the rights of indigenous persons in the Americas and West Indies from attack by Spain, Grotius wrote his own *De Indis* (which was not published until 1868 though the ideas within this early book of Grotius serve as the basis for much of his later thought) in defense of the rights of the Dutch to retain cargo seized from Spanish ships returning from the new world.

beginnings of a new social imaginary. As a starting point, note that they both wrote from a natural law perspective (and Vitoria's thought influenced Grotius' writings), and they both appealed to an understanding of human beings as rational and social to justify their respective turns to natural law. This is not to say that their turns to natural law are identical; there are differences in their natural law approaches that are the result not only of their training but also of their ecclesial commitments. So, for instance, whereas Vitoria wrote in broad continuity with the Aristotelian/Thomist vision of natural law, Grotius' version of natural law dispenses with the metaphysical infrastructure of the Aristotelian/Thomist approach, including its intrinsic relatedness to a theory of virtue and its teleological assumptions.[2] For Grotius, the natural law is evident through rationality alone and would remain in place *etsi Deus non daretur* – "even if God did not exist."[3] Too much, though, has been made of a religious/secular distinction between Vitoria and Grotius. Though Vitoria's religious language is closer to the surface and more obligatory than is Grotius', *The Rights of War and Peace* is suffused with religious perspectives and language. The changes in natural law thinking in the period between Vitoria to Grotius are those of degree, not category.[4]

Note, also, that both Vitoria and Grotius used their natural law perspectives to birth the area of study that would become international law and both wrote international law as a way to address significant sites of violence: Vitoria wrote to respond to the violence of colonization and Grotius wrote to respond to the violence of the wars of religion in Europe. As abstract as their writings on international law could be, both are clearly responding to events of their times in those writings. Indeed, their writings would serve as bases for adjudicating very concrete moral, political, legal, and martial questions raised by those events: Vitoria's writings shaped the New Laws and Grotius' shaped the Peace of Westphalia. Again, the difference between the two writers is one of degree, not kind. The sort of thinking that shaped the New Laws (e.g., recognition of the rights of others, concerns about the stability of

[2] The shift away from virtue ethics is noteworthy and will be picked up again in greater detail in Chapter 5.

[3] The phrase is closely associated with Grotius' work *De jure belli ac pacis*.

[4] Vitoria and Grotius are not the only just war thinkers of their age(s) to write from the perspective of natural law. Beyond Suárez and Gentili (perhaps the best known of the thinkers in the period between the two), there are also Matthew Sutcliffe, William Fulbecke, and William Ames – and after Grotius, there are Locke and Vattel. On these figures, see James Turner Johnson, *Ideology, Reason, and the Limitation of War: Religious and Secular Concepts, 1200–1740* (Princeton, NJ: Princeton University Press, 1975), 150ff.

the current system, admission that conflict inhibits well-being and economic growth, and, above all, opposition to religious causes for war) are amplified in Westphalia, not new to it.

Note, moreover, that both Vitoria and Grotius birthed international law as a way of describing and defending a legitimate authority for addressing interstate conflicts during a period when the established authorities of the Middle Ages were falling away. During Grotius' life especially, the sway of both emperor and pope was much diminished at the nascence of the modern nation-state; Protestantism, the humanist turn, and Cartesian thought followed from and reshaped literature and philosophy; new technologies were leading to the diffusion and localization of economic, political, and social power; the rapid expansion of the "known world" was introducing Europeans to cultures and systems of authority quite different from their own; even the shape of war was changing.[5] Faced with the challenges of finding stable authorities through which to establish and justify moral codes and practices, Vitoria and Grotius both found promise in the impartiality and universality of natural law.

Note, finally, that both Vitoria and Grotius simultaneously used and transformed the resources of Christian just war thinking to address the new sites and causes of violence arising out of the larger transition from one social imaginary into another. Toward that end, they took up the tradition's fundamental questions about self, state, and other and responded to those questions with new ways of thinking about self, state, and other. As seen in the last chapter, Vitoria began with questions about how to understand and engage a new "other" but, in the

[5] As Johnson notes,

> certain fundamental changes in the face of war … appeared early in the modern period and marked war as Grotius knew it. These changes rank alongside the dissolution of a common religious bond as contributors to a new historical context to which the ideas of justification and restraint in war needed to be adapted. There were three such fundamental changes. First was the use of religious-ideological justifications for intra-cultural wars across European peoples: previously such reasoning had been used only across major cultural lines, notably in the wars with Islamic powers. Second was a change in the character of armies as they grew larger and were increasingly made up of common men, not the professional warriors of the knightly class. (The further expansion of warfare to include huge armies fed by nationwide conscription would not come about for another century after Grotius.) The third change was associated with this new kind of army: the replacement of an internalized chivalric morality by an externally imposed discipline as the form of restraint on the behavior of individual soldiers.

Johnson's summary is incisive, as far as it goes. He spends very little time, though, examining the forces that shape these changes. See James Turner Johnson, *Ethics and the Use of Force: Just War in Historical Perspective* (Burlington, VT: Ashgate, 2011), 77.

process of doing so, linked questions about the treatment of these others to questions about how to understand self and state. Grotius, meanwhile, began with questions about the self, recognizing that answers to those questions connect to questions about the state and the other at the start of modernity.[6]

These points of overlap between Vitoria and Grotius also offer genealogical resources for those who would understand the transformation of Christian just war thinking into international law. While that transformation was incomplete (Christian just war thinking stays both Christian and just war-oriented even if the tradition is largely overshadowed by international law), it is nevertheless the case that the transformation helped bring about the shift in just war thinking from tradition to theory and, in the process, into a mode amenable to the secular age.[7] As different as they may seem at first blush, the continuities between Vitoria and Grotius are notable not only because they undermine dualistic narratives about the period (Catholic vs. Protestant; religious vs. secular; medieval vs. modern) but also because they point to the complicated processes through which societies shift from one social imaginary into the next – processes that those dualistic narratives only partially express by focusing on one religious, cultural, economic, technological, or political force to the exclusion of others.

[6] While the term "modernity" is so encompassing and used in so many different ways as to inhibit an agreed-upon starting point, one possible starting point, as suggested in the first chapter of this book, would be the publication in 1641 of Rene Descartes' *Meditations on First Philosophy*. While this date comes near the end of Grotius' life, the social changes and forces that helped give rise to that text are also apparent in Grotius' work during the preceding three decades.

[7] David D. Corey and J. Daryl Charles have also made the argument that the just war tradition became a theory and, in the process, lost not only the range of voices that constitute a theory but also its malleability in addressing conflicts. I worry that they simplify both "tradition" and "theory" in the process and that their argument doesn't adequately locate that transition in history, though these concerns could be the product of their choices about the audience they are trying to reach. See David D. Corey and J. Daryl Charles, *The Just War Tradition: An Introduction* (Wilmington, DE: ISI Books, 2012). Chris Brown makes an even more provocative claim about the consequences of the just war tradition becoming a theory. He wishes to "defend the position that while the categories associated with just war *thinking* may help us to exercise judgment in particular cases, we should avoid just war *theorizing* altogether … [since] those who look to just war theory expect to be given *answers*, those who prefer to talk of just war thinking hope to discover good *questions*." See Chris Brown, "Just War and Political Judgment," in Anthony F. Lang, Jr., Cian O'Driscoll, and John Williams, eds., *Just War: Authority, Tradition, and Practices* (Washington, DC: Georgetown University Press, 2013), 35–48. As much as I find consonance with Brown's language here (his concern about theory, his emphasis on questions), I worry that his conceptions of both are reductive.

Though much scholarship in the history of just war thinking and international law has focused on Grotius' approach to the use of natural law and the implications of that use, of deeper interest are the causes for that approach.[8] In this chapter, I will argue that Grotius' work displays the consequences of a range of forces operative at the time that shaped revised understandings of the self and that those revised understandings not only express themselves in his writings but drove revisions to the Christian just war tradition.[9]

THE RICH LIFE OF HUGO GROTIUS

Hugo Grotius (or Huig de Groot) was born on Easter Day in Delft in 1583, the oldest child of Jan de Groot and Alida van Overschie. At the time, the Dutch Republic was in its second decade of a war intended to free it of Hapsburg rule, and the Republic's economic power – particularly on the seas – was growing rapidly. Trained in humanist and Aristotelian thought by Protestant (Calvinist) parents during his childhood, Grotius' prodigious intellectual abilities soon revealed themselves: He entered Leiden University at eleven; accompanied leading Dutch politician Johan van Oldenbarnevelt to Paris on a diplomatic mission at fifteen;[10] published his first book, on the work of Martianus Capella, at sixteen; was appointed as an advocate to The Hague at about that same time; and

[8] So, e.g.,

> [Grotius'] crucial contribution is usually perceived in his specifically modern conception of natural law, freed from the theological shackles of the scholastics who had lost their credit with the breakdown of medieval Christendom; this offered a neutral ground on which the struggling confessional parties could agree and find their way back to peace in the midst of the Thirty Years' War: Grotius' system was thus to serve as a theoretical blueprint for the so-called Westphalian order supposed to have been set up soon after his demise. Yet this piously nurtured foundational myth, which smacks of 19th-century hero worship and anticlericalism, is hardly borne out by the text of Grotius' masterpiece [*De iure belli ac pacis*]. Neither does it propound a new "secular" conception of natural law, nor is it really a system of international law.

Peter Haggenmacher, "Hugo Grotius," in *The Oxford Handbook of the History of International Law*, ed. Bardo Fassbender and Anne Peters (Oxford: Oxford University Press, 2012), 1098–1099.

[9] I should note early that in this chapter, at least, I have no interest in taking sides in a contemporary debate about whether just war theory or international law should have priority in adjudicating contemporary conflicts (indeed, from the perspective I am developing here, I think the whole debate is misplaced). For more on this debate, see Alex J. Bellamy, "When Is it Right to Fight? International Law and *Jus Ad Bellum*," *Journal of Military Ethics* 8.3 (2009): 231–245.

[10] Where King Henry IV referred to him as "The Miracle of Holland."

was selected as the official Latin historiographer of the United Provinces (over the full professor Domenicus Baudius) at eighteen.

As an adult, Grotius continued the ambitious and intellectually prodigious path he had begun in childhood. He worked as a lawyer, politician, diplomat, historian, philologist, classicist, poet, and theologian. His passions in these areas are visible in his writings and his life. As the Pensionary of Rotterdam (akin to its governor), his pursuit of ecclesial moderation in the Netherlands led him to write both *Ordinum Hollandiae ac Westfrisiae pietas* (1613) and *Resolution for Peace in the Church* (1614), thereby incurring the wrath of Prince Maurice of Nassau. These works offered Arminian responses to conflicts with harder-core Calvinists (including the prince), and the prince's wrath at those works, especially when linked to Grotius' long friendship with the prince's opponent Oldenbarnevelt, led to Grotius' imprisonment in the castle at Loevestein. After escaping prison hidden in a box of books (an escape engineered by his wife, Maria, and her maidservant),[11] Grotius settled in Paris, where he would expand a book begun in prison, *De Veritate Religionis Christianae* (published in Dutch in 1622 and Latin in 1627) and write *De Jure Belli ac Pacis* (1625).

His peripatetic life continued after Paris: He returned to Holland in 1631 after the death of Prince Maurice though was then almost immediately banished. From there, he spent time in Hamburg before returning to Paris in 1635 as the ambassador of Queen Christina of Sweden after King Gustavus Adolphus expressed admiration for *De Jure Belli ac Pacis*. While in Paris, Grotius' work as ambassador focused on pursuing reconciliation between Catholics and Protestants during politically fraught times, including through writing *Via ad Pacem Ecclesiasticam* and *Votum pro Pace Ecclesiastica* on the unity of the church as revealed in Scripture. In 1645, he was recalled to Sweden for relief and the royal family offered him and his family a place to live. Grotius chose, instead, to return to France. His ship foundered off the coast of Germany and Grotius became gravely ill as a result. He died at Rostock on August 28, 1645, and his body was returned to Delft to be buried shortly thereafter.[12]

[11] Charles Butler, a nineteenth-century biographer of Grotius, describes this escape and, in an especially felicitous turn of paraphrase from Cicero, writes that, especially in this instance, "[l]iterature is an ornament in prosperity and a comfort in adversity." Charles Butler, *The Life of Hugo Grotius with Brief Minutes of the Civil, Ecclesiastical, and Literary History of the Netherlands* (London: John Murray, 1826), 140.

[12] Writings about Grotius' life and work are voluminous. For that matter, so are the writings he left us. Probably the best (English) biography is Henk Nellen, *Hugo Grotius: A Lifelong Struggle for Peace in Church and State, 1583–1645*, trans. J. C. Grayson (Leiden: Brill, 2015).

While he wrote widely and engaged in an epic life, Grotius would be best known for two things: among Protestants, he would be known for his defense of Arminianism against Calvinism and his pursuit of unity across Roman Catholic/Protestant divisions; within the wider world, he would be known for *De Jure Belli ac Pacis*, the first great text of international law and one of the great texts within the canon of Christian just war thinking. Before diving more deeply into *De Jure Belli ac Pacis*, it is worth noting that, for Grotius, these two foci were connected. Both are responses to violence and reveal Grotius' commitment to humanism. Both so mix religious and secular language as to reveal the lie that those two terms are antonyms. Both are premised on the ability of human beings to make wise and moral decisions in light of thoughtful observations about the world around them and recognition of their connections to others. Both struggle with questions about the appropriate authorities in which to ground such wisdom. Both are products of a period of transition from one social imaginary to another and show the seams of that transition, including in their ability to be used to justify the very types of behavior (violence, ethnocentrism, the hollowing out of religious perspectives) they nominally oppose. And both are the products of a changing conception of the self that began arising in the sixteenth century. Indeed, whatever else his defenses of Arminianism reveal, they also display the thought of a scholar confident in the capacities of individuals to think and act on their own.

THE SHAPE OF *DE JURE BELLI AC PACIS*

The breadth of *De Jure Belli ac Pacis* (hereafter *De Jure Belli*) is evident in its subtitle: "The Law of Nature and Nations and the Principle Points Relating to Government." Dedicated to "His Most Christian Majesty Lewis XIII, King of France and Navarre," *De Jure Belli* begins with a preliminary discourse in which Grotius offers a brief philosophical foundation for the body of the text, primarily through interaction with a wide range of classical philosophers and poets. In that discourse, he argues for the priorities of right to might and of reason to force, suggesting that such priorities are evident not only through attention to the wisdom of great thinkers and observation of the world around him but the conviction that "God, as being our Creator, and to whom we owe our Being, and all that we have, ought to be obeyed by us in all Things without Exception, especially since he has so many Ways shewn his infinite Goodness and Almighty Power"[13] – which set of claims may feel especially audacious when one thinks of the context of religious strife and national upheaval in which they were written. The human capacity to reason together that arises out of these sources shapes societies and the laws that govern them to the benefit of those who participate in them: Justice is a good intrinsic to human society and necessary for the flourishing of individuals within society as well as for states themselves and their relations with each other (so justice has to do with human flourishing rather than, for example, fear of punishment, the presence of wealth, or the exertion of brute force).

Yet in a world filled with persons with such capacities, the very existence of war is a problem, as, especially, are some of the causes for which wars are fought. As Grotius notes,

> I observed throughout the Christian World a Licentiousness in regard to War, which even barbarous Nations ought to be ashamed of: a Running to Arms upon very frivolous or rather no Occasions; which being taken up, there remained no longer any Reverence for Right, either Divine or Human, just as if from that Time Men were authorized and firmly resolved to commit all manner of Crimes without Restraint.[14]

This tension between the human capacity to use reason to shape just and peaceful states and the human willingness to go to war for frivolous reasons sits at the heart of Grotius' reflections in *De Jure Belli* and motivates his search for a way of thinking about war and peace that can be grounded in human rationality and sociality. Can war be a rule-governed activity? If so, what rules apply to questions about when to go to war and how to fight? How are these rules revealed through natural law? How does a particular tradition of thought, rich in its own history and shaped by the distinctive voices of that tradition, express itself in universal terms such that those not a part of the tradition will still recognize its wisdom and adhere to the laws it produces? And what is revealed

[13] Hugo Grotius, *The Rights of War and Peace* (3 volumes), edited and with an introduction by Richard Tuck from the edition by Jean Barbeyrac in the Natural Law and Enlightenment Classics series, Knud Haakonssen, general editor (Indianapolis: Liberty Fund, Inc., 2005), 89–90. This text maintains the awkwardness of peculiar capitalizations throughout. Partly because I will not be walking through the text in detail, I will maintain its format, awkward as it may be. Typically, *The Rights of War and Peace* is cited by book, chapter, section, paragraph (e.g., "1.2.1.2"); I will follow this method (using the edition named here) after concluding this paragraph on the Preliminary Discourse.

[14] Grotius, *The Rights of War and Peace*, 106.

about the power, purpose, and structures of states and their interactions with each other via these rules?

The three volumes that compose *De Jure Belli* can be read as Grotius' answer to that first question: "Can war be a rule-governed activity?" (which is, for him, tantamount to asking whether war can be just) and an exploration of the implications of his answer. That Grotius' lifelong and preoccupying quest for peace and unity are evident throughout a text about war signals both the significance of his quest and the realism with which he approached it.

Toward answering the question, Grotius began by recognizing the range of opinions that have been given to it and its related questions (including those by near-contemporaries such as the pacifist Erasmus and the just warrior Vitoria as well as poets and philosophers of the classical period, histories, the Old and New Testaments, and many in between). Given the range of resources to which he refers in the preliminary discourse, it is noteworthy that his

> first Care was, to refer the Proofs of those Things that belong to the Law of Nature to some such certain Notions, as none can deny, without doing Violence to his Judgment. For the Principles of that Law, if you rightly consider, are manifest and self-evident, almost after the same Manner as those Things are that we perceive with our outward Senses, which do not deceive us ...[15]

As familiar as he was with a wide range of theological, historical, legal, and literary sources upon which he might ground his arguments, his primary authority would be the "manifest and self-evident" laws of nature. His focus on these self-evident and unchanging natural laws is the reason for his oft-quoted claim that these rules would apply even if there was no God (or if God did not care about such things). To be clear, however, Grotius was quick to reject either of those possibilities; his concern was not whether there was a God who cared (he was sure there was), but how that God related to the world via the natural law.[16]

In Book One, Grotius explores answers to the question of whether war can be just by attending to answers that had been given from

[15] Ibid., 110–111.

[16] "And indeed, all we have now said would take place, though we should even grant, what without the greatest Wickedness cannot be granted, that there is no God, or that he takes no Care of human Affairs. The contrary of which appearing to us, partly from Reason, partly from a perpetual Tradition, which many Arguments and Miracles, attested by all Ages, fully confirm; it hence follows that God, as being our Creator, and to whom we owe our Being, and all that we have, ought to be obeyed by us in all Things ...": Grotius, *The Rights of War and Peace*, 89.

antiquity forward. Yet rather than beginning in history or Scripture, he begins with definitions and descriptions of, for example, "war," "right," "justice," and the "law of nature," as well as descriptions of the possible sources of laws, of human rights, and their limits. Giving explicit definitions of terms from the beginning is generally helpful but was also doubly necessary as Grotius advanced ideas that he intended to carry weight regardless of the traditions and perspectives of his readers. From there, he outlines arguments for war as a rule-governed (and therefore potentially just) activity, the implications of which clarify who can engage in war and the limits of subjects to make laws against their sovereigns. While human beings have rights (to property, to be repaid outstanding debts or be compensated for damages, to punish those who transgress natural law), these rights do not, for the most part, extend to a right to rebel or to engage in private wars.

The "for the most part" of that previous sentence, though, is significant. Given (a) a vision of natural law shaped by the human capacity for reasoning and sociality; (b) a theory of sovereignty that simultaneously recognizes the authority of the state over its people and claims that no stable state could exist in which its population uniformly disagrees with the way that the state's leaders have claimed authority; and (c) the reality of many forms of governance operative in Europe and around the world, Grotius recognizes "there are several Ways of Living, some better than others, and every one may chuse which he pleases of all these Sorts; so a People may chuse what Form of Government they please: Neither is the Right which the Sovereign has over his Subjects to be measured by this or that Form, of which divers Men have diverse Opinions, but by the Extent of the Will of those who conferred it upon him."[17] Book One, then, simultaneously settles some questions ("Can a war be just?") and names new questions. Given the capacity for human beings to rationally discern the natural law, the (limited) reciprocity between the rights of persons and the rights of the sovereign state, and the range of forms of governance that follow, what are legitimate causes for war and how will conflicts be adjudicated?

Book Two takes up those questions. It begins with arguments about just causes, plowing territory familiar to the tradition (self-defense, to punish offenses, in defense of life, and – more interestingly – the recovery of property, about which I will say more later), though Grotius orders those just causes in a way that is more coherent with his vision of

[17] Ibid., 1.3.8.1.

natural rights, including property rights, than a classically Augustinian vision ordered them. It ends with chapters on unjust causes for war (fear, desire, religion), dubious causes for war, and justifications for soldiers to engage in acts of war when commanded to by superiors. Between beginning and end, the book is wide-ranging and not always focused on war. Because his understanding of sovereignty blurs the boundary between subjects and the state (with the result being that the book weaves back and forth between rules about war between states and rules about civil governance within states), and because even the adjudication of just wars can be complicated by context (e.g., what to do about conflicts on the seas when they cannot be "owned"), Book Two serves as the primary basis for the widespread claims that Grotius is the father of international law and that *The Rights of War and Peace* is the first great text of modern international law.

What makes this work different from, for example, Vitoria's *Relectiones* or Alberico Gentili's *Law of War* is the systematic and thorough process through which Grotius links answers to one set of questions of law to answers to other sets of questions in light of the changing understandings of human beings and the state that are arising in his day. Although it is hard to find precedent in the Christian just war tradition for adjudicating questions about, for example, the disposition of property at the death of someone without heirs, sales contracts, or rights to burial, Grotius applies the same reasoning processes to such matters that he has applied to questions about war and peace. When just warriors such as James Turner Johnson claim that international law arose out of the Christian just war tradition, this is what they mean. What Grotius wrote about war in Book Two, "*ubi iudicium deinit, ibi bellum incipit*" ("For where the Methods of Justice cease, War begins")[18] apply transitively across the range of issues Grotius raises in Book Two: Just processes prevent conflict and unjust processes (or the absence of processes) lead to conflict.

Book Three revolves around *jus in bello* questions. For war to be a rule-governed activity in a world with multiple legitimate sociopolitical forms, Grotius admits a wide range of behaviors (including plunder and enslavement) while also restricting some forms of behavior because they undermine the possibilities of rule maintenance (e.g., violating oaths, taking property that cannot legally be claimed in war). It follows that Book Three reads especially harshly: The range of acceptable behavior is wide, the punishment for violating rules severe, and Grotius' treatment

[18] Ibid.,1.2.1.2.

of punitive wars normalizes them. While some have argued that Book Three reads as it does because Grotius was trying to maintain a sense of neutrality in engaging the norms of other cultures, a more likely reading is that his writing really does represent the biases of his age and sometimes the pursuit of "neutrality" manifests as bias. The book closes with an exhortation to preserve faith and pursue peace that goes out not only to rulers and lawyers, but to everyone:[19]

> May the ALMIGHTY then (who alone can do it) impress these Maxims on the Hearts of Christian Powers; may he enlighten their Minds with the Knowledge of every *Right*, Divine and Human, and inspire them with the constant and dutiful *Sense* of their being the *Ministers of Heaven*, ordained to govern *Men*; *Men*, for whom, of all his Creatures, *GOD* has the greatest *Regard and Affection*.[20]

Whether the preservation of faith and the pursuit of peace are more likely to come out of wars fought according to the rules Grotius lays out in *The Rights of War and Peace* is an important, if larger, question than can be dealt with here.

Collectively, the books do not so much initiate international law as, for the first time, read like international law; Vitoria was, after all, advancing a kind of international law. Not only is the tone juristic but Grotius' consistent neutrality reveals itself in a kind of universality if not abstractness. Religion, for instance, plays a role throughout – in the sources he relies on, in his own thought processes, in the way he orders the books – but primarily as the basis for the type of common denominator morality to be expected of a natural law approach in an obviously pluralistic world. As a result, the particularities of religious expression that have arisen in discrete faith communities dealing with violence are smoothed over if not rejected. Sovereignty expresses itself at the level of the state and the individual; the tension between those two expressions of sovereignty is never worked out and the common denominator morality actually undermines any attempt to reconcile the two forms of sovereignty because it prevents Grotius or anyone following in his steps from appealing to a higher authority from which the two expressions of sovereignty can be derived – a problem with which Protestant natural lawyers, and later political theologians,[21] have struggled. The books are full of quotes and commentary on earlier texts

[19] Well: Men, at least.

[20] Grotius, *The Rights of War and Peace,* 3.25.8.1.

[21] See, e.g., the attention given to Ernst Kantorowicz, *The King's Two Bodies: A Study in Medieval Political Philosophy* (Princeton, NJ: Princeton University Press, 1957), a text to which I will return in Chapter 4.

but the earlier sources do not so much determine Grotius' arguments as provide support for them: The power of the sources lies in their synthetic connections more than their unique insights.[22]

Barrels of ink have been spilled on Grotius' life and works, perhaps especially on *The Rights of War and Peace*. Even the briefest summary of these assessments of Grotius' work would expand this chapter beyond all viable reckoning. Here, I want to focus on one particular aspect of that work; namely, the description of the self as agent in the world that underlies his just war writings. This description not only provides an important lens into Grotius' just war thinking – including shaping and undergirding his notable emphasis on defense of property as a just cause for war – but it also has far-reaching implications for just war thinking after the seventeenth century, offering a paradigmatic instance of how a changing conception of the self would shape the Christian just war tradition even as it contributed to the shape of modernity.

GROTIUS' (ALMOST) MODERN SELF: (ALMOST) UNIQUE AND UNIVERSAL

In *Sources of the Self: The Making of Modern Identity*, Charles Taylor argues:

> [t]he disengaged subject is an independent being, in the sense that his or her paradigm purposes are to be found within, and not dictated by the larger order of which he or she is a part. One of the fruits of this is the new political atomism which arises in the seventeenth century, most notably with the theories of social contract of Grotius, Pufendorf, Locke, and others. Contract theory as such wasn't new in this century, of course ... Nevertheless, there was something importantly new in the seventeenth-century theories. Previously the issue of consent had been put in terms of a people establishing government by contract. The existence of the community was something taken for granted in all earlier versions ... For a post-seventeenth-century reader, an obvious question arises: How does the community get started? Where does it get its authority to determine the nature of political authority over its constituent individuals? ... The big innovation of contract theorists from Grotius on is that they do address [that question] ... The new theories add to the traditional contract founding government a second one, which precedes it: a contract of association.[23]

[22] After its first publication in 1625, Grotius would regularly return to *The Rights of War and Peace* in order to add further support, by way of additional references, though without substantially changing his arguments therein.

[23] Charles Taylor, *Sources of the Self: The Making of Modern Identity* (Cambridge, MA: Harvard University Press, 1989), 192–193.

Without dismissing the significance of Pufendorf (1632–1694) or Locke (1632–1704), it is worth noting that Grotius was publishing more than a half-century before either of the others began publishing and that the first edition of *The Rights of War and Peace* predates the birth of either Pufendorf or Locke by the better part of a decade. Even if that text played only a minor formative role in their thought,[24] lumping the three figures together as Taylor does in order to describe the larger phenomenon of natural law-based contract theory threatens to lose sight of two important matters.

First, Grotius' work, pedantic though it was, was also groundbreaking in the shape of its scope: Grotius the synthesizer of information interpreted a wide range of issues through the prism of his natural law-based social contract approach. The fact that theologians read him with an eye toward his religious foundations, political philosophers (like Taylor) read him with an eye to his social contract theory, and legal historians read him with an eye to his founding of international law gives some sense not only of the range of materials with which Grotius was working but also of the synthetic qualities of modern thought that Grotius helps to initiate. Indeed, paradoxically, the "disengaged subject" of modernity – what Taylor will later call the "buffered self" – must be a synthesizing thinker if only because all the streams of a society's thought must pass through him or her on the way to (often idiosyncratic) expression in society. Where the self is embedded in society, it plays a role among other role-playing selves and the responsibilities of selfhood – including in the constructions of social organizations – are distributed among selves because they may be. Subsidiarity reigns in the construction of the self. However, when the self becomes atomistic (to use another of Taylor's words), it must either take up the range of those responsibilities or volunteer them to other selves. The self is sovereign. Said differently, the modern self is a product of individual actions (including the action of not engaging), not the consequence of social assumptions. We live in the societies we shape with each other rather than in societies whose shape has been designed by powers beyond us and we find order in those societies because our powers of discernment grow in tandem with our powers to make choices.

Grotius' contemporary, Rene Descartes, offers a paradigmatic expression of this atomistic self when he begins his *Meditations on*

[24] Jeremy Waldron recognizes Grotius' influence on both Pufendorf and Locke, though he does not explore its depth or range in *God, Locke, and Equality: Christian Foundations in Locke's Political Thought* (New York, NY: Cambridge University Press, 2002), 208.

First Philosophy with a process of systematically doubting everything he could doubt (his parents, his teachers, the church, his habits, his senses ...) in order to arrive at an indubitable foundation upon which he could build the philosophical architecture with which to frame his life. In the end, the indubitable thing he discovered is that someone was there, doubting things. *Cogito ergo sum.*[25] That individual self becomes a funnel through which the streams of sense-making on matters political, economic, scientific, familial, religious, and so forth all pass in the projects of self- and society-construction.

Where Descartes focused on the individual, Grotius was preoccupied with the role of the modern self in the construction of societies, whether those societies are ecclesial or secular. His Arminian self can will the creation of a peaceful church through the power of free choice; his political self can will the creation of a peaceful nation through the power of humanist thought; and his juridical self can will the creation of peace between nations by emphasizing the sovereignty of persons and states. As James Turner Johnson notes, "Grotius' *De Jure Belli ac Pacis* has exerted from its very first appearance a profound influence on the development of international law in the modern period. Grotius drew together the thought of the preceding age, reworked it in light of changed assumptions about men and their interactions, and produced a new conception of the relations among nations for the age on whose threshold he stood."[26]

Johnson helpfully highlights the connection between the (re)construction of societies and the developing conception of the human being at the beginning of the seventeenth century. It is not simply that a new way of thinking about the self leads to a new way of thinking about the state and the rules that will govern its interactions with other states; it is that selves create states that mirror their vision of themselves (hence the language of "sovereign" applied to both state and individual) and advance the authority of those selves to make their own distinct way in the world. Questions about what authorizes or determines the legitimacy of these projects of self- and state-construction become central to the tensions between social imaginaries during transitions from one to another.

[25] To be clear (and to avoid a cul-de-sac of unhelpful criticism): Descartes did not use the phrase "*Cogito ergo sum*" in *Meditations*, though he describes the doubting process there and arrives at the conclusion *ego sum, ego existo* (I am, I exist). He did use the phrase four years later in *Principles of Philosophy* in a nearly identical context.

[26] Johnson, *Ideology, Reason, and the Limitation of War*, 209–210.

Perhaps as a result, *The Rights of War and Peace* displays a distinctive breach in its thought. On the one hand, its range is impressive, particularly as its various focal points all grow out of Grotius' preoccupying question about the proper shape of law in a pluralistic and violent world. This range reveals Grotius' ability not only to reorder but to reorient his predecessors' work by processing it through a vision of the self as discrete and the interactions of selves as voluntary. On the other hand, Grotius continually supports his conclusions through appeals to other scholars.[27] As much as his subject may be new and his method novel, the authorities to which he appeals in defense of his work are not. His reliance on such appeals sits in tension with the work's subject matter and his treatment of that subject matter. To generalize, his focus and method are modern but his process in authorizing/justifying that method is premodern. He pours new wine into old wineskins.

As a result, *The Rights of War and Peace* leaks: It regularly feels cumbersome and unnecessarily digressive, dripping fresh insights in seemingly random ways rather than pouring them into the text systematically. That is, his arguments do not so much build upon each other as array themselves next to each other. Contrasting Grotius' work with Locke's, as Taylor implicitly does, displays that leakiness.[28] Whether this leakiness is the product of the complexity of the subject matter and his approach to it, a revelation of the limits of his genius, the all but inevitable consequence of living in the liminal space between premodern and modern thought, some combination of these, or some other cause, *The Rights of War and Peace*'s expansiveness becomes both the source of its brilliance and a cause for its limitations.

GROTIUS' (ALMOST) MODERN SELF: THE BASIS FOR A JUST WAR THEORY

The second important matter that Taylor's lumping of Grotius, Pufendorf, and Locke under-regards is the way *The Rights of War and Peace* reveals that the synthesizing done by the modern self is not a

[27] Even Butler, in 1826, notices that one of the chief criticisms of *The Rights of War and Peace* is that it defers too little to principle and too much to the authority of other voices. See Butler, *The Life of Hugo Grotius*, 140ff.

[28] I suspect that scholarly complaints about the work's pedantic qualities are expressions of frustration about the way he organizes authorities to justify his arguments more than criticisms about his treatment of his subject matter.

wholly conscious project and modern selves do not develop plans on how to interpret (let alone integrate) the various streams of thought that pass through them. Its social contract theory is voluntaristic but that voluntarism is shaped by an emphasis on agency more than understanding. As with human beings throughout the ages, much that modern selves take in registers primarily at subconscious levels and then shapes unconscious preferences, commitments, and prejudices for how to engage the world around them. The modern self's distinctive way of interpreting the world, though, is to prioritize the buffered self when giving expression to such preferences, commitments, and prejudices, to synthesize in light of them, and to act in ways that are consistent with and promote such a vision of the self. So, for example, while I cannot consciously and thoughtfully justify a claim that I have a right to be happy, when I am unhappy and wish to be happy, my desire for happiness is processed through a pattern of engagement with the world in which individual persons have rights, those rights are a means of assuring the welfare of individuals, they trump whatever the surrounding community can demand of me, and the community must therefore have some obligation to shape conditions that make my happiness more likely. Consciously, I know better than that: The distinctive moral lexicon of modernity (rights-talk) and its vision of the relation between self and community that have helped to constitute me as a modern self means that my conscious awareness of the problems with thinking that I have a right to be happy means I have to deal with the problems of the modern, subconscious filter through which I engage the world.[29]

Grotius was a product of his age even as his work drove Western society in new directions. His self's sovereignty is far more limited than it would later become. His basis for defending his arguments is to align them with earlier arguments rather than to promote their novelty by arguing that persons should have "the courage to think for themselves," to paraphrase Immanuel Kant's definition of enlightenment. Grotius' theological convictions may differ from those of Calvinists and Roman Catholics, but they imbue his corpus because nontheological arguments were not so much heretical as nonsensical at the time. That is, not only his resources but also his perspective reveal a worldview that is still

[29] To be clear: Communities may have obligations to promote individual happiness (at least at some level) and those obligations may be grounded in a variety of moral and political bases. The modern self grounds such an obligation in a vision of rights and the relationship between self and community that follow from that vision.

deeply informed by premodern convictions. Indeed, part of what makes Grotius so interesting is that he is so clearly a transitional thinker.

Nowhere is this more clearly the case than when Grotius is thinking about matters of war and peace. If the first problem with Taylor's lumping Grotius, Pufendorf, and Locke together as new natural lawyers is that amalgamating them conceals the degree to which *The Rights of War and Peace* is a groundbreaking work, the second problem, restated, is that it under-regards the significance of the text as a marker of a transition in just war thinking. With *The Rights of War and Peace*, the Christian just war tradition begins its shift into becoming a just war theory. Recognizing that Taylor's project is focused elsewhere and that it would be unfair to criticize him for not writing about something else, it is nevertheless the case that *The Rights of War and Peace* both initiates and becomes a paradigmatic example of the very transition of Western cultures into modernity that is Taylor's focus. As James Turner Johnson has noted, "[i] n the sixteenth and early seventeenth centuries a host of theorists of various European nationalities contributed to the metamorphosis in just war thought that led it away from a system of rules importantly connected to the value system of Christian religion and toward the secularized system of values and rules that is modern international law ... Of these, Victoria and Grotius are the most significant benchmarks of change."[30] Disaggregating Grotius, Pufendorf, and Locke helps to reveal this transition (and could have strengthened Taylor's arguments by giving him a text that would nuance his arguments about the way that the transformation of natural law thinking contributes to the arrival of a secular age).

More centrally for the purposes of this chapter, the shift from a just war tradition to a just war theory only becomes possible as the distinctness of human voices recedes. Paradoxically, the very emphasis on the uniqueness of individuals making free choices that marks the rise of early modern thought, once located in a world where people will not see the world in the same way or make the same choices, mandates deemphasizing their uniqueness in contexts where selves will interact with each other. All selves will have the same rights in political contexts. All selves will rely on the same processes for discovery in scientific contexts. All selves will submit themselves to larger processes of production

[30] James Turner Johnson, *Just War Tradition and the Restraint of War: A Moral and Historical Inquiry* (Princeton, NJ: Princeton University Press, 1981), 174. Though the quote from this text is convenient, the argument Johnson makes there has been central to his work from its beginning. See, for example, chapters 3 and 4 of his earlier *Ideology, Reason, and the Limitation of War*.

and consumption in economic contexts. Just war thinking, viewed as a theory, protects the integrity of states (and individuals) by removing from it the caprice of individuals making decisions for idiosyncratic or self-promoting reasons; importantly, it must do this because they have the power to do so.

So, for instance, where questions about the legitimacy and/or morality of war had been addressed by appeal to great thinkers prior to the sixteenth century, after Grotius, those questions would increasingly be addressed by appeal to the very type of law that Grotius began laying out. The wisdom of the individual (or, more accurately, the individuals) is replaced by the authority of the law. "How would Augustine answer this question?" or "What does the Roman Catholic Church say about this type of act?" – the types of questions that made Gratian so important for the medieval Christian just war tradition – start to be replaced by questions such as, "How would this (broadly agreed upon and contextually abstract) law apply to this context?" and "What do standardized codes of military conduct say about this sort of behavior?" When Charles Taylor describes the transition from a teleological to a social contract-oriented version of natural law, this is, in part, the shape of the transition he describes. And this transition is deeply linked to the wars of religion in sixteenth- and seventeenth-century Europe.

SOCIAL CHANGE, EARLY MODERN WAR, AND GROTIUS' (ALMOST) MODERN SELF

The conventional narrative of the wars of religion treats religion as the primary source of that violence even as it treats 1648's Peace of Westphalia and the "privatization" of religion as the solution to that violence. However, not only were the roles religion played in those wars – including the Thirty Years War – far more complex than that conventional narrative allows, but also there were many other drivers of both the conflicts and the peace that followed. Likewise, those wars and the privatization of religion that followed were only some of the causes of the changes that the Christian just war tradition underwent during the period – and as it changed, so did the understanding of the self that helped make it coherent.

A first change to note was with war itself. As Johnson has pointed out, medieval armies were structured around knights that, at least in theory, adhered to codes of chivalry as part of the maintenance of their lofty positions within European societies – codes that maintained hierarchical structures while allowing victorious knights to devastate the commoners who served in support roles to the knights that were defeated. Yet as Johnson notes, there "was a change in character of armies as they grew larger and were increasingly made up of common men, not the professional warriors of the knightly class," one consequence of which was "the replacement of an internalized chivalric morality by an externally imposed discipline as a form of restraint on the behavior of individual soldiers."[31] Simultaneously, weaponry was changing: Firearms were introduced to the battlefield (including affixed bayonets), as were hand grenades. Such weapons do not require the same amount of training as swords or pikes. They speed up conflicts, diminish opportunities for escape or withdrawal, expand the range of enemy combatants, and reduce the possibilities of "noble combat." By the time Grotius wrote, war "was not chivalric combat but a struggle involving armies of commoners, often inspired by religious zeal."[32]

The changing shape of war contributed to revisions to the evolving concept of the modern self. It undermined established medieval hierarchies, empowered commoners to think of themselves as capable of bringing about political changes through their own use of deadly force (while also further legitimizing them as targets under the *jus in bello* principle of discrimination), and reinforced developing ideas about the disciplining of masses of ordinary people to pursue complicated political ends. As it destabilized the *jus ad bellum* principle of legitimate authority, it also shifted questions about authority toward individuals and away from established political and ecclesial systems.[33] Combined with expanding visions of humanism and pacifism and exposed to the horrors of entrenched and prolonged war, it even introduced these developing selves to the idea that they had the power to choose not to participate in conflicts. The changing just war tradition – via Vitoria, Suárez, Gentili, Grotius, and others – reinforced the significance and expanded the range of individuals making choices about when and how to fight. Paradoxically, the inhumanity of wars in the early modern period also lent support to the ideas of the humanity and agency of the modern self.

[31] Johnson, *Ethics and the Use of Force*, 77.

[32] Ibid., 78.

[33] See Anthony F. Lang, Jr., Cian O'Driscoll, and John Williams, eds., *Just War: Authority, Tradition, and Practices* (Washington, DC: Georgetown University Press, 2013) for a provocative collection of essays about the (deleterious) impact of modernity on the just war criterion of legitimate authority.

One place this is clearly visible is in Grotius' shift in the priorities that establish just cause: "[R]ather than finding the justification of resort to armed force in the need to establish or maintain justice, Grotius reasons upward from the individual's right to self-defense to that of the state. After some discussion, he reaches his definition of just cause: 'Three justifiable causes for war are generally cited: defense, recovery of property, and punishment'."[34] This shift in priorities demonstrates both how Grotius equated the pursuit of justice with the establishment of juridical processes and also how, paradoxically, the modern focus on the individual could threaten the well-being of persons in war by absenting any larger teleological conception of justice from those processes.

GROTIUS, SELFHOOD, AND THE CHANGING SHAPE OF EUROPEAN POLITICS

A second change had to do with the changing political structures of early modern Europe. Even rulers in nominally Roman Catholic countries were freeing themselves from obligations to the Roman Catholic Church. Royal leaders in Western, Central, and Northern Europe were pursuing their own purposes in either initiating new communions (as Henry VIII did), protecting those who did (as Fredrick the Wise did for Martin Luther), or offering an audience to founders of new communions (as Francis I did for John Calvin). These purposes included attempts to accrue greater power in their respective regions, squelch rebellions, promote national interests during a period of exploration and colonialization, step into the power vacuums that were developing as the papacy and emperor lost hegemonic control, pursue their own shares of the growing wealth of Europe, institute political systems that followed from new visions of political philosophy, raise their own armies to defend themselves in the face of new types of war and the armies that fought them, and be willing to bind themselves to the beliefs coming out of the new communions being formed at the time. Over time, all these motives shaped what would become the modern nation-state.

Even hundreds of years after it came into existence, getting a good and tight definition of the modern nation-state is not an especially easy thing to do. The definition could be grounded historically: The Peace of Westphalia (itself at least partly the product of Grotius' thought) established the

[34] Lang et al., *Just War*, 26–27.

idea, at least in Europe, that all nation-states have sovereignty over their own territory. Certainly, this claim about sovereignty has been hugely influential – so much so that it continues to undergird global political structures to this day, including that most obvious of transnational political bodies, the United Nations. The definition could be grounded philosophically: The idea of state sovereignty only became possible as the result of a set of fundamental ideas about autonomy and its concomitant notion of freedom from external authority; universalizability and the extension of rights-language into international law; and instrumentality and its attendant emphases on process and usefulness that demarcates bureaucratic systems. The definition could be grounded genealogically: The Renaissance's commitment to immanence and the Reformations' oppositions to the Holy Roman Empire laid the groundwork for the wars of religion (which were really the extended death rattles of the earlier empire in the face of the insistence on the part of state leaders that they could go their own way). The exhausted conclusion was Westphalia and a tacit agreement in which religious bodies would surrender their own use of violence to achieve their ends provided the state, which would then hold a monopoly on the legitimate use of violence, would protect them and grant them a kind of independence. It could be grounded functionally: Grotius, Hermann Conring, and others who had lived through the wars of religion, who were shaped by the Reformation, and who were both reading in and shaping the early philosophical moves of modernity were also the ones establishing the rules of international law and interaction, and it was the willingness on the part of states to agree to international law that made the modern nation-state viable. It could even be grounded culturally: The modern nation-state is a peculiarly Western idea that became global only because of the success of Europe in colonizing the rest of the world and that would only remain globally significant while the effects of colonization continued. Prudence suggests taking all these sources of a definition of the modern nation-state seriously without being bound to any one of them.

With the rise of modern nation-states, the religio-political connections through which people identified themselves were changing. No longer could it be the case that the majority of Europeans (and everyone else they knew of in Europe, if not the world) shared a common connection through the church and the emperor. While these connections may not have been central to their life experiences, they nevertheless lent a sense of stability and continuity across time and space to personal identity: At least conceptually, persons existed in a vast community that ordered their existence and oriented them vis-à-vis each other. People fit into the cosmos and, as Taylor suggests, their "primary agency was the social group as a whole, or some more specialized agency recognized as acting for the group."[35]

In the absence of a unified Europe, identity does not suddenly shift from the group to the individual. Instead, identity increasingly becomes mediated through other units, the most obvious of which are those associated with faith and the state. Tribal loyalties become less assumed and more matters of choice. To be a Lutheran, for example, not only meant not being a Roman Catholic; it also meant being capable of choosing one's faith. This capacity to choose – including, eventually, to choose not to attach oneself to any religious community or perspective – is the basis for what Taylor has called a secular age. Religion no longer functioned simply as a background condition – an assumed reality – of human existence. Instead, it became something about which one must think and make choices. As religious faith became an expression of conscious reflection and commitment, it also lost its hegemonic power to shape assumptions about causation in the natural and political world: The multiplication of communions drove disenchantment and participated in shaping modern scientific worldviews.[36]

Of course, these changes in choice, secularization, and disenchantment were occurring at the very point when states were ascendant. In a world of *cuius regio, eius religio*, persons would not only find their faith and their national identity linked but would also discover that *regio* determined *religio* while the inverse was not necessarily (or at least comfortably) so. Public loyalty to religious faith only worked if one adhered to the faith of one's prince; otherwise, it was superseded by loyalty to the state (and if one did not adhere to the faith of one's prince, one either fled to a place where one's faith could be publicly expressed or acceded to pressure to keep one's religious faith out of one's public engagements, which are just other ways that *regio* determined *religio* in nonreciprocal ways). Grotius' translation of the Christian just war tradition into

[35] Charles Taylor, *A Secular Age* (Cambridge, MA: Harvard University Press, 2007), 148.

[36] My agreement with Taylor's thesis in *A Secular Age* does not imply that I agree with the secularization thesis, which argues that religion fled (or was pushed out of) the public sphere – including where political and economic transactions occur – leaving that sphere a place where people of different religious convictions could gather to shape society, provided they left their religious convictions behind (or at least translated them into nonreligious language). The next two chapters will explore the continued presence of religion in the nominally secular public sphere.

international law both exemplifies and reinforces this change through its social contract-based version of natural law and the vision of secularity that sits behind it. As much as Grotius may have thought about international law in theological terms, the theological terms through which he thought were those that reinforced state sovereignty, national identity, and the capacity of religious people to choose the shape of their faith far more than the shape of their state.

GROTIUS, SELFHOOD, AND THE SOCIAL CHANGES THAT SHAPED INTERNATIONAL LAW

A third driver of this changing understanding of the self was the interlinked developments in technology, economics, and the projects of colonization that they funded. Even as the conceptual boundaries between states were being solidified via a particular understanding of state sovereignty, European states' geographic boundaries were dissolving as they raced for control of the rest of the world. Developments in science and technology (including developments pertaining to maps, seagoing vessels, and the ability to locate oneself in the world) and expanding economic resources made exploration of the world possible. Those explorations, in turn, reinforced the value of these scientific and technological advances and introduced both new sources of wealth and new causes for conflict. Having exhausted many of the natural resources of Europe, state-funded explorations were designed to gain control over resources from Asia, Africa, and the Americas (including land, foodstuffs, precious metals, fibers, and people), thereby dramatically extending the power of the merchant classes in Europe and driving conflicts between European states onto the oceans and into other continents. Colonialization also provided a relief valve for those fleeing conflict in Europe and those seeking a better life with access to abundant resources even as it reinforced the political power of European states and the ethnocentrism of European peoples. Conquest and missionization reshaped political configurations and alliances while establishing bragging rights for successful European countries. Encounters with indigenous persons led not only to conflicts and genocide but also to modern conceptions of race and their attendant practices of racism.[37]

[37] To be clear: Race and racism existed prior to modernity. Modern understandings of race and racially ordering the world, though, were new. See Geraldine Heng, *The Invention of Race in the European Middle Ages* (Cambridge, UK: Cambridge University Press, 2018).

In the conventional narrative about the establishment and expansion of international law, Grotius and others shaped international law to deal with conflicts in Europe and, over time, expanded it in order to deal "justly" with the shape and impact of colonialization. That is, the Westphalian model of state sovereignty would gradually extend from Europe to the rest of the world, nominally for the non-European world's benefit. Yet as Antony Anghie has argued, "colonialism was central to the constitution of international law and the sovereignty doctrine … Sovereignty is formulated in such a way as to exclude the non-European; following which, sovereignty can then be deployed to identify, locate, sanction and transform the uncivilized."[38] The concepts of self and state sitting behind and being shaped by Grotius' international law may have imagined themselves to be universally applicable and culturally neutral. Much as we saw in Vitoria's approach to indigenous persons in the Americas and West Indies, though, in both shape and impact these concepts of self and state displayed Eurocentric and imperialist characteristics that continue to impact sociopolitical systems to the present day.

The sovereignty of states is only possible where the forms of state fit models that emphasize secularity, a focus on self- and property protection, and the priority of national identity. Those political systems that do not fit such models are deficient and need to be treated as such until they can grow into "mature" forms and attain true sovereignty. Until then, the thinking went, they can be controlled by others; indeed, they must be controlled by others in order to guide their maturation. And the sovereignty of selves is only possible where those selves fit models that emphasize autonomy, favor transactional relations, prioritize choice, can control their own destinies, and determine their own tribal connections. Those persons that do not fit such models are deficient or not even fully human and need to be trained until they mature or used until they cease to exist.

The Eurocentric models of state and self-sovereignty that grew out of Grotius' vision of international law have produced important benefits for the world: human rights, national and international stability, prosperity that reaches to more and more people over time, scientific developments that make life easier and reduce suffering, and possibilities for cross-cultural interactions, to name just a few. Yet these models have also come at great costs: race-based chattel slavery, massive wars, economic inequality driven by the control and abuse of natural

[38] Antony Anghie, *Imperialism, Sovereignty and the Making of International Law* (New York, NY: Cambridge University Press, 2005), 310, 311.

resources, increasingly dangerous and indiscriminate weapons, cultural appropriation and insensitivity, and planetary despoliation, to name just a few. International law offers a mixed bag of costs and benefits.[39]

As the Christian just war tradition shaped international law and was, in turn, shaped by it, that tradition offered its own mixed bag of costs and benefits. As noted in the previous chapter, Christian just war engagement with indigenous persons in Africa, the Americas, and the West Indies shaped patterns of engagement with "the other" that could both inhibit and stimulate violence against such persons. Just as importantly, though, international law's emphases on secularism, universality, neutrality, and proceduralism shaped the transition of Christian just war thinking from tradition (framed by connection to a past that expresses itself in the particularity of earlier persons and their ideas) to theory (which would be framed by abstract and secular laws that could be universally applied by and coherently defended toward any person or group of people in order to make determinations about the morality of war). Grotius sat at the crux of this transition.

JUST WAR CRITERIA, JUST WAR THEORY, AND MODERNITY

As described in the Introduction, the primary expression of this transition from Christian just war tradition into just war theory was the coalescence of the *jus ad bellum* and *jus in bello* criteria into an ordered and regularized framework that was functional regardless of the various criteria's theoretical/theological origins, ordered according to a kind of *a priori* rational framing, and applied via procedurally neutral systems across a wide range of conflicts. That is, these criteria, nominally universal in scope, neutral in form, and regularized in implementation, ceased to be embedded in the frameworks that had given them meaning and could therefore stand in for those frameworks without appeal back to the traditions from which they had arisen. In their juridical form, the criteria themselves became the center of just war thinking, disconnected from the values and virtues that sat behind them.[40]

39 For an extended assessment of the problems of a Grotian approach to international law with regard to matters concerning colonization and imperialism, see Eric Wilson, *The Savage Republic:* De Indis *of Hugo Grotius, Republicanism, and Dutch Hegemony within the Early Modern World-System (c. 1600–1619)* (Leiden: Martinus Nijhoff, 2008).

40 See, again, my "Changing the Rules: Just War Theory in the 21st Century," *Theology Today* 54.1 (January 2003).

Unsurprisingly, as the tradition became a theory, it cemented the very *jus ad bellum* and *jus in bello* criteria that favored those patterns of military engagement visible at the beginning of modern Europe. Defense of self and property grew in prominence as a just cause. Legitimate authority became harder to ascertain or value. Moral discernment via a Eurocentric social contract version of natural law blurred rather than clarified differences between what is permitted of all persons and what is permitted of Christians.[41] Possibilities for conscientious objection grew even as the preconditions for what could count as "conscientious" favored Europeans, thereby allowing persons from other parts of the world to be pressed into military service regardless of their wishes.

There is, then, an irony at work in the way just war thinking shifts from being tradition-based to being theory-based: Even as it surrenders particularity (of sources, authorities, and expressions) in favor of universality (of laws, *a priori* forms of reasoning, and global reach), it implicitly accepts a particular way of thinking about law, reasoning, and utility that favors European modes of thought, war, technologies, politics, and interests. While it may have played a role in limiting wars in Europe – and particularly "religious" wars – after Westphalia (at least for a time), the only way just war theory could limit wars outside Europe would be if European countries could so exert power in their own projects of colonization and conquest that they could force such wars to be fought according to European values (private property), structures (nation-states), sensibilities (standing armies), and stratagems (distinguishing combatants and noncombatants). While the conventional narrative about the history of just war thinking tells a story in which it seemed irrelevant between Grotius and Ramsey, a more finely tuned narrative might suggest that it was the very success of just war theory to reinforce particular reasons for and ways of fighting at political, legal, and conceptual levels that shaped the wars between Westphalia and Vietnam, particularly those associated

[41] As Johnson notes, "Grotius also differs from earlier just war doctrine in distinguishing between what is known by nature, which binds all men, and what is known by charity, which though good is binding only upon Christians ... The result in his thought is a minimal concept of just war, a moral floor on the right of man in nature to make war as he likes, with another set of standards superimposed, which define what Christians may do. This separation has an ironic result. Since European nations were by definition 'Christian,' and since the customs and mutual agreements in the *jus gentium* were also the products of interactions among European nations, it was difficult during the early modern period to distinguish between standards drawn from Christianity and those drawn from the *jus gentium*." Johnson, *Ideology, Reason, and the Limitation of War*, 219.

with colonization, conquest, and their effects. And Grotius played an important role in shaping such implications.

THE SHIFTING PLACE OF CHRISTIAN THOUGHT IN THE MOVE FROM TRADITION TO THEORY

The various changes in war, political structures, technology, and economics described above were only some of the reasons that the Christian just war tradition became a secular just war theory. As important as they all were, a fourth driver for the shift from tradition to theory was the failure of the tradition to adjudicate conflict between Christian nations in Europe in the sixteenth and seventeenth centuries. While the wars of religion are sometimes portrayed as conflicts between Catholics and Protestants over doctrinal matters, such a portrayal ignores the frequency with which Catholics and Protestants on one side fought against Catholics and Protestants on another side as well as the causes of those conflicts. These conflicts were much more likely to be the product of states assuming powers previously accorded to the church and going to war in the absence of any superior adjudicating authority like the pope or the wisdom of dominant voices within a tradition like the Christian just war tradition.[42]

That the Christian just war tradition struggled to relate to conflicts across religious perspectives was nothing new: The challenges of shaping a moral response in conflicts with Muslims, for instance, were long-standing. The tradition consistently asked questions about how to engage the "other." Yet in sixteenth- and seventeenth-century Europe, the stability of defined opposing sides dissolved, and with it clarity not only about one's opponents but about one's allies and whether they fought according to the same rules and with the same vision of war and peace. Lutherans, Calvinists, Anglicans, and Roman Catholics may all have drawn from the same broad tradition but they did not all draw from the same parts of it, and their varied approaches to conflict not only were the products of those differences but also exacerbated such conflicts by introducing greater levels of chaos into political systems.

So, for instance, if the question of private armies had been an acutely troubling one for the Christian just war tradition in the Middle Ages, it

[42] See, e.g., William Cavanaugh, "'A Fire Strong Enough to Consume the House': The Wars of Religion and the Rise of the State," *Modern Theology* 11.4 (October 1995): 397–420.

was now replaced by the question of what to do with rebels – whether rebels against the Holy Roman Empire, against the new princes of Europe, or against civil governance in general – and that question was answered differently by representative thinkers from the various communions of Europe. Catholic Spain fought against independence-seeking groups in the Netherlands; Martin Luther railed against peasant rebellions; John Calvin wrote about the obligations of magistrates to punish the religiously deviant and others who disrupt public peace even as John Knox encouraged Scottish rebellion against a Catholic queen; and the English Reformation led to the English Civil War. And throughout Europe, alliances – including some unlikely ones – were created and dissolved to address the increasingly chaotic continent even as the varied visions and resources used by allies and enemies threatened to exacerbate that chaos.

Grotius' location at a transitional point in Christian just war thinking is evident in the fact that he turns to the question of what to do with rebels very early in *The Rights of War and Peace*, treating it as the first practical matter to address after completing his argument that it can be lawful to wage war (even before self-defense or defense of property). As much as it is the central question of Book One, though, he frames his arguments around the distinction between public and private wars and the varied cases of wars made by subjects against their sovereigns. Again, the language he uses to make his case reflects a medieval lexicon even as the arguments he makes do not: While rebellion is not allowed by nature, the Scriptures, the early church, or even reason itself, there are cases "of extreme and inevitable Necessity" (e.g., when people are understood as free by their own governing systems, when a ruler abdicates his or her responsibilities or behaves as an enemy of the people) where a war of rebellion can be waged. Even here, Grotius argues that there ought to be laws governing the causes for and shapes of rebellions to which those who would rebel can appeal in defense of the legitimacy of their cause.

Indeed, the creation and growth of international law at the time might better be understood as an attempt to found a new adjudicating authority, albeit one that could gain such authority only by affirming and reinforcing the sovereignty of the very states that it would regulate.[43] Grotius' social contract approach may appeal to agreements between persons on the shape of social structures for its focus but it gains whatever power it has through

43 There is much still to be written, I think, about the way Carl Schmitt's understanding of political theology – with some modification – could help make sense of the place and prominence of international law in modern political thought.

the agreements of states to the laws he describes, and these agreements turn on states recognizing their own interests being served in those laws.

This new adjudicating authority gains traction through its theoreticity. In a geographically and politically expanding world where there are not only different religions, but also multiple and seemingly incommensurable ways of being Christian (each with access to political and military power), appeals to mutual religious narratives, agreed-upon ways of reading a common text, or shared experiences manifested as problems with political implications as often as solutions to political crises. The diminished value of appeals to tradition, Scripture, or experience left a particular kind of reason – one framed by abstraction, an awareness of the power of subjectivity, instrumentality, and a tolerance for difference – ascendant. This kind of reasoning meshed with the way Grotius framed international law. Said differently, the arrival of a secular age in the sixteenth and seventeenth centuries shaped both the necessity for and the value of a theory-oriented approach to politics, law, and rules of war.

A GROTIAN ANTHROPOLOGY AND THE SHAPE OF MODERNITY

That Grotius' vision of a social contract-based version of natural law shaped a polity is obvious: That is its point. That it was shaped by and shaped a particular anthropology is less obvious but no less significant. Polities become operative only as they are funded by (and reinforce) a particular way of thinking about those who live within them. The anthropological perspectives that underwrote Grotius' work in *The Rights of War and Peace* reveal a profound shift away from medieval thought even as Grotius used the resources of medieval thought in directing that shift.

That perspective begins to be revealed early in *The Rights of War and Peace*, when Grotius writes:

> But it must be owned that a Man grown up, being capable of acting in the same Manner with respect to Things that are alike, has, besides an exquisite Desire of Society, for the Satisfaction of which he alone of all the Animals has received from nature a peculiar Instrument, viz. the Use of Speech; I say, that he has besides that, a Faculty of knowing and acting, according to some general Principles; so that what relates to this Faculty is not common to all Animals, but peculiarly agrees to Mankind.[44]

[44] Grotius, *The Rights of War and Peace*, 84–85.

At first blush, the sentence does little to distinguish it from earlier thinkers such as Aristotle (human beings as social animals) or Thomas (human beings as able to reason from general principles). In context, though, distinctions begin to emerge. His larger concern with law suggests that the human ability to know and act according to general principles pertains primarily to knowing that there need to be rules when human beings enter society, and, as will emerge later in the text, these rules look like a protean version of the social contract theories that would begin to emerge a half-century after his writings. His sense that such principles are real and discernible via a "faculty" distinguishes him from both the more Machiavellian approaches to political life that downplay the existence of such principles and the more empiricist approaches of, for example, Alberico Gentili. And his linking of knowing and acting, especially once combined with his emphasis on the power of speech, hints at one of the claims central to his larger corpus: that human beings act in voluntary ways shaped by their capacity to independently reason rather than being obedient to larger forces, whether those forces arise from nature or providence, and to convey such reason through the powers of persuasion rather than force.

None of these ideas independently – or even collectively – are unique to Grotius: His strength was not in shaping a new anthropology through the force of his intellect. And neither are the versions of natural law and natural rights that flow from these ideas; as Brian Tierney has written, "Grotius did not create a new theory of natural rights and natural law; but what he achieved was equally important. He made it possible for the old theory to live in the new world."[45] He did so by making this anthropology seem not so much something discerned via some special faculty provided to only a few or accessible to only the most learned but something that arose simply from common sense, regardless of one's political or religious persuasions. In a way, then, he democratized reason across the pluralization of vocations and universalized it across religious and political differences.[46] It was this insistence on a kind of ordinary

[45] Brian Tierney, *The Idea of Natural Rights: Studies of Natural Rights, Natural Law, and Church Law 1150–1625* (Grand Rapids, MI: Eerdmans, 1997), 342.

[46] "According to Grotius, it was possible to receive reasonably certain answers to [the problems of religious conflicts, of disputes over territory, of taking hostages in war or lying to enemies, and of dealing with tyrants] by examining human nature itself; more particularly, the rules and principles that upheld social life … Grotius' law could be understood and employed even (and perhaps especially) by religious or political adversaries – after all, they 'flow [ ] from the internal Principles of Man'." Martti Koskenniemi, "Imagining the Rule of Law: Rereading the Grotian 'Tradition'," *European Journal of International Law* 30.1 (2019): 26–27.

rationality that drove the more explosive qualities of his project: Rulers could claim the obedience of their subjects because both rulers and subjects could see the wisdom of such obedience; people could own things and pursue profits because these projects flowed from a vision of natural rights that simply made sense for those who live with others and benefit personally from doing so; these same people could demand from their rulers the enforcement of laws that gave everyone the freedom to own things and make money; such laws blurred distinctions between natural and positive law because they arose not as expressions or mirrors of divine law but from the kind of reason that springs from human beings simply interacting with each other in a society in which they can live peaceably with each other. For Grotius, human beings are, in short, independently capable actors who can make voluntary choices to shape their religious and economic engagements, who can find their identities within bureaucratic state systems, and who can advance their work by recourse to pragmatic and reasonable efforts to live with others in a world filled with religious, economic, and political differences.

It is the very applicability of this anthropology that empowers it, such that it shows up in the way Grotius thought about religion and religious tolerance, domestic and international law, economics and property rights, and the rules of war, and such that it could undergird nascent modern understandings of human rights, modern science, and religious freedom.[47] It is the kind of anthropology that can fund the transition of just war thinking from residing within particular traditions to functioning as a universally accessible theory; after all, theories rely on particular understandings of the type of human beings who can operate them.[48]

It is, then, this anthropology that propels the distinct directions in which Grotius takes just war thinking and, as he does so, shapes both its novelties and its failings. To make sense of these novelties and failings, especially as they impact our entrance into the wars of the Anthropocene, I turn toward two features of that thinking. The first has to do with the way Grotius forwarded concerns about private property and the right to

[47] My argument here echoes that of Koskenniemi, who writes, "I believe it was the objective of *De iure belli ac pacis* (1625/1632) to consolidate an authoritative form of legal speech that could sustain firm government at home, help orient subjects towards profitable activities and order human relations in general outside the sphere of domestic laws." See Koskenniemi, "Imagining the Rule of Law," 20.

[48] This construct, I think, helps to explain the shift in just war thinking away from the pursuit of the common good to the maintenance of sovereign systems that has been noted by, among others, James Turner Johnson.

defend it through force of arms if necessary. The second has to do with the way Grotius overlooked concerns about refugees and the ways they both lead to and are among the primary effects of war.

GROTIUS, PROPERTY, AND WAR

Grotius' understandings of property and of the justification for defending it through the use of force are broadly familiar and display the degree to which he was simultaneously inheriting old and advancing new ideas in the transition from late medieval to early modern thought. Developed across a range of his works, beginning with *De Jure Praedae* (written for the Dutch East India Company in the first decade of the seventeenth century but unpublished – aside from a single chapter – during his lifetime) and continuing through *De Mare Liberum* (1609), his thoughts on property and war find their culmination in *The Rights of War and Peace*. Throughout, he advanced three foundational claims: that in the beginning, when human beings exist in a state of nature, all things are free, shared, and seemingly abundant; that by seizing particular things and occupying particular spaces, those things and spaces become our own; and that once things are owned, we have the right to fight to keep them as an expression of justice. Because those three arguments are broadly familiar, I will not rehearse them in detail here. However, each of them warrants a bit of explication in order to develop the implications of Grotius' understanding of property and its defense for making sense not only of his justification for war, but also of his understanding of the natural world.[49]

Begin with the first claim: that in the original state of nature, all things are free, shared (or at least equally accessible by all), and abundant. The idea can be traced back at least through Vitoria and Thomas. The starting point is important not only because it funds the idea that things not otherwise acquired can become so (that is, things begin their existence as readily available and not, for instance, already claimed by a god, and, just as importantly, there is – at least in theory – originally no prohibition based on a concept such as race or class that restricts who can gain

[49] For much more detailed accounts of Grotius' understandings of these three points, see, among others, Alejandra Mancilla, "What We Own before Property: Hugo Grotius and the *Suum*," *Grotiania* 36 (2015): 63–77; John Salter, "Hugo Grotius: Property and Consent," *Political Theory* 29.4 (August 2001): 537–555; and Ileana Porras, "Appropriating Nature: Commerce, Property, and the Commodification of Nature in the Law of Nations," *Leiden Journal of International Law* 27 (2014): 641–660.

access to what), but because it initiates a division between human beings and everything else: Nature is reduced to natural resources. Moreover, because things are plentiful, human beings and the beginnings of property start not in conflict but in possibility. Contra Hobbes and, for that matter, European projects of colonization, acquisition begins in neither contest nor conquest.[50]

Second, seizing things and occupying spaces has the effect of changing the status of those things and spaces: They become owned. When they do so, they also become transferable. Because things are not evenly distributed and not everyone can maintain control over everything, human beings, as social animals, enter into acts of trade. For Grotius, trade does more than shape an economy; it is the fundamental action upon which societies and, indeed, civilization are formed. It drives social animals toward each other and orients them vis-à-vis each other as well as vis-à-vis the nonhuman natural world.

Third, then, because acquisition, occupation, and trade are so fundamental to the formation and maintenance of societies, these societies begin the shaping of law around just and unjust ways through which things are acquired, maintained, and change hands. We can and should defend our property – even to the point of taking the lives of those who would take our stuff – because the alternative to doing so undermines not only the legitimacy of ownership but the basis of societies and the foundations of justice. Questions of justice and injustice begin, for Grotius, with a set of assumptions about property: about what can be ours and, therefore, not anyone else's.

The implications of these three points are far-reaching and extend well beyond the scope of this project. Here, though, let me name and explore two implications that are especially pertinent to the project. First, Grotius' arguments at the beginning of the second book of *The Rights of War and Peace* about defense of property as just cause for war build from his third claim: The very idea of justice more generally is bound to matters having to do with the acquisition, ownership, and control of things. And, second, by thinking of nature as a collection of plentiful resources and human beings as expressing their sociality fundamentally through trade, Grotius divides the world between human beings and

[50] Grotius exercised the same inconsistency here that Vitoria had: Even as he recognized that the indigenous persons of the Americas and West Indies had seized some things and acquired some spaces, he still thought of the new world as full of free and unclaimed natural resources in a way that Europe was not – and did so without ever having been there or engaged the way such indigenous persons thought about things and spaces.

everything else – and everything else exists to make civilization possible. It is, moreover, important to note that these two implications are linked such that if the second does not hold, neither can the first.

GROTIUS, JUSTICE, AND A WORLD OF SCARCITIES

Because his questions about justice are so bound to matters of ownership and control and, therein, to law, it is unsurprising that Grotius would link his just war thinking to his ideas about property. Stepping back from that linkage, though, brings a range of assumptions and tensions into focus. The linkage between justice, property, and law treats war as an expression of an individual's right to defend his or her own property writ large. Societies shape laws to clarify and order their basic visions of justice and, for Grotius, societies' laws, (not to mention the *ius gentium* more broadly) grow from the rights of individuals to those of nations without giving attention to the possibility that the laws governing nations (and the visions of justice that underwrite those laws) should be different in kind from the laws governing individuals. Wars over property are tantamount to wars of self-defense because threats about the loss of property put states at risk in a way that is comparable to that in which threats about the loss of property put individuals at risk. Individuals can attack those who would take their stuff because they need that stuff for their own welfare; the same is true of states. By treating the laws governing nations as continuous with the laws governing individuals and orienting those laws around a vision of justice that begins in human sociality expressed in terms of ownership and trade, Grotius not only emphasizes war as punitive but also views permissions to go to war and the range of behaviors manifest in war expansively.

War is punitive because it is a response to a violation (or a threat of violation) of the most basic rules of justice. A nation can wage war not only to recover goods and spaces that it has lost but also to punish the ones who took them even after they are recovered. It follows that weighing out variables within the *jus ad bellum* and *jus in bello* criteria of proportionality includes giving attention to the severity of punishment due an offending party for violating the demands of justice in the first place, therein expanding not only the variables to be accounted for but also the levels of force that can justifiably be used. And permissions to go to war are expansive not only because war can be waged against any marginally organized body that would take a state's stuff (pirates, other states, rebels claiming land, etc.) but also because justification for war turns on the loss (or threat of loss) of goods or spaces regardless of the contextual realities that shape such contestations. Questions about who took these things and why they were taken are secondary, at best, to the prior right to keep what one owns.

Because Grotius starts his understanding of the nonhuman world with a vision of it as abundant stuff which can be acquired or occupied, it is unsurprising that attention to the implications of war for the nonhuman natural world largely go begging. He is not alone in this; as Ileana Porras notes, "nature became visible to the early modern Europeans who contributed to a new tradition of *jus gentium* (the law of nations) only in the moment they articulated it as a material thing subject to appropriation, reducible to property, and capable of entering the stream of commerce."[51] The nonhuman natural world matters only to the degree that it lubricates the interactions of human beings. Nature gains meaning/significance when it is used, not because it exists, even as the fact that it exists and interacts with the human world is essential to the human world. There is, then, a first paradox in the way Grotius engages questions about the nonhuman natural world: Grotius effectively gives more attention to the nonhuman natural world than do many just war thinkers, but the form of that attention reinforces a binary (human vs. everything else) that rejects any valuing of the nonhuman natural world beyond its utility for persons.

At first blush, this should suggest that Grotius could give closer attention to the impact of war on the nonhuman natural world. After all, if a war decimated the nonhuman natural world in any particular area, those losses would redound upon the human beings fighting the war and living in that area. That Grotius shows so little concern with this matter reveals just how significant ideas about abundance and unequal distribution are to his thought: The loss of resources in one place just increases the possibility of trading for them with someone from another place. Recall that, for Grotius, the essential function of the nonhuman natural world is conditioned by its abundance, on the one hand, and its unequal distribution, on the other: If everyone had access to the same stuff, nobody would need to trade, and if there was not enough stuff, nobody would be able to trade. Violence used to unjustly take stuff therein not only violates appropriate and essential expressions of human sociability (and therefore warrants punishment); it also violates Grotius' vision of a world in which things are abundant and unequally distributed (and

[51] Porras, "Appropriating Nature," 642.

therefore warrants rejection as an idea). There is, then, a second paradox involved in connecting his vision of the nonhuman natural world to his understanding of war: Wars may damage natural resources in one place, but they also make the nonmartial forms of human interaction more meaningful by driving people toward trade in order to get natural resources from one another.

One piece of evidence for this is found in Grotius' conviction, beginning in *De Jure Praedae* and continuing through his writings to *The Rights of War and Peace*, that things that are abundant, that cannot be divided up because they cannot be controlled, and that make trade possible – things such as the oceans – cannot be owned. Because they cannot be owned, they can be fought upon but not fought over (to be clear: false claims to own such things can be justification for war, but it is the claim, not the things themselves, that is at the heart of such a *casus belli*).

A second, related, piece of evidence for this is found in Grotius' references, throughout his writings, to Seneca's belief that "the supreme blessing conferred by nature resides in these facts: that by means of the winds she brings together peoples who are scattered in different localities, and that *she distributes the sum of her gifts throughout various regions in such a way as to make a reciprocal commerce a necessity* of the members of the human race."[52] His adoption of stoic ideas about natural law, providence, and a hierarchical ordering of the universe (with human beings above other animals and things) reinforced his ideas not only about abundance and distribution, but also about the insignificance of any part of the nonhuman natural world that is not involved in trade.[53]

And a final piece of evidence arises out of Grotius' conviction that rights are transferable things – that they do not inhere in persons but can, if a person wishes, be given to another. Indeed, Grotius is so convinced of this that he uses the argument (problematically) to support slavery: Persons can volunteer to sell themselves and, so long as they are volunteering to do so freely, enter into slavery. These three bits of evidence – that human beings cannot own or trade things that are universally abundant, that they live in a world that has been ordered in such a way as to promote this way of thinking, and that even many of the things that

[52] Ibid., 653. Emphasis hers.

[53] See Teun Tieleman, "Stoicism and the Natural World: Philosophy and Science" in Paul T. Keyser and John Scarborough, eds., *The Oxford Handbook of Science and Medicine in the Classical World* (New York, NY: Oxford University Press, 2018), 677–698.

we find most inherently human such as rights are transferable – clarify the way Grotius thought about the nonhuman natural world, the extent to which he framed that world through economic lenses, and, therein, the way Grotius understood the self.

Note, though, one implication of Grotius' approach: the foundation of his claims about justice broadly and just war thinking more specifically pertain *because* human beings are traders and they are traders because the world has been organized – both geographically and conceptually – in a way that favors trade. The Grotian moral self is, fundamentally, an economic self: *Homo sapiens* is *Homo economicus*. If that framing and understanding of the organization of the world does not hold, then neither will Grotius' vision of justice. If scarcity is a reality in a way that abundance is not and/or if the distribution of resources exists in a way that inhibits or prevents trade, then a Grotian vision of justice (and just war thinking) suffers, as does his understanding of the moral self. For much of modernity, that Grotian vision seemed to hold, at least to those of European ancestry and those who adopted or were forced to adopt Western ways of thinking during and immediately after colonialization. As we enter further into the Anthropocene, however, that vision is taking increasingly severe body blows.

While it is true that technological advances have increased abundance in some parts of the world (one thinks of the green revolution in India in the latter half of the twentieth century), it is also true that those same technological advances have contributed to the systems leading to climate change and, therein, to increasing levels of scarcity at global levels. Further technological advances may help address some of the problems of scarcity that are arising and will grow more severe in the coming decades, but those advances now need to be assessed in much more forward-thinking ways. Increased attention to the precautionary principle – including its use in critiquing proposals for geoengineering our way out of the climate crisis – is one piece of evidence in this regard: *principiis obsta et respice finem*. That precautionary principle is also shaping contemporary thinking about war.[54]

[54] See, e.g., Richard E. Berkebile, "New Generation Warfare and the Just War Tradition," *Interagency Journal* 9.3 (2018): 17–32; John Forge, "Proportionality, Just War Theory and Weapons Innovation," *Science and Engineering Ethics* 15.1 (October 2008): 25–38; Elizabeth L. Chalecki and Lisa Ferrari, "Engineering the Climate – or Deploying Disaster? Applying Just War Theory to Geoengineering," *New Security Beat: The Blog of the Environmental Change and Security Program* (June 25, 2018), www.newsecuritybeat.org/2018/06/engineering-climate-or-deploying-disaster-applying-war-theory-geoengineering/ (accessed April 22, 2020).

While it is true that North Atlantic visions of modernity (including visions having to do with politics, economics, cultural forms, and metaphysics) now manifest globally, it is also true that those visions are meeting increasing resistance from people and groups around the world who are either recovering or creating alternative ways of living in and understanding the world that they believe better address life in the Anthropocene.[55] They are, moreover, becoming increasingly organized and politically engaged at global levels (that is, they are revealing their sociality without binding it in a Grotian fashion to trade) and among their concerns are what Rob Nixon has called "slow violence," which is the violence shaped by climate change.[56] They want justice and they are justifiably suspicious of Western expressions of justice because such expressions have been linked to the very patterns of domination (of others and of the nonhuman natural world) that they believe have driven the creation of injustice in the first place.

While it is true that variables involved in assessing future needs and capacities (e.g., those linking population growth to education, access to previously unattainable or underdeveloped resources to changing markets, and efficiencies driven by the exponential rate of technological change to environmental concerns) remain vague, it is also true that even optimistic scenarios still cannot overcome the costs associated with ocean acidification, desertification, flooding, drought, wildfires, deforestation, and the range of other forces that are increasingly driving problems of scarcity with regard to access to basic goods such as food and water. Though there are a variety of schools of thought in environmental security studies (e.g., neo-Malthusians arguing that the fundamental problem driving environmental conflicts is resource scarcity, neoclassical economists arguing that the problem is the lack of diverse and open markets, and political ecologists arguing that the problem is unjust distribution patterns), all of them forecast a bleak future in which current patterns of overconsumption will bear dramatic long-term costs with regard to access to basic goods, including costs associated with political violence and war.[57] The future is, simply, viewed as less stable and less hopeful.

55 See, e.g., Willis Jenkins, *The Future of Ethics: Sustainability, Social Justice, and Religious Creativity* (Washington, DC: Georgetown University Press, 2013).

56 See Rob Nixon, *Slow Violence and the Environmentalism of the Poor* (Cambridge, MA: Harvard University Press, 2011).

57 See, e.g., Simon Dalby, *Security and Environmental Change* (Cambridge, UK: Polity, 2009) and Colin Kahl, *States, Scarcity, and Civil Strife in the Developing World* (Princeton, NJ: Princeton University Press, 2009) for overviews on environmental security studies.

Further entrance into the Anthropocene, then, reveals that scarcity is an expanding problem, that conceptions of justice – especially as they relate to environmental justice – are multiplying, that uncertainty about the shape of the future is justifiably growing, and that a vision of moral selfhood shaped by economic sensibilities cannot overcome such problems. Each of these revelations undermines the Grotian conception of justice that funds his visions of law, war, and engagements with the nonhuman natural world. The implications of these blows to Grotian ideas not only undermine Grotius' vision of the world and with it his understandings of justice, law, and just war thinking; they also undermine his understanding of the moral self engaged in the pursuit of justice, law, and just war thinking as they intersect with the nonhuman natural world. To get at just how deficient that understanding of the moral self is, especially as we enter further into the Anthropocene and the types of climate-shaped conflict arising therein, I turn to the way Grotius thought (or, rather, did not think) about refugees.

GROTIUS, WAR, AND REFUGEES

The absence of attention to refugees in Grotius' thought is shocking; after all, sixteenth- and seventeenth-century Europe was awash in refugees. Jews fled the Spanish Inquisition. Huguenots fled France. Members of the historic peace churches fled Germany. Waldensians fled Italy. Catholics fled lands where the authority of Rome was no longer recognized. Geneva tripled in size in the course of a few decades as it took in Protestant refugees. The establishment of colonies around the world coincided with (and was sometimes directly due to) the movement of peoples seeking safety and stability. A sociological version of the ideal gas law was operative in early modern Europe: As the pressures of social change increased, the temperatures of conflict and the volume of Europe changed as well, and refugees are manifestations of changing volume. As Greta Grace Kroeker has noted,

> [t]he experience of exile, with its profound sense of displacement, loss and alienation, formed a central feature of early modern life. While local religious upheavals, violence and dislocation certainly existed before the fifteenth century, the religious changes and challenges posed by the Reformation, the centralization of political power characteristic of many parts of the continent, and the first era of globalization transformed Europe's religious landscape and disrupted traditional patterns of religious practice and settlement on an unprecedented scale.[58]

[58] Greta Grace Kroeker, "Introduction," in Timothy G. Fehler et al., eds., *Religious Diaspora in Early Modern Europe: Strategies of Exile* (New York, NY: Routledge, 2015), 1.

Early international law and early modern expressions of Christian just war thinking remained nearly silent on refugees. Grotius, himself a religious refugee, attends to them only briefly in Book Two, primarily in a section on the communication of punishments and the obligations of the state to protect refugees while their cases are pending.[59] And while he recognizes a limited right to resettlement (provided refugees abide by the laws of the place they resettle), he can do so only by ignoring a primary reason that many persons were refugees in the first place; namely, that their faith convictions did not align with those of the state that was persecuting them.

Refugees constitute a problematic category for any sort of international law premised on state sovereignty while recognizing the rights of persons.[60] They are not just crossing borders; they are doing so to escape their treatment at the hands of one sovereign country by entering another. When they cross those borders, they carry their beliefs and cultures with them, and it is an open question whether such beliefs and cultures fit much better in their new space than their old one. Their commitments to whatever state they find themselves in are treated with suspicion, given the fact that those very commitments did not seem to fit in whatever state they left. They are dependent on others for aid and support and can threaten to swamp the social services that are available in their new location. In sufficient numbers, they reinforce the very distinction between being part of a state and being part of a nation that the notion of the modern nation-state attempts to erase. Four hundred years into the development of international law, not only the status but even basic conceptions of refugees remain a thicket of competing values,

Surprisingly little theoretical research has been done on refugees during the early modern period, including those moving to flee conflict and religious persecution. See, though, Fehler et al., *Religious Diaspora in Early Modern Europe*; Susanne Lachenicht, ed., *Religious Refugees in Europe, Asia, and North America (6th–21st Century)* (Hamburg: Atlantic Cultural Studies, 2007); Dagmar Freist and Susanne Lachenicht, eds., *Connecting Worlds and People: Early Modern Diasporas* (New York, NY: Routledge, 2016); and Susanne Lachenicht and Kirsten Heinsohn, *Diaspora Identities: Exile, Nationalism, and Cosmopolitanism in Past and Present* (Frankfurt: Campus Verlag, 2015). The advantage of these books is that they are case-based, offering rich descriptions of particular communities and movements. The disadvantage is the comparative lack of the larger contexts and social forces driving dis- and relocation, let alone a sustainable theoretical construct about those movements.

59 "Rights of Suppliants or Refugees, and the many Precedents of *Asylums*, affect not our last Conclusions; for they are intended only for the Benefit of them who suffer undeservedly, and not for such whose malicious Practices have been injurious to any particular Men, or to human Society in general": Grotius, *The Rights of War and Peace*, 2.20.5.1ff.

60 And religious refugees during a secular age occupy an even more peculiar space.

opinions, and political possibilities.[61] One wonders how refugees would be understood and their conditions addressed in international law had Grotius paid more attention to his own situation in shaping early international law. Almost undoubtedly, the trajectory of international law toward state sovereignty would be different.

Refugees also create a problem for just war theory and modern identity. Refugees clearly have the capacity to make choices, but those choices are not framed by identification with their own state; indeed, they regularly make choices to identify with a nonstate community as they decide where to go. In the process, their existence raises questions about who or what constitutes legitimate authority in war. Their choices simultaneously reveal their independence from others and put them in situations where they are likely to be dependent on others. In the process, their movements not only can put them in harm's way but also can lead to suspicions that they are in the service of an enemy state, both of which complicate the maintenance of the *jus in bello* principle of discrimination. Flight from state-sanctioned violence is tantamount to a rejection of the modern nation-state's claim of a monopoly on the legitimate use of force, and flight from nonstate actors (e.g., religious groups, gangs, tribes, families) is tantamount to an accusation that the modern nation-state does not have such a monopoly. In the process, refugees reveal the possibility of a resort to something other than violence. They can economically and politically stress places taking them in, thereby threatening to undermine the receiving state's ability to defend itself, putting a just cause at cross-purposes with a reasonable hope of success. They make the weighing of costs and benefits for purposes of proportionality almost impossible.[62]

[61] "The refugee in international law occupies a legal space characterized, on the one hand, by the principle of State sovereignty and the related principles of territorial supremacy and self-preservation; and, on the other hand, by competing humanitarian principles deriving from general international law (including the purposes and principles of the United Nations) and from treaty. Refugee law nevertheless remains an incomplete legal regime of protection, imperfectly covering what ought to be a situation of exception. It goes some way to alleviate the plight of those affected by breaches of human rights standards or by the collapse of an existing social order in the wake of revolution, civil strife, or aggression; but it is incomplete so far as refugees and asylum seekers may still be denied even temporary protection, safe return to their homes, or compensation": Guy S. Goodwin-Gill and Jane McAdam, *The Refugee in International Law*, 3rd ed. (New York, NY: Oxford University Press, 2007), 1.

[62] A scan of the indexes of fifteen major and recent texts on just war thinking coming from both religious and secular writers (including books by, e.g., Walzer, Elshtain, and Ramsey) reveals only two sentences – one each in two different books – that even mention refugees. Indeed, it is worth noting that arguments about duties to refugees are becoming clearer only as ideas about a *jus post bellum* are arising out of just war thinking – and that such ideas are only just over a decade old at the time of this writing.

That so many refugees are women and children suggests that one reason for the absence of discussion of refugees in just war thinking and early international law is that the modern self is a masculine self. That so many colonizing projects are the product of refugees seeking new space in the new world (and regularly oppressing indigenous persons in the process) suggests that one reason for the absence of discussion of refugees in just war thinking and early international law is that the modern self is white. Both these matters deserve far more attention than I have given here; I hope to return to them in the future.[63] However, that so many refugees are not women or children suggests that even masculine selves, when they are refugees, do not fit the modern conception of the self as expressed in just war thinking and (at least early) international law. And that so many refugees went to a "new world" where they oppressed other persons suggests that the patterns of violence that make just war thinking and international law necessary do not recede just because the locations and victims of violence change.

REFUGEES, AGENCY, AND CAUSATION

Why raise the topic of refugees in this chapter? At an obvious level, questions about refugees are central to a larger project of which this chapter and this book are a part: addressing the types, causes, and understandings of war in the environmental age. Current estimates suggest that the twenty-first century faces the prospect of hundreds of millions of "climate refugees": persons who are forced to move because of rising sea levels, drought, fire, desertification, ocean acidification, crop collapse, hurricanes, and torrential rains and floods.[64] And these estimates do not include the number of refugees who are fleeing violence that has

For explorations of duties to refugees see, e.g., Jennifer Kling, "Who Owes What to Refugees?" and Jovana Davidovic, "What Do We Owe Refugees? *Jus ad bellum,* Duties to Refugees from Armed Conflict Areas and the Right to Asylum," both in *Journal of Global Ethics* 12.3 (2016): 327–364.

63 As a starting point for thinking about the masculinization of just war thinking at the beginning of the modern age, see Laura Sjoberg, "The Inseparability of Gender Hierarchy, the Just War Tradition, and Authorizing War," in Lang et al., *Just War*, 81–98.

64 The term "climate refugee" is, itself, contested and used – where it is used – in such broad ways as to be relatively unhelpful in the context of shaping international law to deal with persons forced to move because of climatic events. See Silja Klepp, "Climate Change and Migration" in *Oxford Research Encyclopedia of Climate Science* (New York, NY: Oxford University Press, 2016), https://oxfordre.com/climatescience/view/10.1093/acrefore/9780190228620.001.0001/acrefore-9780190228620-e-42?rskey=M62RCF&result=1 (accessed March 18, 2020).

been caused, in part, by changes in the natural world. That is, there is an ethical and political urgency to thinking about how to deal with refugees as the modern age comes to a close even as modern resources give inadequate attention to them. Indeed, the very questions of how to compare a climate refugee to a refugee fleeing political violence and how the presence of climate refugees may induce political violence turn out to be both practically and conceptually complicated.[65]

At a subtler level, I raise the topic of refugees in the context of the transition of Christian just war thinking into international law at the beginning of the modern age in order to get at one of the significant lacuna of modern thought: the reductivism of its conceptions of causation. Wars between states cause refugees, but not in direct ways. There are, after all, options for responding to war that do not involve fleeing across a border. One can hide or fight or subvert or simply hunker down and wait. And not all refugees are the result of wars between states. Refugees can be fleeing nonstate violence, economic oppression, or natural disasters. Modern nation-states and their corresponding international law struggle to know how to deal with such refugees. While all social imaginaries contain lacunae, aporias, blind spots, anachronisms, and so forth, the modern age's vision of the self and the understandings of just war that follow from it express themselves in at least this distinctive absence.[66]

The modern understanding of cause tends to be mechanical, orderly, measurable, diagnosable, and tamable. Gone are Thomistic understandings of formal, material, efficient, and final causes (not to mention John Calvin's primary and secondary causes); the disenchanted world finds no more need for them than it has for a teleological vision of the universe or a doctrine of providence. In their place, one finds Newtonian physics, Leibnitzian mathematics, and Darwinian science (not to mention Clausewitzian war, Lockean politics, Smithian economics, Grotian law,

[65] See, e.g., USAID, "Navigating Complexity: Climate, Migration, and Conflict in a Changing World," a discussion paper from the Office of Conflict Management and Mitigation (November 2016), www.wilsoncenter.org/publication/navigating-complexity-climate-migration-and-conflict-changing-world (accessed October 12, 2017). Indeed, the very term "climate refugee" is contested, and alternative terms (climate migrant, environmental refugee, etc.) are also politically fraught.

[66] To be clear: the Christian just war tradition also stays nearly silent on refugees. This is its own failing. That said, the Christian just war tradition assumes that it is part of a larger social and ecclesial context in which treatment of refugees is addressed within some other part of that same context by reference to, for example, Christian obligations of hospitality and care for strangers and outsiders. International law before the twentieth century could not make the same assumption; it meant to be encompassing in a way that the Christian just war tradition did not.

etc.). Even where human beings can only make sense of the universe without being able to change it, human beings retain their agency as observers capable of understanding that universe without divine aid or reliance on nontransferable insights. And where they can change the universe, they not only do so, but they are also the only actors who can give rational accounts of the purposes for their actions. As "man" becomes the measure of all things, all things increasingly submit to his insights, bend to his will, and conform to his projects.

The transition into a secular age in Europe carried within itself a contraction in the range of forces that people perceived could cause things to happen. Left behind was a social imaginary in which supernatural powers, including divine and demonic ones, are simply assumed to be acting in the world. While many people in Europe and its colonies would continue to see the world as a site of supernatural activity – including in a variety of complex ways that manifest as reacting against modernity, romanticizing either a past or a place in which the supernatural continues to exert itself, segregating the realms in which natural and supernatural forces operate, or shaping narratives in which natural and supernatural forces align – the shared assumption that supernatural forces were operative no longer held. Appeals to supernatural causes increasingly mandated appeals to authorities that were not necessarily recognized by the wider public and suffered in comparison with the growing adequacy of appeals to naturally observable and defensible causes, which causes simultaneously grew in authority as their capacities to explain the world grew. F = ma solved problems that teleological narratives did not.[67] The late medieval Christian just war tradition aligned itself with a metaphysically complex universe in which supernatural forces were among those that needed to be taken into account when assessing whether and how to fight (e.g., Thomas' four causes shaped his sense of what makes an authority legitimate; Augustine's emphasis on the telos of love as found in God shaped his sense about what should motivate just warriors when they fight, and Honoré Bonet looked to the natural world for evidence of divine purpose). Unsurprisingly, just war theory in early modern thought aligned that theory with the notions of causation prominent in a secular age: Measuring likelihood of success did not need to include questions about whether "God is on our side" or even, "What is God telling us right now?"

[67] I would note that the literature on the rise of scientific ways of thinking and the relation of those ways of thinking to other ways of thinking (including religious ways) is rich and reveals fascinating complexity to which this paragraph hardly does justice.

The transition into a secular age in Europe also carried within itself a contraction in the ways that forces can cause things to happen. The growing reliance on modes of thought that emphasize measurability, order, and control and can be used to describe events and diagnose consequences (e.g., mechanical and scientific modes of thought) shaped understandings of how things happened. Animals behaved like machines, weather followed patterns described in almanacs, metals were changed through smelting rather than alchemy, the properties of gases (not to mention the idea of gases) were discovered, and science advanced through a method described in Francis Bacon's 1605 *The Proficience and Advancement of Learning.* One consequence of this emphasis on the measurable, orderable, and controllable was the increasing tendency to assess complex effects by separating out the various forces that caused them. Where such effects are the consequence of natural forces measured by the laws of physics, such assessment is coherent and effective. Where such effects are the consequence of a wider mix of forces, such assessment grows more complicated. And where such effects manifest as an interplay of a wide range of forces combined with the variable effects each of those forces is having on the others, such assessment increasingly leads to the exclusion and/or undue reduction of some variables in the attempt to lock in certainty.[68]

Modern just war traditions in all their varieties, including Christian and secular international law varieties, build on and benefit from this vision of causation. Just causes are premised on the possibility that someone or some group could have done something different and thereby avoided the sanctions of justified violence. Right

[68] The transitions in the sciences since the end of the nineteenth century begin to reveal the inadequacies of such a simplified vision of causation. Albert Einstein demonstrated the relativity of time and the connection between gravity and space-time. Werner Heisenberg highlighted the impossibility of measuring both the location and the velocity of an object because the very act of measuring introduces a nonreducible variable. Barbara McClintock did groundbreaking research in genetics on the mobility of transposons. Studies in epigenetics revealed the way environmental and biological structures interact in complex ways. Neuroscientists have discovered that people act and then decide to act. And two particles that are linked by quantum entanglement can be simultaneously affected across light-years of distance by a force applied to only one of them. Beyond changing understandings of cause and effect in science – and one consequence of such changing understandings – has been the rise of attention to what Jane Bennett calls "vibrant matter": the quasi-agentic participation of nonhuman (and even nonbiological) forces in events. See Jane Bennett, *Vibrant Matter: A Political Ecology of Things* (Durham, NC: Duke University Press, 2010). Matters of causation still warrant attention; they have just grown much more complex.

intent presumes the capacity to understand the motivations of agents. Legitimate authorities are those capable of and empowered to act and to be held responsible for actions. Last resort, likelihood of success, and proportionality are all premised on the capacity to do complex calculations of measurement. *Jus in bello* principles of proportionality and discrimination only function where military personnel can make choices about how to fight and whom to target. Laws, including those that are premised on structures that favor state sovereignty, only function in contexts of accountability. And while just war traditions and law have existed both before and outside the Western modern context, it is nevertheless the case that modern conceptions of causality have given these traditions added weight and depth around the world. In a secular age, the causes that are of political interest are those made possible through the powers of human agency.

Yet the paragraphs above start to get at a complexity of causation that threatens to transcend the powers of human agency to describe or create. To begin with, they suggest that the conventional narrative of the transformation of the Christian just war tradition into international law at the beginning of the secular age (one in which the wars of religion lead to a commitment to a secular world in which religious motives should not drive violence and religious reasons should not count in arguments on public matters) is inadequate. That narrative ignores technological changes (new kinds of weapons), demographic shifts (armies made up of commoners), changing political structures (the nation-state), expanding and interdependent economies (via trade), the continued impact of religious thought in shaping nominally secular polities (see Chapter 4), the geographic transformation of the known world (exploration and colonization), and climatic changes (the beginning of the Little Ice Age), among many others – not to mention the interactions of these causes. Had any one of these drivers of change not been in place, the transformation of Christian just war thinking into international law might have looked very different. And many of these drivers cannot be attributed to the insights or actions of a particular person or group of people.

Complicating matters further, as all these causes interacted, they increased the rate of change and degree of impact of each. New weapons shaped new armies, which made new political structures viable, and those structures required new sources of wealth and philosophical justifications for their existence. This, in turn, contributed to the reordering of religious categories, which further reshaped political structures, armies, and weapons. Human interactions with the nonhuman natural world drove the exhaustion of some resources and the search for new ones even as changing climates complicated European economies and shaped exploration, colonization, conquest, and so forth. Indeed, one piece of evidence for the shift from one social imaginary to another is the degree to which many spheres of human existence are all simultaneously in flux.

Moreover, complexities grow exponentially in the context of interacting causes because the rapidity and impact of change leads to the interpenetration of causes and effects. Like an Escher drawing, the effects of a changing system loop back on themselves and drive unanticipatable further change. Where a new nation-state established a successful colony in some other part of the world, not only did it draw resources from that colony to enrich itself and fund further colonization, but also its foothold in that part of the world became a base from which to establish further colonies – even though those new colonies faced new struggles that would not be resolved through the resources of earlier patterns of colonization. Where new ways of thinking about political structures arose, the success of such ways of thinking invited further scholarship, including backlashes against earlier ideas. Arming commoners for battle shaped conflicts and military strategies that relied on larger armies, required more commoners, and led to the politics of suspicion about state motives among empowered commoners. The interpenetration of cause and effect, especially when combined with the interactions of causes, makes measuring them or predicting their consequences all but impossible (and enormously subjective at the very least). Rather than using the language of "cause and effect," one might better speak in terms of probabilities, potentialities, feedback loops, and the political equivalent of chaos theory.

This, then, is one of the central paradoxes of the developing early modern self in the context of just war thinking and international law: He is gaining agency – indeed, being required to assume agency – and recognizing his power to bring about change even as the possibility to actually measure or understand that power is being diminished in the face of the complexity of causes.[69] And both the Christian just war tradition as it changed and international law as it came into existence relied on accounts of human agency that, in turn, downplayed the complexity of the causes that led to those changes/origins. Attending to the lacuna of reflection on refugees is one place this paradox expresses itself. Making refugees visible, then, can serve as a model for making other absences visible.

[69] The gender-specific pronoun is not accidental in this instance.

MISSING THE NONHUMAN NATURAL WORLD: A LACUNA IN EARLY MODERN JUST WAR THINKING

An even larger lacuna of the early modern self acting in the context of just war thinking in international law – one that has been hinted at above – is the absence of any recognition of human interactions with the nonhuman natural world. That war has an impact on the nonhuman natural world is apparent; that the nonhuman natural world (especially with regard to climate change) has an impact on war is less apparent. Yet changes in the climate during the sixteenth and seventeenth centuries, especially due to the Little Ice Age, played a role (that is, served as a cause for) the movement of peoples, conflicts over resources, technological innovations, and the reshaping of economic and political systems. As Brian Fagan describes it,

> Climatic stress, when it does not bring complete collapse, often acts as a spur to social reorganization and technological innovation. In the fourteenth century, Europe was a rural continent with only the most rudimentary infrastructure of roads, harbors, and local mills. Kings and queens reigned over kingdoms and crowded cities haunted by the constant threat of insufficient food. Nine of every ten workers were engaged in growing food, and still the entire continent lived only from year to year. But the exigencies of the Little Ice Age helped bring about an agricultural revolution, which began during the fifteenth and sixteenth centuries in the Low Countries, then spread to England a hundred years later. Many English landowners embraced the new agriculture as larger, enclosed farms changed the landscape. Innovative crops like turnips and clover provided insurance for herds and people against winter hunger. Britain, Flanders, and the Netherlands were self-sufficient in grain and livestock by the onset of the Industrial Revolution in the late eighteenth century. A new scale of agricultural production, combined with an ever-more efficient infrastructure, fed burgeoning cities and rising rural and urban populations.[70]

Fagan's description reveals the difficulties of linking climate to cause. The changes climate brought about were open to a wide range of variables, occurred over significant swaths of time, and only indirectly impacted political structures. Indeed, as much as connections between climate changes and conflict may seem self-evident across such wide sweeps of time, causal connections prove surprisingly difficult to name. The causal relationship between climate and armed conflict has been so tenuous that as recently as the late 1990s Nils Petter Gleditsch could argue that "[d]espite numerous pronouncements on the relationship

[70] Brian Fagan, *The Long Summer: How Climate Changed Civilization* (New York, NY: Basic Books, 2004), 249.

between conflict and the environment, there is no consensus on causal mechanisms. Indeed, several writers have questioned the overall argument [that environment can play any causative role in conflict]."[71] There were simply too many challenges to the research to justify such claims.

Or such was the loose consensus of historians, scientists, and environmental security experts until a 2011 study by Solomon Hsiang and two colleagues lent statistical support to the claim that climate change and civil conflict can be causally connected, including in the current day. Examining the oscillations in the El Niño/La Niña weather patterns and their correlation to violence in the tropics over the past sixty years, Hsiang and colleagues were able to "directly associate planetary-scale climate changes with global patterns of civil conflict [revealing that weather patterns] may have had a role in 21% of all civil conflicts since 1950."[72]

As significant as Hsiang's study has been, it is only the start – a toe into rising waters, as it were – of understanding the causal connections between climatic changes and conflict, and it arrived hundreds of years after the other forces and structures that have shaped modern war were described and defended. In later chapters, I will take up the project of connecting changes in the nonhuman natural world to the changing social and political structures during the early modern period, as well as exploring how those changes manifest themselves in the transition of the Christian just war tradition into international law. For now, though, it is enough to note that too-simple understandings of causation at the beginning of modernity have not only had the effects of ignoring the multiplicity and interactions of many causes, but also have biased contemporary just war thinking and international law against attention to the very causes and effects that may be most significant to wars in a warming world.

More broadly, it is important to note that these understandings and effects are the products of a particular way of answering the question, "How should we understand ourselves?" where answers to that question established the preconditions not only for shaping effective modern polities but also for shaping the Christian just war tradition into a just war theory capable of being transformed into international law.

71 Nils Petter Gleditsch, "Armed Conflict and the Environment: A Critique of the Literature," *Journal of Peace Research* 35.3 (1998): 383.

72 Solomon M. Hsiang, Kyle C. Meng, and Mark A. Cane, "Civil Conflicts Are Associated with the Global Climate," *Nature* 476 (August 25, 2011): 438.

More immediately, it is important to think through a third manifestation of how new answers to the central questions of the Christian just war tradition – namely, those of theological anthropology – led to significant changes in that tradition. Having explored the early modern answers to the questions of how to understand the other via the work of Francisco de Vitoria in the previous chapter and how to understand the self via the work of Hugo Grotius, I now turn to the third interrelated question of theological anthropology: how to understand the state. Toward this end, I turn to the Constitution of the United States and the answers it offers to just war questions about the construction of political systems.

3

Shaping the State

The US Constitution as a Christian Just War *Document*

No Soldier shall, in time of peace be quartered in any house, without the consent of the Owner, nor in time of war, but in a manner to be prescribed by law.

The US Constitution, Amendment 3

Limited wars are as old as the history of mankind, as ubiquitous as armed conflict. In the history of international conflict the wars that have been truly momentous and rare are those that were fought to annihilate, to completely defeat or completely dominate the adversary. These wars have been the principle stimulus of theories and doctrines of how to limit and fight wars as rational instruments of national policy.

Robert Osgood, *Limited War Revisited*

The promise of constitutional politics is that worthwhile gains such as newly won human rights are taken out of political play for a period of time. Activists could use the respite. But the risk is that we rest on our laurels and are overtaken by new events. Another risk is that we forget that even entrenched rights depend for their meaning and power on their vulnerability, on the fact that we (or our forebears) took them and that we must take them again and again.

Bonnie Honig, "Dead Rights, Live Futures"

At first blush, treating the Constitution of the United States of America as an important source for thinking about early modern Christian just war thinking is manifoldly odd. It isn't a person. It isn't an explicitly Christian document. It isn't primarily about war. It recognized in the literature of just war thinking. It comes about comparatively late in the construction of early modern just war thinking; indeed, it comes about during that time period that most Christian just war thinkers simply skip past on their way from Grotius to Ramsey.

I treat it as a source for thinking about early modern Christian just war thinking not in spite of those oddities but because of them. In choosing a document rather than a person, I want to continue to undermine the "great man" myth of modern just war thinking and undergird my claim that understanding changes to Christian just war thinking involves giving attention to technologies, practices, events, and even artifacts such as documents. Artifacts don't simply represent just war thinking; they shape it (so, for instance, Augustine's *City of God* has probably had an impact on the tradition in ways that Augustine's more explicit writings on war have not). And as much as the US Constitution had authors, its power comes not from its ability to convey their thoughts but from its capacity to shape a political community in ways that its authors would not have anticipated. In choosing a nominally secular document rather than an explicitly Christian one, I want to undermine understandings of secularism that treat that term as the opposite of "religious" and push forward a claim that theological concerns shape the modern social imaginary in its construction of the state. In choosing a legal document that lays out the shape of a particular modern polity, I want to undermine the separation of questions about war from larger questions about self, state, and other. In choosing a document that other just war thinkers have ignored in their descriptions of the tradition, I want to undermine the accretionary stories they tell about the movement of the just war tradition through time and undergird my claim that attention to changing social imaginaries uncovers previously hidden sources for just war thinking. In choosing a late eighteenth-century document, I want to undermine the idea that social imaginaries have firm and narrow boundaries during which the revolutionary work of giving new answers to old questions occurs and offer evidence for the idea that periods of seeming silence within the Christian just war tradition are anything but.

More broadly, I've focused on the US Constitution because it puts on full display those questions about the construction of a political system – a "state," broadly construed (and, within that system, a people) – that shape the movement of the Christian just war tradition through time. And, most broadly, I've focused on the US Constitution because it not only displays the marks of modern approaches to Christian just war thinking (almost paradigmatically so) but, in so doing, it also conceals from view arguments, values, resources, and persons that are of import for the Christian just war tradition, especially as it moves into the environmental age. Since the larger purpose of this book is to reveal the significance of the interactions between the Christian just war traditions and the nonhuman natural world, and because modern political thought has disregarded the natural world and concealed the interplay of nature and politics to its own detriment (and ours), getting clearer on the processes through which that tradition hides things is of paramount importance.

Each of those claims warrants further development in order to justify the larger arguments of this chapter, the next, and the larger project that is this book. Because there are so many claims – and because at least two of those claims are so big (that the US Constitution is a *just war* document and that it is a *Christian* document), I have broken these claims into two chapters. In this chapter, I take up the first of these large claims. Seeing the US Constitution as a just war document will mandate attending to not only the ways that it attempts to address the perennial questions of the state asked by just war thinking (and, more concretely, how to construct a state that is attentive to the same types of just war considerations that produced the *jus ad bellum* and *jus in bello* criteria) but also how the answers it gives to those questions display a particular kind of just war thinking that is characteristic of modern just war thought.

Prior to attending to such matters, some background is in order.

SHAPING A STATE OF STATES: ORIGINS AND STRUCTURES OF THE US CONSTITUTION

After the American Revolutionary War came to an end in 1783, the now formerly British colonies needed some way to organize and order themselves. They did not begin *de novo*. The First Continental Congress (September 5–October 26, 1774) had been organized with representatives from twelve of the thirteen British colonies in attendance in order to determine the appropriate response to a series of punitive measures that were taken by the crown in response to the Boston Tea Party.[1] After their appeals to the crown went unrecognized, they called the Second Continental Congress, which began meeting in May 1775. By July 1776, they had decided the only way forward would be to establish their independence from Great Britain. Though there had been skirmishes between colonists and British soldiers prior to this, the signing of the Declaration of Independence formalized the call to war. The Second Continental

[1] Georgia did not send representatives.

Congress functioned as a provisional government during the war and the early years that followed.

While the proceedings of the American Revolutionary War took priority in their considerations, the Second Continental Congress had begun to think through the shape of the nation that was created by the fiat of that signing. It was a heady and uncertain time. Not only was triumph not guaranteed, the representatives in the Second Continental Congress had no formal power to act as they did: They had not been elected in an open process, their opinions were not universally shared in the colonies, and they were a distinctly unrepresentative group, composed only of white male landowners who, while agreeing with each other on a few basic claims, also disagreed with each other on a wide range of issues.

They were, though, thoughtful and far-sighted. While shaping the direction of the war during their gatherings, they unilaterally did things such as appoint ambassadors and military generals, sign treaties, take on loans, issue money, and disperse funds. They also began to think about how their assortment of colonies should relate to each other. Between July and November 1777, articles of confederation were drafted that clarified the loose affiliations between separate colonies. These articles were argued over, revised, rerevised, and eventually ratified by representatives of all thirteen colonies; as of March 1, 1781, the Articles of Confederation took force. These articles were the set of agreements that the representatives reached about how the states and their representatives would govern themselves. Unsurprisingly, the agreements in the Articles of Confederation were few and weak.

After the war, those Articles stayed in force but in the years immediately following, their limitations and weaknesses became increasingly apparent. They did not help build alliances with foreign nations, nor did they do enough to clarify domestic relations between the states, which regularly chose to act in their own interests even when those interests had been prohibited by the Articles. By 1786, their utility was openly questioned by most representatives. So, on February 21, 1787, a new congress was created for the purpose of shaping a functional government to order interstate and international relations by revising the Articles of Confederation.

The committee selected to revise those articles was influenced by contending streams of thought about the shape of governance. John Locke's emphases on rights to "life, liberty, and property" and rule by the consent of the governed played roles, as did Edmund Burke's whiggish approach to history, John Witherspoon's suspicion of pooled power,

Roger Williams' priority of religious liberty, and Calvinism's federal theology, which saw public life as a place where God's purposes were being worked out. These contending streams of thought, combined with the political machinations of representatives seeking to further the power of their own states and extended debates between Federalist and Anti-Federalist forces, shaped a document that would prove to be remarkably awkward in its manifold seams and compromises and surprisingly stable over the decades that followed. The original seven articles of the Constitution gave broad definition to the three branches of government, their relation to each other, the relation of states' governments to federal government, a process for its amendment, the supremacy of the Constitution over other governing documents, and the means for its ratification.

The process of ratification by the states in 1788 extended the debates but also allowed for the creation of a secondary literature[2] especially *The Federalist Papers,* that defended, clarified, and expanded on the document and its various articles. Even more importantly, the debates shaped the first ten amendments to the US Constitution, the Bill of Rights, which collectively expanded on the individual rights of citizens over against the power of the government; these were approved by Congress in 1789 and ratified in 1791. Over the centuries since its initial ratification, twenty-seven of thirty-three proposed amendments have made their way into the document, including the Eighteenth Amendment prohibiting the manufacture or sale of alcohol (ratified in 1919), which was then repealed by the Twenty-First Amendment in 1933.

That the US Constitution can be and has been amended reveals the peculiarities of its strengths. Including a process for amending – imperfect though that process may be – displays a degree of humility on the part of its framers, who recognized that they could not produce a perfect document capable of addressing all the problems of the future. Its revisability allows it to be a living and durable document, capable of organic change and resilient in the face of the challenges that it has faced since its writing. As *The Federalist Papers* and the addition of the Bill of Rights hint at, though, its framers were not so much thinking about shaping a document that could weather the future as providing one that could deal with the storms of their present. Attempting to answer questions about the rights of the various states vs. the rights of the nation, the rights of individuals vs. the rights of state and nation, who

[2] Excluding Rhode Island and North Carolina, which ratified the Constitution in 1790.

should count as a person (which led to the Three-Fifths Compromise, wherein slaves were treated as three-fifths of a person for matters having to do with the numbers of legislators from southern states and taxation), the appropriate structures of governance for moments of crisis vs. those for maintaining order during long stretches of mundane living, and honoring the forces that motivate corporate action (such as religion) without favoring one of those forces or letting them determine national vision, not only drove its creation but also the process of creating and ratifying the Constitution mandated pliancy just to get it passed. Rather than resolve all these questions in their details during the intense debates about their possible answers, the framers created a way for those debates to continue without allowing the debates to prevent ratification. On June 21, 1788, the Constitution was ratified, having been approved by nine of the thirteen states. Not incidentally, the fact that they were writing and ratifying with the memory of the Revolutionary War in their heads and deep concerns about the possibilities of impending wars also shows throughout the document, which set a framework for addressing questions about resort to war and the shape and limits of a national military, thereby ordering matters of war within the context of domestic law.

SHAPING MODERN STATES IN AN AGE OF CONSTITUTIONS

The United States was not alone in creating a constitution. The end of the eighteenth and beginning of the nineteenth centuries was an age of constitutions. Some 1,600 constitutions, pieces of constitutions, declarations of rights, and drafts of constitutions came into existence at the time.[3] Sweden passed a constitution in 1772, the Polish–Lithuanian Constitution came into existence in 1791, and, under the influence of Jean-Jacques Rousseau, Corsica created a constitution in 1755 that existed until Corsica was absorbed by France in 1769. Then, in 1794, Corsica wrote a second constitution that lasted only a few years. France put together a series of constitutions in 1791, 1793, and 1795, each of which superseded the previous one and none of which endured past the nineteenth century. Great Britain even put together an "unwritten" constitution that contained the sum of its laws as they came into existence after the Glorious Revolution of 1688. Norway, the Netherlands,

[3] See Horst Dippel, ed., *Constitutions of the World from the Late Eighteenth Century to the Middle of the Nineteenth Century: Sources on the Rise of Modern Constitutionalism*, 5 vol. (Berlin: De Gruyter, 2008).

Belgium, Spain, and Brazil all passed constitutions in the first decades of the nineteenth century.

Distinct as each one was, they collectively demonstrated the changing visions of Western politics arising during the seventeenth and eighteenth centuries. Subjects were beginning to think of themselves as citizens.[4] Early versions of claims to human rights began to carry weight, including the appeal to rights offered in the US Declaration of Independence.[5] Positive forms of law increasingly found favor among persons disillusioned with the laws of kings and suspicious of turns to natural law. The defining ideas of modernity (described in much greater detail in Chapter 5) increasingly shaped political sensibilities.

The pressing questions that shaped the US Constitution grew out of this changing vision of politics and the instabilities that it promoted. Who are these citizens that have rights they can exert against the state and each other? Where did these rights come from? What is the power of law to enable and constrain their actions in the world? What kind of individual lives are made possible in a world where (some) people are empowered to and charged with shaping their own destinies? How will states filled with such persons engage each other? What opportunities for new ways of living are available to these persons? As of the eighteenth century, questions about self and state increasingly blur into and out of each other as state governments begin to be "of, by, and for the people."

The rhetorical brilliance of the US Constitution – or at least its Preamble – is that it creates the very audience it would speak for: "We the people of the United States ... do ordain and establish this Constitution for the United States of America." Author and audience are the same. Unity not only trumps division, but it also promotes a politics of shared purpose that carries both sacred ("do ordain") and secular ("and establish") undertones. Lacking the Preamble, the imperatives that constitute the Constitution could sound like outside impositions and the powers given to each branch of government could seem arbitrary. Indeed, lacking a society of persons that think of themselves as the "We" of the Preamble, the rest of the document lacks the force to build or maintain any of the structures it describes. The nation's sovereignty is a collective loan from the banks of its citizens' sovereignty. As legal scholar and literary theorist James Boyd

[4] See Michael Walzer, *The Revolution of the Saints: A Study in the Origins of Radical Politics* (Cambridge, MA: Harvard University Press, 1982).

[5] For a provocative reading of the origins of human rights, see Lynn Hunt, *Inventing Human Rights: A History* (New York, NY: W. W. Norton and Co., 2007).

White has argued, "[t]his Constitution – like other such instruments – is thus in a literal sense a rhetorical constitution: it constitutes a rhetorical community, working by rhetorical processes that it has established but can no longer control. It establishes a new conversation on a permanent basis."[6]

As much as the ideas that shaped it may have come from earlier thinkers, it is a peculiarly self-creating and self-authorizing document. Thomas Hobbes may have provided a vision of the sovereign body of the nation as composed of the actual bodies of its people; John Locke may have offered a vision of individual rights to life, liberty, and property that are the center of an implied social contract between citizens; Montesquieu and Madison may have gifted it with a vision of checks and balances in order to constrain power with power; Thomas Jefferson may have provided a vision of church–state relations via his *Statute of Virginia for Religious Freedom;* the *Magna Carta* may have supplied a vision of due process; the English Bill of Rights may have inspired trial by jury; and the Iroquois Confederacy may have modeled a way for multiple nations to unite. But these and other sources of wisdom, even thoughtfully combined, wouldn't necessarily shape both the community and the laws to which that community would be beholden. The authority of the US Constitution is a product of rhetorical sleight of hand, creating the very thing it is nominally describing.

This suggests that the foci of autonomy, universalism, instrumentalism, and immanence that define modernity (discussed in greater detail in Chapter 5) don't pertain only to human beings and their capacities. In its own limited way, the Constitution is an autonomous thing: It relies on its own instrumental qualities to create a universal community (or at least a state community that will not be defined by any particular cultural characteristic or national idiom). At the same time, it authorizes itself by assuming that this community's existence and support will be adequate to justify its continued existence by reference to its continuing usefulness for the previously fictive "We the people" that it creates.

PEOPLE CREATE LAWS AND LAWS CREATE A PEOPLE

In a series of important books, Roman Catholic ethicist and legal scholar Cathleen Kaveny has argued that "law functions as a moral teacher, as a teacher of virtue, even in pluralistic Western democracies like the

6 James Boyd White, *When Words Lose Their Meaning: Constitutions and Reconstitutions of Language, Character, and Community* (Chicago: University of Chicago Press, 1984), 246.

United States. In such societies, ... it is particularly important that the law strive to support autonomy and solidarity, which can be understood as aspects of the virtues of prudence and justice, respectively."[7] Kaveny's argument, clearly implied through reference to the virtues of prudence and justice, is that in a classic (read: Thomist) understanding of the law, the law functions not only as a "police officer" but also as a teacher and an engineer in the construction of a particular type of community. In the American context, "autonomy and solidarity are key virtues for American law to foster, precisely because they enable both individual and communal flourishing in pluralistic liberal democracies."[8] Moreover, it is vital that Kaveny specify the American (or, rather, pluralistic liberal democratic) context in order to advance her argument that autonomy and solidarity are primary virtues that the law teaches because, "[i]f it is to be pedagogically effective, law must key its moral message to the actual culture and customs of the people it purports to govern."[9] Kaveny's argument about the pedagogical function of law – and, therein, the need to attend to the virtues that the law teaches – helps make sense of the founding legal document of the United States. As much as the Constitution attempts to frame rules for the nation's political structures, the inclusion of the Preamble as part of the Constitution reminds us not only that there is a teleological function inherent in the document (for, on Kaveny's terms, all systems of rules will at least implicitly have such a teleological function), but that this particular document has a particular teleological function, namely, the construction of a people that can grow in autonomy and solidarity.

Kaveny also makes a closely related but distinct argument: that human beings are the type of creatures that not only can grow in autonomy and solidarity but also are created to grow toward autonomy and solidarity. Here, her Roman Catholic perspective registers strongly. So, for example, when she takes up solidarity, she notes that

> [t]he virtue of solidarity, as understood in contemporary Catholic thought, has three fundamental components that correspond to and supplement ... conditions for true human autonomy ... First, solidarity requires us to meet the basic

[7] Cathleen Kaveny, *Law's Virtues: Fostering Autonomy and Solidarity in American Society* (Washington, DC: Georgetown University Press, 2012), xi. See also her *A Culture of Engagement: Law, Religion, and Morality* (Washington, DC: Georgetown University Press, 2016) and *Prophecy without Contempt: Religious Discourse in the Public Square* (Cambridge, MA: Harvard University Press, 2016).

[8] Kaveny, *Law's Virtues*, 18.

[9] Ibid., 2.

needs of each person inasmuch as that person is not only an abstract chooser but also a being who is an integrated unity of body and soul. Second, solidarity requires us to recognize the nature of each person as essentially social ... Third, solidarity requires us to provide vehicles through which all persons can contribute to the community through the expression and development of their own talents and interests.[10]

By arguing from a particular anthropology, Kaveny's Roman Catholic perspective about human nature helps fund her arguments about the law's pedagogical function and its promotion of virtues (though some other Christian communions, such as Calvinism, could do the same, as could other religious traditions such as Judaism). However, in this instance, her perspective also partially conceals something vital to the genius of the Constitution: Its implicit claim that no particular vision of human nature and no preceding anthropology is necessary for it to create "We the people." Kaveny's argument might better be framed as a way of revealing how a Roman Catholic anthropology can align with the Constitution's project rather than how it can underwrite it.

This distinction is important because the Constitution treats the law that comes from the people that it creates as free from any obligation to be bound to a larger set of claims about an intrinsically moral universe that can be known through natural law (which distinguishes it from its sister document, the Declaration of Independence, and that document's "self-evident truths"). That is, it evinces a form of positive law that not only doesn't need to grow out of natural law (making it more functional for a pluralistic society in which there may not be agreements about that natural law), but also doesn't even need to attend to natural law (making it more distinctively liberal in form).

Moreover, by creating the people that will both uphold and submit to it, the Constitution is reflexively self-validating. Indeed, it must be so because this reflexivity mirrors the way that the people it teaches think of themselves: as both sovereigns and subjects. The modern self (at least in liberal society) is the type of creature whose autonomy is both a law arising from the self and a law adequate to order the way many selves interact. The Constitution creates the conditions in which those who shape it and those who live under it can be the same people.

Whether the positivism implied in this vision of the law and its relation to its people is adequate to the tasks that law takes up is a question that moves far beyond the scope of this work to address. Whether

[10] Ibid., 28.

autonomy and solidarity can really be promoted without grounding them in a prior vision of human nature, or whether law can be both stable and free-floating, or whether positive law shorn from natural law can endure over time, or whether the promotion of individual freedom and equal justice can happen together in a pluralistic society: These are questions for another day and for persons far wiser than I to take up. I am not offering a defense of this vision of law so much as describing a way of understanding that vision because that vision has significant ramifications for how we understand the Constitution as a just war document. More importantly, as I will contend shortly, it underwrites my argument that this document (rather than a particular person or even group of people) reveals its location in an axial point in the development of Christian just war thinking. That is, the US Constitution functions as a sign mandating our interpretation of the way the Christian just war tradition moves through a particular point in time that a preoccupation with persons within that tradition conceals.

This is one of the reasons that I have chosen to focus on a document in a book on the history of the Christian just war tradition. When that tradition is described in terms of a history of great thinkers, it not only disregards the degree to which powers, practices, artifacts, events, and technologies beyond the capacities of any individual to control drive the tradition, but it also wrong-foots those who would use the tradition to address enduring questions about whether, when, why, and how to engage in acts of violence. In this instance, debates about the receding place of "legitimate authority" in modernity have missed the way that authority in modernity is not so much diminished as diffused:[11] The process, rather than a particular person, authorizes. As unsettling as such an approach may be (Who manages the process? How open to manipulation is that process? How is a process trained to make decisions? How can a process be held accountable?), and as much as this diffusion of authority shapes new debates, getting into the right debates is preferable to winning the wrong ones.

Having now claimed that the US Constitution functions as an important signifier about the way the Christian just war tradition moves through time, the next steps are to take up four linked questions of "state" that arise out of that claim: What does it mean to think of it as a just war document? How is it interestingly "Christian?" What do

[11] See Anthony Lang, Cian O'Driscoll, and John Williams, eds., *Just War: Authority, Tradition, and Practice* (Washington, DC: Georgetown University Press, 2013).

answers to those questions reveal about its significance for the tradition? And what are the implications of these questions for climate-shaped conflict in the twenty-first century? The rest of this chapter and the next one address these questions.

THE US CONSTITUTION AS A JUST WAR DOCUMENT

Though there is a significant body of literature on just war thinking in the Constitution of the United States, it is almost entirely restricted to the realm of law.[12] As such, it is primarily caught up in questions about the war powers provided in the Constitution. Its arguments have to do with those matters that have preoccupied legal scholars and been the sources of debates about proper interpretation of the Constitution (such as debates about originalism). What most of this literature does recognize, though, is that the framers were familiar with the writings of just war thinkers such as Grotius, Vattel, Pufendorf, and Locke, and that they attempted to incorporate insights from such thinkers into the document they were creating.[13] Outside of US jurisprudence, though, attention to the Constitution and just war thinking is much rarer. Anthologies of just war thinking of both religious and secular varieties do not include it. Just war theorists do not discuss the Constitution apart from, for example, treating it as a resource for addressing questions about conscientious objection. And scholars of religion and law that do attend to the Constitution tend to be preoccupied with the First Amendment.

Yet the document – and especially the original seven articles and first ten amendments – is suffused with claims about and the concerns of war.

[12] There are arguments about how antecedents to the Constitution of the United States of America (including the Declaration of Independence) and documents arising concurrent with the Constitution (e.g., the Federalist Papers) are either shaped or not shaped by just war thinking. These arguments are complicated by the tendency of contemporary authors treat modern just war theory as a kind of moral calculator: one punches in data and a result (a war is either just or not) comes out. However, arguments about war and just war thinking were far less formalized in the eighteenth century than they have become since. For purposes of simplicity and clarity, I will only note the existence of these arguments here. Among such arguments – and to get a sense of the literature – see, for example, Eric Patterson and Nathan Gill, "The Declaration of the United Colonies: America's First Just War Statement," *Journal of Military Ethics* 14.1 (2015): 7–31.

[13] See, for example, Louis Fisher, *Presidential War Power*, 3rd ed. (Lawrence, KS: University of Kansas Press, 2013); Charles Lofgen, "War-making under the Constitution: The Original Understanding," *Yale Law Journal* 81.4 (March 1972): 672–702; or Daniel J. Hessel, "Founding Era *Jus Ad Bellum* and the Domestic Law of Treaty Withdrawal," *Yale Law Journal* 125.8 (June 2016): 2394–2444.

Those claims and concerns not only reveal the impact of earlier thinkers such as Grotius and Locke, but they also display a series of novel arguments about fundamental *jus ad bellum* and *jus in bello* questions. That these are novel arguments both helps to explain the larger silence of just war thinkers about the Constitution and also is unsurprising, given the degree to which the framers were trying to create a new kind of state and people. There were not only technical politico-legal questions they were trying to answer as they argued over it, but also there were questions about how this new nation would understand itself. Toward that end, the document mixes form and vision in a rhetorically complex document.

JUST WAR THINKING IN THE ORIGINAL ARTICLES

Nowhere is the significance of the rhetorical power of the US Constitution to shape a people and a vision of how that people might morally engage a world of conflicts more important than in those matters that are most quintessential to the modern nation-state and its monopoly on the legitimate use of physical force. All states maintain some interest in their own defense; the modern nation-state, constituted by its citizens, must describe that defense in terms that recognize the complex interplay of the sovereignties of state and the citizen while also describing and maintaining the state's internal workings and external relations. So, for example, where the powers of Congress are described in Article I, the Constitution gives to the legislative branch the powers:

> To define and punish Piracies and Felonies committed on the high Seas, and Offenses against the Law of Nations;
>
> To declare War, grant Letters of Marque and Reprisal, and make Rule concerning Captures on Land and Water;
>
> To raise and support Armies, but no Appropriation of Money to that Use shall be for a longer Term than two Years;
>
> To provide and maintain a Navy;
>
> To make Rules for the Government and Regulation of the land and naval Forces;
>
> To provide for calling forth the Militia to execute the Laws of the Union, suppress Insurrections and repel invasions; [and]
>
> To provide for organizing, arming, and disciplining, the Militia, and for governing such Part of them as may be employed in the Service of the United States, reserving to the States respectively, the Appointment of the Officers, and the Authority of training the Militia according to the discipline prescribed by Congress.[14]

[14] The Constitution of the United States of America, Art. I, Sect. 9.

Then, in Article II, the Constitution says that "[t]he President shall be Commander in Chief of the Army and Navy of the United States, and of the Militia of the several States, when called into the actual Service of the United States."[15] And Article III extends the power of the judicial branch

> to all Cases, in Law and Equity, arising under this Constitution, the Laws of the United States, and Treaties made, or which shall be made, under their Authority;—to all Cases affecting Ambassadors, other public Ministers and Consuls;—to all Cases of admiralty and maritime Jurisdiction;—to Controversies to which the United States shall be a Party;—to Controversies between two or more States;—[between a State and Citizens of another State;—] between Citizens of different States,—between Citizens of the same State claiming Lands under grants of different States, [and between a State, or the Citizens thereof, and foreign States, Citizens or Subjects.][16]

Through just its first three articles, the Constitution has already connected all three branches of government to matters of war and the impacts of war. And as much as it has attempted to separate out various wartime powers (including pre- and postwar matters) among those three branches, these separations are more principled than practiced since the three branches of government interact with each other in myriad ways. By way of hypothetical contrast, for example, it is fully possible to imagine either a constitution in which the military serves as its own branch of government (thereby maintaining its independence from the processes and structures of civil life – which is, roughly, what John Locke recommended) or a constitution in which control over the military and all war-related matters is given to one of the branches of government for purposes of mobilizing efficiency and clarity of authority (as Blackstone recommended). This complexity is purposeful; when *jus ad bellum* questions of legitimate authority arise, deliberation is paramount and best done in groups. As David Gray Adler has noted, it is "the cardinal tenet of republican ideology that the conjoined wisdom of many is superior to that of one."[17]

That all three branches of government have military-related responsibilities, including responsibilities related to oversight and governance, manifests the complicated relationships between the state and the state's

[15] US Const., Art. II, Sect. 2.

[16] US Const., Art. III, Sect. 2. Oddly, Article III also includes discussion of treason, limiting how the term should be understood and establishing standards for conviction while also giving to Congress the "Power to declare the Punishment of Treason" (US Const., Art. III, Sect. 3).

[17] David Gray Adler, "Foreign Policy and the Separation of Powers: The Influence of the Judiciary," quoted in Fisher, *Presidential War Power*, 1.

military that must have preoccupied the Constitution's framers. Theirs would not be a military state but it would be a state in which the military would play an important role. Theirs would not be a state in which military expertise was a precondition for public office but it would be one in which all federal public officers had to pay attention to the military. Theirs would not be a military freed to pursue its own political purposes but it would be one in which the very complexity of the military's relationships to the state could leave it bogged down in the politics of checks and balances (and, over time, one in which the complexity of checks and balances could either inhibit its clear functioning or serve to buffer it from responsible civilian leadership). It is as if the framers felt ambivalence about the military – perhaps the consequence of their experience of the glories and horrors of revolution, of their convictions about the freedoms and sovereignty of citizens given as a bulwark against state violence even as their young nation needed to express its own sovereignty on the international stage, or of the concerns that a young and vulnerable country would have about European empires with standing armies in an age of colonization – and that ambivalence expressed itself in the overlapping responsibilities of distributed governance. This posture of ambivalence characterizes just war thinking, distinguishing it from the postures of those who either think only about beating swords into plowshares or only about beating Mars' drums.

Having delineated responsible parties, the Constitution proceeds, in Article IV, Section 4, to note that the government has the responsibility to protect states from invasion. Even prior to the inclusion of a single amendment, then, the Constitution addresses *jus ad bellum* questions about just cause (self-defense, the punishment of wrongdoers such as pirates), legitimate authority (diffused through the three branches of government), and, at least implicitly, proportionality (by limiting the appropriation of money for wars) as well as *jus in bello* concerns such as discrimination and proportionality (via its attention to traitors and the disciplining of militia). And it does this within the context of a brief series of relatively short articles intended to give shape to the wide range of concerns that come with defining the basic political structures of a new, sizeable, and diverse nation composed of many smaller states.

Moreover, the US Constitution addresses such questions in novel ways. The idea that a civilian – and a temporarily elected one, at that – should be in charge of the military assumes not only that someone outside the military chain of command should ultimately be answerable for questions about when and whether to fight but also that persons without military training or combat experience are as capable of making wartime decisions and shaping military strategy as those who have such things. Not only are *jus ad bellum* questions of legitimate authority diffused through the various branches of government, but they are also reframed by modern presumptions about universality (anyone in such an elected position can make these decisions) and instrumentalism (the data for making such decisions are publicly available and separable from the decision-maker's experience).

JUST WAR THINKING IN THE BILL OF RIGHTS

What begins in the seven articles of the Constitution continues in its first set of amendments, starting with the Second Amendment. The occasion of much controversy in the latter half of the twentieth century and beginning of the twenty-first, the Second Amendment links two ideas that don't necessarily need to come together: the establishment of a "well regulated Militia" and the "right of the people to keep and bear Arms."[18] Not only does the amendment reveal the changed shape of modern war (i.e., one in which the masses, rather than a particular class, fight, therein mandating the imposition of disciplines for those masses), but it also reveals the changing understanding of those masses: as people with rights to keep and bear their own arms. More than simply an expression of a "frontier spirit" in which weapons might be necessary for personal self-defense, the back half of the amendment affirms the sovereignty of citizens who can not only own their own weapons but also be trusted to be responsible for them.[19] Whether those citizens will adhere to a principle of discrimination in any particular conflict is a legitimate concern addressed to the Second Amendment; the changing shape of war during

[18] "A well regulated Militia, being necessary to the security of a free State, the right of the people to keep and bear Arms shall not be infringed." US Const., Amend. II.

[19] While it has since been interpreted to mean that the people can keep their own weapons in order to protect themselves from the government, I find no indication of that interpretation in the Second Amendment – and have trouble imagining a mindset in which the framers would have been so concerned with the oppressive reach of government that they kept an armed constituency around to constrain that government. Restraints to government are built into their ordering political system (via, e.g., checks and balances and an expansive list of rights); for them, revolutions should happen in voting booths. Moreover, the prospect of collections of citizens being capable of organizing and arming themselves into effective military opposition to government forces seems a far-fetched supposition. And far from undermining that supposition, the American Civil War reinforces it because it was states using their respective militias, not individuals, that revolted.

the eighteenth and nineteenth centuries would reveal the need for something like Lieber's "Code for the Government of Armies in the Field." The answer to the question of whether those citizens had the capacity to adhere to such a code, though, is affirmed and based in a right.

Of the first ten amendments to the Constitution, the Third Amendment is most obviously connected to just war thinking; it is also, though, most clearly outside the frame of reference of the just war theory that had been developing in modernity. "No Soldier shall, in time of peace be quartered in any house, without the consent of the Owner, nor in time of war, but in a manner to be prescribed by law."[20] The amendment names and links two kinds of rules about the behavior of soldiers: a rule about peacetime behavior and a rule about where to get guidance for soldiers' behavior during wartime.

These two parts to the amendment make it doubly odd from a modern just war theory perspective. The first rule is the easiest to apply as it doesn't point toward some other adjudicating source and it clearly applies to what soldiers can and cannot do. However, since it applies during peacetime and regulates nonmartial behavior, it doesn't fall within the sphere of *jus in bello* categories either temporally or conceptually. Its existence presumes not only the existence of mass armies, but also that these will be large standing armies. Further, it presumes an understanding of law in which persons have a right, among other things, to own and control their own property. These presumptions complicate the direct application of either a premodern Christian just war tradition or a modern just war theory. To what sources in the tradition would the first part of the Third Amendment go? Perhaps John Locke's writings on, especially, civil government might be of help, or Emmerich de Vattel's writings on noncombatant immunity, but those writings would offer tenuous foundations at best for some of the convictions that sit behind the Third Amendment. The first part of the Third Amendment should also, then, serve as an important reminder that just war thinking doesn't arise out of rules so much as just war rules arise out of more basic sets of values and virtues that are operative in particular societies. In both war and peace, respect for the other and the refusal to let right be adjudicated by might pertain.

Where the first part of the Third Amendment doesn't fit modern just war criteria but clearly fits just war thinking, the second part of the Third Amendment could, at least by extension, fit just war criteria but

[20] US Const., Amend. III.

may or may not fit just war thinking. Pertaining to acts during wartime, it could be understood as derived from the *jus in bello* criterion of discrimination: Distinguishing between combatants and noncombatants extends to the treatment of noncombatants' stuff. Yet by deferring decisions about whether and how to quarter soldiers in private houses to the law reigning in that time and place without naming that law, it at least has the potential to be derived from values and virtues foreign to just war thinking (e.g., property rights). And especially coming where it does in the Constitution, the second part of the Third Amendment doesn't even fit international law categories as those were shaped by the just war tradition. Moreover, by the Third Amendment, it's clear that these rules apply domestically since the Constitution shapes laws that pertain to the behavior of the US government and its citizens, not to foreign states.

All of which makes the Third Amendment especially interesting as an expression of just war thinking. The failure of either part to neatly fit within either premodern just war thinking or modern just war theory reveals the contextuality of any expression of just war thinking on the one hand and the cramped quality of a just war theory to address the full range of issues that just war thinking might take up on the other. Little wonder that just warriors haven't paid attention to the US Constitution, given the failure of the Third Amendment to fit in to either category.

Think, for instance, what closer attention to the Third Amendment as a piece of just war thinking might do. It could expand the range of just war considerations beyond those that admit to *jus ad bellum* and *jus in bello* calculations as manifest in modern just war theory. Provided that the language of the current just war criteria continues to function, it could remind us that the various criteria are interrelated: The costs to noncombatants' property might need to be expressed not only in calculations of proportionality but also in matters of discrimination as well. It could foreground attention to just war thinking not only in international law but also in domestic law. It could point toward the question of what else has been improperly and unhelpfully left out of the narrative that the Christian just war tradition tells about itself at a point when that narrative may be changing in discontinuous ways. And it highlights the importance of seeing artifacts as well as people as having the capacity to shape the movement of the Christian just war tradition through time.

Aside from a brief though provocative comment about excepting military cases during time of war from the rights guaranteed in the Fifth

Amendment,[21] just war thinking then recedes from view in the rest of the Bill of Rights. This is unsurprising; the remaining amendments must cover a wide range of basic rights for this society of citizens. Yet the Fifth Amendment further reveals how complicated just war thinking proves to be when instantiated in (domestic) constitutional law. What does it mean for military personnel to be excepted from due process and the other rights guaranteed therein during times of "War or public danger?" To what degree are they excepted from these rights? Why are military personnel a distinct category, especially in a military composed of persons who come out of and return to the wider civilian population? Why are these exceptions only to occur during "War or public danger," how broad a category is "public danger," and who determines the answers to those questions? And why is the military not excluded from other basic rights or parts thereof enumerated in the Bill of Rights, whether in wartime or not? While these questions don't obviously spring out of just war thinking, they do imply it, at least to the extent that they draw a clear difference between patterns of acceptable behavior during wartime and/or by soldiers and those that are acceptable outside such times or by such persons and they reinforce just war notions about the relation between soldiers and civilians that have already been shaped by the earlier parts of the Constitution.

JUST WAR THINKING AND THE CHRISTIAN JUST WAR TRADITION AFTER THE BILL OF RIGHTS

After the American Civil War, questions about how to deal with those engaged in insurrection or rebellion came to the fore. Sections Three and Four of the Fourteenth Amendment touch on these matters, though like the military exceptions in the Fifth Amendment, the statements made in the Fourteenth Amendment do not obviously arise exclusively out of just war concerns (that is, those who are not just warriors might affirm them). Then, from the back half of the nineteenth century onward, no other ratified amendments address war issues (nor do any failed amendments,

[21] "No person shall be held to answer for a capital, or otherwise infamous crime, unless on a presentiment or indictment of a Grand Jury, except in cases arising in the land or naval forces, or in the Militia, when in actual service in time of War or public danger; nor shall any person be subject for the same offence to be twice put in jeopardy of life or limb, nor shall be compelled in any criminal case to be a witness against himself, nor be deprived of life, liberty, or property, without due process of law; nor shall private property be taken for public use without just compensation." US Const., Amend. V.

though debates about the Equal Rights Amendment did lead to questions about women in combat).

The absence of just war considerations in the bulk of the constitutional amendments does not negate my larger argument that the US Constitution is a just war document any more than the absence of discussion of war during long sections of Grotius' *The Rights of War and Peace* discounts that text from being a significant just war document. Indeed, one might take the absence of just war considerations in amendments after the Fifteenth as evidence of a larger shift in society in which war was increasingly viewed as an abnormal aspect of civilized human existence. As I hope the arguments above have made clear, though, the US Constitution, especially in its earlier years, is a document framed by just war thinking.

As a just war document that shapes the trajectory of the movement of the Christian just war tradition through time, the US Constitution is worth analyzing in terms of the changes it brings to the tradition. What are the continuities it shares with the tradition from which it came? What is novel about it? And how do the answers to those questions clarify the way the just war tradition asks questions about the shape of the state? That is, what is the US Constitution's formative power for the tradition?

Let us begin with an argument already named: It is a legal document about the basic structure of a state. That is, it is one way of answering questions about what a state is and does. There are others. Some states (or their rough equivalents) existed long before they were created through law. Some states do not have a written legal document that defines them. Some states existed and then wrote documents about their shape and purposes later. There are, moreover, written laws that define communities that do not necessarily shape states (one thinks of the Torah). The Constitution reveals one particular and modern way of answering questions about the state by inviting us to take our questions to a text of laws rather than, for example, a collection of contemporaneous scholarly works, a particular tradition, or the *habitus* of persons, communities, and their behaviors.

When modern just war thinking answers the basic questions of the tradition about the state via a founding legal document, it does so by referring them to structuring laws, with all that such reference entails. It does not begin in the pursuit of wise men and women (though it may invite wise men and women to help it reflect on the law's claims about what makes wars just and unjust). It does not begin via recourse to a longer tradition of thought (though it may discover something about the law by situating it in a tradition of law). Nor does it begin with an initial explication of a particular set of experiences, events, or cases (though it may seek narratives about how this set of laws came to be). Addressing the law to discover answers to questions about the state implies accepting the abstract, the generalizable, and the ahistorical as built into the starting point of the conversation because good and just laws apply widely, maintain impartiality, and function across multiple contexts. Small wonder that just war thinking is increasingly understood as a just war theory in modernity: the forms of reasoning used by a theory mesh with those used by a state that has been framed by a legal document.

Similarly, addressing the state to discover answers to questions about war implies accepting the way states think. A state defined by its constitution will see questions of war through law-tinted lenses and will address such questions not only through law but also according to its own concerns, especially those having to do with its governing structures, its own continuance and legitimacy, and its defense of its citizens – including, in this instance, against unjustified incursions by the state against its own citizens. A state formed by a constitution will answer just war questions about just cause, legitimate authority, proportionality, discrimination, and other just war matters by reference to law and it will shape those laws according to its own needs, not according to any supposed mandates from a tradition of thought or a particular person that speaks for that tradition. Constitutional and just war frameworks overlap, but they don't align. As such, some just war questions (e.g., is there a probability of success in this case?) will go unanswered in the constitution even as some of the state's just war concerns as they are expressed in the constitution will go unanswered by the just war tradition (e.g., how should soldiers behave when they are not at war?).[22]

[22] Perhaps a contrast is instructive. Immanuel Kant published *Perpetual Peace* four years after the Bill of Rights was ratified. In that book, Kant simultaneously criticized and argued for a reshaped just war tradition according to his larger political vision. So, on the one side, Grotius, Pufendorf, and the rest are "sorry comforters" of the tradition that are useful in justifying military aggression and useless in constraining such aggression because they offer only recourse to toothless international law. On the other side, if one could build a legal framework in which it was obvious that states' legitimacy turned on their participation in an international law framed by mutual recognition, then rules about whether, when, and how to fight might follow. Building from these claims, Kant advanced a series of preliminary articles that he thought would be defensible in the context of mutual recognition and that meshed with his deontological/categorical imperative approach to morality.

One implication of seeing the US Constitution as a just war document is that such seeing recognizes that the Christian just war tradition didn't translate itself only into international law in the modern age; it also translated itself into law more broadly – including domestic law – and positive law specifically. The broader translation is unsurprising. On the one side, positive law of the sort revealed in the Constitution provides an immanent and instrumentalist framework for doing political philosophy. On the other, growing questions about the stability of a pluralistic state that also honors the autonomy of its citizens mandates prioritizing that minimum set of commitments that a wide range of people could support and these commitments tend to group around acceptable processes rather than agreed-upon conclusions.

A second implication follows. As just war thinking becomes increasingly institutionalized in law, clarity about a just war tradition's independent existence and function diminishes. Laws about whether, when, and how to fight are fitted into a larger set of occupying concerns about how states understand themselves and relate to each other. And when states – especially states framed by constitutions – take up just war thinking via their laws, they do so thinking as states. As states determined whether and when to fight according to their own interests, *jus ad bellum* criteria wouldn't cease to function so much as be reconceived according to positive law frameworks that states would advance in the interests of defending their own moral legitimacy. Ultimately, this transition would shape products such as the Hague and Geneva conventions and the United Nations, all of which are premised on the priority of the state in jurisprudence about war. It also may help to clarify why twentieth-century just war theorists such as Paul Ramsey and Michael Walzer tended to give most of their attention to *jus in bello* considerations: Such considerations

Both Kant and the US Constitution think in terms of law. Both emphasize the sovereignty of persons as citizens in modern society. Both give only limited attention to classical just war thinking and offer to address novel concerns in an age of mass armies. And both frame their arguments immanently, without reference to a larger moral framework derived from prior agreed-upon convictions about the existence and work of God and/or a moral order to the universe. But Kant's project can give voice to a particular perspective, offer reasons behind his conclusions, explicitly engage traditions of thought, and are guided toward a particular teleological conclusion (the "perpetual peace" of the title). All these go missing in the US Constitution because it is the product of negotiated agreements, serves a single particular purpose, and treats the state itself as the basic good with which it is concerned. See Immanuel Kant, *Perpetual Peace* (New York, NY: MacMillan, 1903). See also Brian Orend, *War and International Justice: A Kantian Perspective* (Waterloo, ON: Wilfred Laurier University Press, 2000).

were what was left to individual just war theorists after *jus ad bellum* matters became the purview of the modern state.[23]

Finally, a third implication of the turn toward state-oriented thinking is that the forces that drove constitutional thought also undermined early modern just war thinking. Political philosophers such as Immanuel Kant and Jean-Jacques Rousseau ridiculed the just war thinking of Vitoria, Grotius, and others and advanced visions of political systems based on social agreements among enlightened persons. Legislators faced questions about reconciling a commitment to the rights of all persons with the existence of slavery, which led to debates (and, eventually, wars) about whether and how to end slavery that early modern just war theorists were ill-equipped to answer. Militaries grew in size and strength as classes leveled out, and as this happened rules governing actions in battle such as those of the Lieber Code were formalized and enforced by these new types of militaries. And wars increasingly took on Clausewitzian flavors as states interpreted conflicts as political events via their language of positive law. Just war thinking never disappeared in modernity; paradoxically, though, it did make a fairly abrupt turn away from the very early modern sources that had shaped modern thought (a turn, I would note, that is characteristic of modern thought more generally).

FROM THOUGHT TO TRADITION TO THEORY AND BACK: THE US CONSTITUTION AS A JUST WAR DOCUMENT

That the US Constitution is a modern just war document that advances its own version of just war thinking is less surprising once one recognizes that anyplace one finds a relatively stable and durable state (or its political equivalent), one finds just war thinking. States (or their equivalents) have an interest in continuing to exist, including by exercising rights to national self-defense through force of arms, and stable and durable states have a matching interest in couching their potential or actual recourse to arms in terms that lend moral legitimacy to their arguments. Whatever else it is, the US Constitution is a document about the structuring of a state that wishes to continue to exist (after all, for all its failings, the US Constitution offers the framework for a remarkably

[23] Among the best accounts I have read for this aspect of the transition of just war thinking into a kind of legal thinking is Alex J. Bellamy, *Just Wars: From Cicero to Iraq* (Malden, MA: Polity Press, 2006), esp. ch. 4.

durable state: Powers stay divided and amendments are always possible). And by grounding itself in law (rather than, e.g., the divine right of a king), the US Constitution claims the moral authority offered it by citizens who recognize its moral legitimacy in shaping their political structures and interactions.

To this extent, the US Constitution can function synecdochally for any modern law-based framework for shaping states or their political equivalents through the lenses of just war thought. That is, thinking of the US Constitution as a just war document explodes the range of sources that shape just war thinking. When Michael Walzer declared that the just war theory had proven itself victorious,[24] he wasn't arguing that just war thinking had finally broken through to defeat its rivals so much as arguing that a particular vision of just war thinking – a theory that arose in modernity and was strongly shaped by premodern religious (primarily Christian) influences – was victorious. Just war thinking more broadly has always played a significant role in world affairs, and the US Constitution is just one more piece of evidence in this regard.

More importantly, linking the US Constitution to just war thinking contributes toward exposing the way the just war tradition, as it moved from its medieval to modern forms, not only mirrored but also shaped modern notions of sovereignty. Though a more complete explication of the influence of the tradition on sovereignty will occur in the next chapter, for the moment note the way just war thinking was expressing itself in the face of technological, economic, political, social, and other changes – and the way those expressions are revealed in the Constitution. The sovereignty of persons grows in significance as the capacity for a wide range of persons (rather than simply the elite) to influence historical events also grows; simultaneously, new and (comparatively) widely accessible weapons reshapes soldiering and military practices in order to discipline mass armies. In its first articles as well as its second and third amendments, the US Constitution displays this vision of individual sovereignty. The sovereignty of states grows in significance as the proliferation and independence of states, especially through revolutions, continues: The Peace of Westphalia solidifies a particular vision of *jus ad bellum* concerns by emphasizing the rights of states to control their own borders and internal workings. In its first several articles – and especially Article 4 – the US Constitution displays this emphasis. The mirroring

[24] Michael Walzer, "The Triumph of Just War Theory (and the Dangers of Success)," *Social Research* 69.4 (Winter 2002): 925–944.

of citizen sovereignty and state sovereignty is made manifest in social contract-based positive law to avoid the war of all with all (Hobbes) or to protect life, liberty, and property (Locke). From its Preamble forward, the US Constitution promotes this attitude. Said differently, the concerns that preoccupy just war thinking influenced how early modern states and citizens were engaging the Christian just war tradition as they reimagined themselves and shaped just war theory – but in the process, they were also reshaping that tradition and the framing questions of just war thinking. Attention to the Constitution as a just war document helps to reveal that influence. State and citizen sovereignty do more than answer *jus ad bellum* questions about legitimate authority; they bear witness to the influence of just war thinking on modern thought.[25]

ENVIRONMENTAL BLINDNESS: CONSTITUTIONAL JUST WAR THINKING AND THE LIMITS OF SOVEREIGNTY

The US Constitution has almost nothing to say about human relations with the nonhuman natural world. Why not? The facile answer – that modern nation-states, by virtue of being modern, draw a hard distinction between the natural and political worlds and focus only on the latter – only pushes the question back: Why did modern thought advance such a distinction in the first place? The answer to that question has to do with modern conceptions of sovereignty that arose out of the transition of the Christian just war tradition into its early modern form, that shaped the US Constitution, and that reinforce contemporary blindness to the interactions of the human and the nonhuman natural world.

There were reasons for the modern notion of sovereignty to arise as it did. That notion solved important problems, many of which were contributing to violence. So, for example, the emphasis on human rights shaped both structures and sensibilities that opposed the casual use of force against people – which is to say, it aligned with and shaped modern just war thinking. That emphasis also reinforced the division between the human and the nonhuman natural world. As important as the emphasis on human rights has been in the development of a global

[25] James Turner Johnson develops this idea in much greater detail than I am offering here in *Sovereignty: Moral and Historical Perspectives* (Washington, DC: Georgetown University Press, 2014). See also Luis E. Lugo, ed., *Sovereignty at the Crossroads? Morality and International Politics in the Post-Cold War Era* (Lanham, MD: Rowman and Littlefield, 1996).

ethic, the advancement of human beings out of oppressive systems, and modern just war thinking's *jus in bello* criteria, though, its starting point in the sovereignty of individuals is not well situated in addressing environmental crises. The causes of most major environmental crises are so diffused that it is difficult for individuals to connect their actions to those crises: One commuter's SUV has a negligible impact on increasing CO_2 levels; one omnivore's choice to eat meat has a negligible impact on growing aquatic dead zones; one farmer's slash-and-burn method has a negligible impact on deforestation; one homeowner's leaky faucet has a negligible impact on access to clean water.

Complicating matters even more, the modern focus on individual morality that has followed from individual sovereignty is ill suited to respond to these crises. It has shaped patterns of moral behavior focused on authenticity, personal growth, purity, and catharsis as moral action: marchers in climate protests proclaim "people power" with little understanding of the way actual power (including, in this case, literal sources of power such as oil) flows through political systems.[26] Green voters assume that voting for the right leader will resolve problems without recognizing that elected leaders are still constrained by laws and obligations that limit their capacity to effect radical change. Even as environmental activists urge us to change our patterns of personal behavior, they inevitably leave themselves open to charges of hypocrisy when they must rely on the use of fossil fuels for their work (as if pursuing personal purity has any actual bearing on the actions of complex and interlocked corporations and governments).[27] How can an emphasis on the sovereignty of individuals help bridge conceptual divisions between the human and the nonhuman natural world, anyway? Is extending rights-talk to other animals or to the earth really the best way to recognize the radical interdependence of all things?

If a focus on the sovereignty of the individual is inadequate (at best) for addressing environmental crises, what about focusing on the sovereignty of the modern nation-state? Might such a focus, given its connections to just war thinking (especially as they manifest themselves in

[26] "No matter how many people take to the streets in massive marches, they cannot put their hands on the real flows of power because they do not produce it. They only consume." Roy Scranton, *Learning to Die in the Anthropocene: Reflections on the End of a Civilization* (San Francisco: City Light Books, 2015), p. 60.

[27] See, for example, Alexis Shotwell, *Against Purity: Living Ethically in Compromised Times* (Minneapolis: University of Minnesota Press, 2016).

the form of constitutions such as that of the United States) be a better starting point for addressing climate-shaped conflict?

When just war thinking is seen through the lens of the modern nation-state – and especially when it is seen through the lens of a modern nation-state that is defined by a constitution such as that of the United States – its priorities will express themselves in forms amenable to the ways that modern nation-states think. And perhaps no characteristic of modern nation-states is more central to their identity than that of sovereignty. When modern nation-states define themselves against each other, they do so via reference to their borders and their right to determine the laws that pertain inside those borders. When they engage other nation-states in international matters, they do so via reference to laws that support sovereignty and resist claims to the legitimacy of interference by other nation-states. When they engage in shaping internal economic and cultural matters, they do so as singular entities with some power to enforce laws on those matters. When they take up projects having to do with internal punitive functions, they do so claiming a monopoly on the legitimate use of force. At least in theory, they manage their own territory, function autonomously on the international stage, exert political power over economic and cultural matters, control violence, and use domestic and international laws to do so. Sovereignty is hardwired into the circuitry of modern states.

An international polity shaped around state sovereignty will struggle to address environmental issues. Such issues – whether pollution, desertification, drought and flooding, changing weather patterns, the extinction of species, or climate change – neither respect borders nor admit to simple solutions when causes and effects are geographically separable. States, after all, are obligated to pursue acts and policies that advance the interests of their citizens. Not only do those interests always potentially conflict with the interests of persons in other states, but also they are, themselves, multiple and potentially at odds with each other even within states. Where not only are the consequences of environmentally deleterious patterns of behavior unequally distributed but also the effects of some such patterns of behavior in one country can be felt in entirely different countries, states pursuing their own interests will stand at loggerheads to each other. And even where the consequences of environmentally deleterious patterns of behavior are felt within the same state that is causing them, there are likely to be disagreements about how to weigh the costs of environmentally deleterious patterns of behavior against the benefits of those patterns of behavior in the pursuit of other interests.

At domestic levels, states such as the United States address these issues, in part, through the system of representative democracy advanced in the Constitution. At federal levels, representatives are elected to serve their constituencies, but their oath is to the Constitution and, therein, to the whole nation and its basic governing structures. In principle, at least, debates about, for example, water rights, payments for storm damage, the protection of wildlife, the costs of resource extraction, or the impact of rising sea levels can be dealt with through established laws, the creation of new laws, and the willingness of "we the people" to carry each other's burdens. In practice, such debates can be contentious: Established laws are imperfect, new laws are framed by the biases of the legislators that pass them, and there are limits to citizens' neighborliness. But the imperfections of such systems have not wholly undermined the capacity of those systems to pursue strategies that avoid or mitigate some of the costs of environmental changes – and the more apparent and clear those costs are, the more likely it is that they will be addressed. Democracies have sizeable political carrying capacities.

Things are trickier on the international stage. For a variety of reasons (e.g., geographic variation in the distribution of natural resources, histories of colonization, inequities in the distribution of technologies, and the complexities of international relations), political, economic, and environmental resources are not only unequally distributed, but are seemingly an intractable condition of sovereign state-based international affairs. Given that states are obligated to pursue the goods of their own citizens on the international stage, the sovereign state-based platform for adjudicating disputes between nations is notoriously weak. And where there is a collective desire to resolve such disputes (as is visible, for instance, in actions recommended by the United Nations Security Council), such resolutions tend to focus on immediate and obvious crises with limited ability to address diffused, prolonged, or slow-moving crises or the problems that weaker nations face when stronger nations benefit from the causes of those problems. That the United Nations Security Council convenes in New York City and the United Nations Environment Programme is headquartered in Nairobi is a provocative metaphor for the distance between matters of security and matters of climate within the UN – and the UN, which already depends on the willingness of all parties to continue to participate in its various programs, is constrained by both its member states' voluntary participation and its sovereign state-based Charter.

Unsurprisingly, then, international state-based actions meant to address environmental crises (including those that make conflict more likely) are almost certainly inadequate. The modern priority of state sovereignty assures that. No wonder that hopes arising from the Kyoto Protocol (1992), the Copenhagen Agreement (2009), the Paris Agreement (2016), and other international climate conferences have foundered in the face of the behaviors and postures of individual states.

An exclusive focus on the sovereignty of the modern nation-state is, therefore, also unlikely to be of much aid in addressing climate-shaped conflict. What about attending to such conflict by prioritizing the link between individual and state sovereignty in modern thought? After all, for many modern nation-states (and certainly for liberal constitutional ones) the emphasis on state sovereignty mirrors the recognition of the sovereignty of their individual citizens. A collection of basic and (largely) inviolable human rights prevents the state from reaching too far into citizens' lives or unduly interfering with their interactions. They guarantee these rights through the creation of a bureaucratic infrastructure that both orders and constrains the behavior of those within their borders. The modern emphases on autonomy, universalism, and instrumentalism manifest themselves at both individual and state levels, with the sovereignty of the state mimetically repeating (and thereby reinforcing) the sovereignty of the individual.

MODERN SOVEREIGNTIES, DIFFUSED AUTHORITY, AND THE CONDITIONS FOR ENVIRONMENTAL CRISES

This link between state and individual sovereignty in modern political thought has been provocatively explored in a wide range of books.[28] The consistent argument across these books is that whereas in medieval thought the king had two bodies (one his natural, biological body and one his political body such that for all intents and purposes, the king was the bodily representative of the state on earth), with the dawn of

[28] See, among others, Ernst Kantorowicz, *The King's Two Bodies: A Study in Medieval Political Theology* (Princeton, NJ: Princeton University Press, 1957); Michel Foucault, *The History of Sexuality*, vol. 1, trans. Robert Hurley (New York, NY: Vintage, 1990); Bernard Bourdin, *The Theological-Political Origins of the Modern State: The Controversy Between James I of England and Cardinal Bellarmine*, trans. Susan Pickford (Washington, DC: Catholic University of America Press, 2010); and Eric L. Santner, *The Royal Remains: The People's Two Bodies and the Endgames of Sovereignty* (Chicago: University of Chicago Press, 2011).

modern thought that former body was distributed to the people: The king's two bodies become many natural ones while also being one political one. According to Carl Schmitt (who will play a leading role in the next chapter), Jean Bodin was the first theorist of the modern state, arguing that the concept of sovereignty arose out of the claims of the French monarchy that they sat atop a divinely instituted hierarchy and derived their authority from God, an authority that could be passed on to heirs because of the continuity of their second (political) body: *Le roi est mort. Vive le roi!* The French revolutionaries did more than decapitate their king in 1793; they decapitated that hierarchy and claimed for the people the king's authority. Divine right began to become human rights. Authority was diffused.

Among the problems with the diffusion of authority is not only that agreement among such diffused authorities is difficult to achieve; it is also that individuals feel a gap between the authority they've been given and the power such authority should provide for them to act: "With democracy the concept of the nation replaced the monarch and sovereignty was dispersed from the king's body to all bodies. *Suddenly every body bore political weight....* With the old sartorial and behavioral codes gone, bodies were less legible, and a person's place in the nation was unclear."[29] The modern state stepped into this gap, exerting a disciplinary power over the people's bodies through systems of signs and values that shaped not only how citizens would treat each other but also the codes through which they determined what would count as normal and what would be treated as aberrant and meriting punishment. Law – an abstraction nominally arising out of and directing the people – replaced royal decree.

It is not necessarily that sovereignty itself is a modern invention; indeed, James Turner Johnson has argued that whereas the idea of sovereignty associated with the modern state has two characteristics,[30] there is also an older a concept that can be understood as "the moral responsibility of the ruler for the common good of the people governed."[31] Instead,

[29] Sara Melzer and Kathryn Norberg, eds., *From the Royal to the Republican Body: Incorporating the Political in Seventeenth- and Eighteenth-Century France*, quoted in Santner, *The Royal Remains*, 4.

[30] "[T]he definition of sovereignty in terms of territory that may be defended and the concept of the people as having one fundamental right, that of self-defense, whose exercise they delegate, within the political community, to their rulers for defense of the entire national territory." Johnson, *Sovereignty*, 1.

[31] Johnson, *Sovereignty*, 2.

sovereignty in modern political thought is framed by the tensive relationship between state and individual as that relationship is negotiated through law.

This tensive relationship offers stability to modern political forms. Law is less arbitrary than decree, the tensive quality of the relationship between individual and state diffuses power and prevents either side from annihilating the other through collapse into anarchy or calcification into totalitarianism, and the relationship favors democratic systems in which individuals can express their power toward shaping representation for the state: "We the people" can be defined by the laws we make. And undoubtedly, the tensive relationship between the two sovereign authorities has shown favor to the greatest production of capital and expansion of cultures in human history.

Yet this tensive relationship also creates problems. What should be done, for instance, when state sovereignty and individual sovereignty sit at cross purposes with each other, as, for instance, when a state oppresses its own citizens while claiming that international laws premised on state sovereignty prohibit outside interference? Or when states demand allegiances from their citizens that trump religious, familial, cultural, ethnic, or geographically oriented allegiances as payment for protecting them? Or when individuals transport their personal sovereignty across state borders? Contemporary debates about whether and when states can violate national borders in the interests of humanitarian interventions (including whether and how to shape international laws around the responsibility to protect or "R2P"), the disruptive power of various forms of nationalism in pluralistic societies (including ethno- and religious nationalisms),[32] and the treatment of refugees all arise out of the modern premises that sovereignty pertains to both states and individuals and that it should be maintained as a good greater than other political goods.

Perhaps, then, the problems we face in addressing conflicts in the Anthropocene are not the product of the modern focus on individual sovereignty, on state sovereignty, or on the way those two sovereignties are joined. Perhaps the problems we face are the product of the modern preoccupation with sovereignty more generally.

[32] Indeed, one of the great risks of the forms of nationalism that arise with this notion of sovereignty is that national identity, as constructed identity, is malleable and open to manipulation: appeals to nationalism are always also appeals to a cipher and therein capable of being arbitrarily shaped by whatever values most benefit those making the appeal.

SOVEREIGNTY AND THE WICKED PROBLEM OF CLIMATE CHANGE

It is worth noting, to begin with, that environmental issues are one genus within a family of problems created by the emphasis on sovereignty in the politics of the modern social imaginary. After all, sovereignty at both individual and national levels complicates prospects for dealing with any problems that are bigger than individuals can manage and that don't respect national borders. If they are one genus, though, they are increasingly becoming the representative genus of that family of problems, and climate change, as a species within the genus of environmental issues, is the most obvious among the genus. The problem with climate change isn't simply that it is occurring on a global scale – although its scale makes it difficult for a system organized around discrete states to manage, as repeated attempts by the international community to address it have revealed (see, again, the Rio Declaration of 1992, the Kyoto Protocol of 1997, the Climate Change Conference in Copenhagen in 2009, the Paris Agreement of 2015, etc.). The problem isn't simply that its causes are diffused – although the global diffusion of its causes makes it far more likely that individuals will not feel personally indicted for their actions in contributing miniscule amounts to the causes of climate change and that some states will pursue free rider strategies that swamp effective global action. The problem isn't simply that its effects are unequally distributed and that those areas with the fewest resources are likely to be hit hardest – although it lays bare and then exacerbates the disparities between developed and developing nations. The problem isn't simply that the disciplinary powers exerted by states in shaping the habits of citizens (especially habits of consumption linked to national economic health) have exacerbated the problems of climate change – although such habits are even harder to break when they are still being reinforced by the disciplinary powers of states. And the problem isn't simply that the overwhelming scale of climate change and the problems it creates make a mockery of the idea that individual persons acting out of their own sovereign power can make any difference in addressing the problem – although the nihilism that accompanies feelings of helplessness reinforces the sense that sovereignty (whether nation or individual) is a useless value. It is all these things, but more.

It is also that climate change exacerbates so many other problems. The most pernicious problem of climate change – the wicked problem of climate change, to use an increasingly common phrase about the matter – is that it exacerbates the very problems that the early modern turn to state and individual sovereignty was intended to resolve.[33] In its own myopic way, the modern state focuses on these other problems because they are both more obvious and more likely to demand our attention than those of climate change. As a result, modern states are persistently focused on symptoms while ignoring the underlying causes of disease and they do so, in part, because attention to those underlying causes would mean confessing their own complicity in the conditions that gave rise to the disease and our individual and current collective helplessness in curing it.

What, though, are the wicked paradoxes of the politics of climate change in a world shaped by modern notions of sovereignty? There are at least three such wicked paradoxes: Where early modern political theory reshaped notions of sovereignty in an attempt to reduce the explosive qualities of violence by quarantining one of the threat- and force-multiplying factors of early modern violence (i.e., religion), it has stimulated another such factor (climate change). Where early modern political theory reshaped notions of sovereignty in an attempt to redistribute power in more equitable ways (including by promoting democracy), climate change in a world of sovereignties is exacerbating global inequities of wealth and political power. And where early modern political theory reshaped notions of sovereignty by privileging the individual and the state as the primary political actors, climate change funds nonstate actors for both weal and woe in ways that personal and interstate politics are ill-equipped to respond to. Take each of these three wicked paradoxes in reverse order.

CLIMATE CHANGE AND THE WICKED PARADOX OF NONSTATE ACTORS IN A NATION-STATE WORLD

First, consider the ways that climate change funds nonstate actors that the international community isn't well situated to deal with because of the modern emphasis on sovereignty. As the conditions of political sovereignty are undermined by the instabilities that come with climate-shaped

[33] The term "wicked problem" comes from work by Horst Rittel and Melvin Webber on urban planning. Rittel and Webber distinguished tame problems (which are soluble) from wicked problems (which are ill defined and not so much solved as repeatedly engaged/resolved, with each solution exacerbating a related problem). For more on the use of the phrase "wicked problem" as it applies to climate change, see, for example, Kevin J. O'Brien, *The Violence of Climate Change: Lessons of Resistance from Nonviolent Activists* (Washington, DC: Georgetown University Press, 2017), 18–19.

conflict and the inequitable distribution of resources and as global political systems built around the ideas of sovereignty continue to prove themselves not well oriented for the tasks of dealing with climate change, nonstate actors play increasingly significant roles. Transnational corporations, most notably those associated with the petroleum industry, exert global power in pursuit of their own ends, collecting yearly profits that dwarf the gross national products of most countries and revealing the lie that political sovereignty can protect countries against economic manipulation from the outside. Nongovernmental organizations (NGOs), often in agonistic relationships to national governments and with limited support, pursue the causes of justice and sustainability around the world. Armed groups affiliated with no internationally recognized formal political organization pursue agendas of conquest, disruption, and personal gain in contexts destabilized by climate change.

On the one side, nonstate actors are often best able to see the effects of climate change, either because they are most affected by the localized impacts of those effects or because they are operating across multiple countries and can see parallels in geographically disparate effects. Indigenous populations in the far northern latitudes see the effects of melting permafrost on animal populations and their own habitations and so band together to raise alarms. Refugee agencies see the way political and climate refugees blur into each other where climate change has exacerbated political instability and warn of the consequences that come with failing to address the causes of climate change. The United Nations' Intergovernmental Panel on Climate Change compiles the results of scientific research from around the world and provides assessments about the causes and effects of climate change that the international community can use to mitigate the effects of such change. The insights such groups and organizations provide are vital for understanding and dealing not only with climate change but also with its political, economic, and cultural effects. These nonstate actors depend on both the goodwill of the nation-states in which they find themselves and their own independence from those nation-states in order to maintain the integrity of their lives and work.

The one realm into which they have not historically been allowed, at least for the most part, is that of serving as a military force. Maintaining their monopolies on the legitimate use of force, states and their proxies are the only recognized bodies that can put military muscle behind attempts to address the destabilizing impacts of climate change. Sometimes this can look like humanitarian relief, as when the United States sends its floating military hospitals to hurricane-ravaged islands. Sometimes this can look like a display of military power, as when developed nations send navy vessels to accompany private craft through contested waters off the Horn of Africa. Sometimes this can look like the use of military force, as when militaries send fighter jets to strafe or bomb the terrorist camps of warlords who have attempted to control access to life-sustaining natural resources. Yet even the restrictions on nonstate actors pursuing military ends is disappearing in the face of local militias, private military contractors, and mercenaries that are called upon to sustain localized sites of violence related to private companies' pursuit of resources.

The tensions here are not only that nonstate actors such as NGOs are likely to have missions that sit at cross-purposes to military force but also that each side (military and NGO) will struggle to understand how to relate to the other even as it relies on that other for information, defense, insight, and/or support. So, for example, where climate-shaped conflicts create both humanitarian and martial emergencies, how will each side not only work with the other toward addressing those emergencies but also maintain distance from the other in order to avoid accusations of collusion and the feeding of propaganda by those who benefit from instability? While it is surely better that nonstate actors do not gain legitimacy in the use of force, the structures of modern sovereignty have established a barrier between legitimate military and all other types of actions, and the very existence of this barrier problematizes responses to the interconnected humanitarian and martial crises brought on by climate-shaped conflict.

On the other side, the growing power of many nonstate actors, particularly transnational corporations, complicates the tasks associated with dealing with climate change and climate-shaped conflicts. Where climate change functions as a threat multiplier in some of the very areas where such corporations are pursuing agendas of resource extraction, we find both the recruitment of private armed forces to protect workers and corporate interests and the animosity of armed belligerents intent on forcing those corporations out of their territory, often in order to claim its resources for themselves. Or, worse, large corporations, working through subcontractors and intermediaries, use armed belligerents to undermine or stage revolts against state governments that have restricted or opposed those corporations' interests. In such cases, the rules of war are likely to be ignored because there are no recognized structures of accountability: Just war thinking hardly seems applicable in dealing with large, if private, conflicts or criminal activity.

Said differently, when just war thinking entered modernity in Europe, it oriented itself around interstate conflicts and the theories that arose out of that thinking were designed to adjudicate those conflicts. State and individual sovereignty are preconditions that guide and sustain the clarity of those theories. Those same theories are not, however, well designed to address nonstate actors that engage in armed conflict. Private military forces are not necessarily trained in or obligated to the rules of conflict embedded in international law. Many terrorist groups (e.g., Boko Haram, ISIS) deny the legitimacy of the states in which they reside; other terrorist organizations (the Taliban, Hezbollah) are given tacit support by weak governments to pursue the types of armed actions for which the state would be condemned. Private militaries and terrorist groups may abide by rules, but the rules they abide by are constructed toward ends other than peace and justice. Just war criteria such as just cause, legitimate authority, proportionality, and discrimination aren't so much violated in these contexts as toothless.

CLIMATE CHANGE AND THE WICKED PARADOX OF GLOBAL INEQUITIES

Begin paying attention to the second wicked paradox of sovereignty intended to promote equality but now driving increasing inequality by noting the way that appeals to sovereignty reinforce political and economic inequities at the beginning of the Anthropocene. One of the ways that modern nation-states inhibited violence was by redistributing power, flattening out, diversifying, and diffusing it through a wider range of locations: Subjects became citizens as they gained individual sovereignty in systems that also emphasized the sovereignty of nation-states. It is not accidental that modern forms of representative democracy, such as the one developed in the US Constitution, are expressions of that redistribution. And these forms of democracy have done enormous good in the world, establishing systems that were more just and served to promote and protect individual rights throughout the world. They move slowly, can bog down in bureaucracy, and are driven by a complex blend of hope about what they might lead to and fear about what they might become, but they work. As Winston Churchill once noted, "[n]o one pretends that democracy is perfect or all-wise. Indeed, it has been said that democracy is the worst form of Government except for all those other forms that have been tried from time to time."[34] Underexamined as the argument may be, the development of modern democracies was itself the product of just war thinking, especially as that thinking shaped particular ways of thinking about law and political stability.

But the political egalitarianism that theoretically marks modern democracies has not kept up with or offered an adequate response to the impacts of climate change. Whether because of widening global economic inequality between states, the priority of economics over politics in interstate interactions, the growing power of transnational corporations, the vicissitudes of geography and demographic changes, patterns of development and militarization that followed colonization, the prioritization of the nation-state as the dominant player in global political systems, the trafficking of cheap weapons in less developed countries, or other proximate causes for global political inequities, global political power has increasingly pooled in a few states (e.g., the permanent members of the United Nations Security Council). The global distribution of vulnerability to the consequences of climate change maps onto global political inequities: Developing states have fewer resources available to apply toward protecting themselves from the impact of climate change even as they're being asked to make sacrifices in their own development as a means of inhibiting the causes of climate change. Small countries, particularly small island countries, lack the megaphones necessary for their voices to be heard, their GDPs are too small for their economies to exert pressure, the insidious powers of racism that followed from colonizing projects continue to exert social pressures that dehumanize and devalue vulnerable populations, and industrialized (and now computerized) economies produce disproportionate quantities of greenhouse gasses.

One of the more troubling results of these inequities has been the tendency to treat those populations that are most vulnerable to the deleterious effects of climate change as dangerous. Farmers in agrarian societies watching fields become deserts, fishermen in poor coastal communities facing ocean acidification and depleted fisheries, city dwellers in poor urban communities losing access to resources basic to human welfare, refugees fleeing the effects of climate change and climate-shaped violence: These are the persons most likely to be branded as terrorists or

[34] Winston Churchill, Speech to House of Commons (November 11, 1947). Quoted in Winston Churchill, *Churchill by Himself: The Definitive Collection of Quotations*, ed. Richard Langworth (New York, NY: PublicAffairs, 2008), 574.

pirates and they live in countries most likely to be branded as failed states, kleptocracies, and/or breeding grounds for terrorism. As Sanjay Chaturvedi and Timothy Doyle note,

> [s]ince there is a highly complex geography to both the materiality and perceptions of "risks" and "catastrophes" ... it is rather difficult to find framings, howsoever dramatic, that could act as common climate rallying points for the US citizens in Manhattan, the slum dwellers of Mumbai and the coastal communities of Somalia. For millions situated on the margins, especially in the global South, whereas the roots of "climate catastrophes" are intricately entangled with deeply entrenched and disempowering social-economic hierarchies, the routes are crisscrossed by trajectories of ecological injustices and acute livelihood insecurities. In a sharp contrast to the anxieties and fears of the affluent urban dwellers, the relevance of climate change is far more multifaceted, multilayered, and much more nuanced for the subaltern ... [As a result, there] appears to be a growing militarization of climate change and the concomitant up scaling of "climate fear" to "climate terror" on yet another site of vital importance: military-defense establishments.[35]

Malthusian and social Darwinian visions of the world have become increasingly prevalent in the (mostly democratic) global North. Thus far, the empathetic social fibers that sustain robust democratic nation-states have been inadequate in understanding the perspectives or addressing the needs of those most vulnerable to the pernicious effects of climate change. Majority world countries have been undermined by antipathies toward their populations, many of which have been inhibited from pursuing stable and democratic political systems by the same powerful countries and corporations of the global North that now scorn them. In an insidious case of victim-blaming, countries that failed to achieve stable democratic systems and vigorous economies during the explosive growth of globalization are now having those failures used against them when they are also buffeted by the effects of climate change and climate-shaped violence. Treated as politically sovereign on the international stage, such countries in the majority world find themselves sacrificing that sovereignty in the face of violence shaped by natural systems that pay no attention to boundaries or rights.

Modern just war theory, largely restricted as it has been to *jus ad bellum* and *jus in bello* questions and bounded as it is by assumptions

[35] On the equating of vulnerable populations with dangerous populations, see Sanjay Chaturvedi and Timothy Doyle, *Climate Terror: A Critical Geopolitics of Climate Change* (New York, NY: Palgrave Macmillan, 2015), 16.

about its universal applicability, finds little purchase in addressing these conflicts. Nor does it offer resources for combatting the underlying assumptions and systems that have helped to give rise to such conflicts and their inequitable distribution in the first place. Indeed, in a context of global political, economic, and environmental inequities where victims are also viewed as threats, some security/defense thinkers are advancing arguments to set just war criteria such as just cause, last resort, and discrimination aside in order to use the military to prevent the political instabilities caused by climate changes from leading to failed states or terrorist activity that metastasizes to other countries. This new form of "humanitarian intervention" would ideally be pursued in cooperative and trust-building ventures, but there is no guarantee of this and should situations degenerate, the resort to armed force wouldn't so much violate as defy the very type of modern just war thinking displayed in contemporary just war theory (not to mention the US Constitution).

Further complicating these matters is the fact that the threat of climate-shaped violence in the vulnerable countries and interventions by countries in the global North is, paradoxically, reinforcing the significance of national sovereignty. Countries that fear climate refugees close and arm their borders. Countries that fear incursions by forces from other nations appeal to the rights of sovereignty within their borders. Countries dealing with armed nonstate actors inside their borders seek out bigger and more powerful weapons in order to regain the monopoly on the legitimate use of force that partly defines the modern state. Again citing Chaturvedi and Doyle,

> [climate terror] is fast turning ... into a largely conservative grand-strategy deployed by faltering sovereign states, at various stages of a neo-liberal embrace, to discipline and regulate various faultlines in statecraft. In the context of the complex and dynamic political geography of climate change, the processes of deterritorialization and reterritorialization, operating in conjunction, do not so much question the system of sovereign spaces as they reproduce it.[36]

Though the early modern emphasis on national sovereignty was intended to spread political power more broadly, its effect at the beginning of the Anthropocene is to reify the very sorts of political inequities that are arising at modernity's end and then cement them into place through appeals to the importance of national sovereignty.

[36] Ibid., xii.

CLIMATE CHANGE AND THE WICKED PARADOX OF SOVEREIGNTY PROMOTING VIOLENCE

Finally, third, consider the paradoxical problem of sovereignty and violence. After the wars of religion that followed the Protestant Reformation, early modern political theorists saw the dangers that came with religious fervor inserted into regional conflicts. The conventional limits on warfare (limits that were related to but not identical to those associated with medieval Christian just war thinking, including restrictions on the number of warriors drawn from the appropriate classes of society, the weighing out of the economic costs and benefits of going to war over resources, and clarity about who had proper authority to declare war in medieval political forms) faded in the face of a new political form: the modern nation-state. These new nation-states drew from wider classes for their armies, proffered religious arguments as a means of recruitment and in defense of their armies' actions, and elevated their own leaders, with their own distinct religious affiliations, rather than fit their leaders into a hierarchy that included emperor and pope. In response, early modern political theorists treated the power of religion in politics like strong medicine: to be used sparingly and only as necessary. Concerns about the violence of these wars and the way religious fervors and arguments exacerbated them helped induce the birth of the secular age and, eventually, the political separation of church and state visible in the First Amendment of the US Constitution. As a result, religion ceased to be an obvious driving force in modern interstate conflicts (at least in Europe) and was increasingly disconnected from the tensive quality of the two sovereignties that both sustained and mandated pluralistic societies.

Climate change increasingly plays a role in conflicts around the world that is analogous to the role religion played at the beginning of modernity: as a force that doesn't so much cause conflicts as make those conflicts more likely and worse. It does not, though, admit to quarantining the way religious impulses did during the secular age. Worse, the tensive qualities of the two sovereignties actually inhibit means for addressing climate-shaped conflicts. Because it was reacting to violence and being shaped by the modern social imaginary, modern political forms favored state sovereignty, in part, for its preoccupation with immanent, crisis-oriented, state-shaped, political threats to political stability. This is visible, among other places, in the shape of just war theory criteria.[37] Yet

37 See, again, my essay, "Changing the Rules: Just War Theory in the 21st Century," *Theology Today* 54.1 (2003): 59.

climate change is the product of slow-building, long-term, global phenomena that arise out of natural systems and are reaching a point of crisis long after passing a point of efficient repair. As a result, the modern political forms that were designed to address one kind of violence are especially ill-equipped to address the new kinds of violence that come with climate change.

Moreover, as climate-shaped violence is also shaped by sociopolitical forces, states find themselves better able to address those other forces while ignoring or downplaying the role that climate change plays in conflicts.[38] States respond to the causes of violence that make sense to them and that they are shaped to address in the promotion of political stability; as a result, climate drops out of the field of view of states responding to violence and pursuing international political stability. To make matters even more complex and worse, state responses to violence tend to exacerbate climate change by producing significant quantities of greenhouse gasses. The forms of agency promoted by the two sovereignties are not well shaped to address a threat multiplier such as climate change, the result of which is its own threat multiplier and violence exacerbator. In sum, the very shape of political sovereignty that arose, in part, as a way of dealing with religiously inflected violence now makes it harder to deal with environmentally inflected violence.[39]

The growing power of nonstate players on the world stage; the growing inequalities of the political haves (who are comparatively uninterested in addressing climate change) and the have-nots (who are most affected by climate change and, therein, climate-shaped conflict); the proliferation of the types and causes of violence in the world: these are the wicked problems – the wicked paradoxes, really – of dealing with

[38] The current conflict in Syria is a case in point. Were it not for repeated years of severe drought in the region, young men from the rural areas in Syria would not have gone to the cities looking for work, where they gathered with other disaffected young men, observed uprisings in Tunisia and Egypt, and initiated a revolt that a corrupt and brutal Assad regime moved to squelch, that quickly drew in religious radicals who reframed the focus of the war, that occurred in an oil-rich area in the eastern Mediterranean, that happened shortly after other destabilizing events in the area (including the war in Iraq), and that drew international powers into the conflict, each with their own contending political goals. The drought did not cause the war. The various factors that have contributed to the war are so complex and interrelated that the international (state-oriented) political community is at a loss as to the best way to address them – and though drought was a significant factor that led to the war, concerns about drought have been all but forgotten in the face of the other factors bringing about and exacerbating violence there.

[39] In a later volume, I will return to this concern and argue that religion itself needs to be reunderstood in the process of responding to climate-shaped conflicts.

climate-shaped conflict in a political context framed by early modern emphases on sovereignty. To the degree that just war thinking shaped that political context (which is a quite significant degree, especially once one begins to recognize the way just war thinking played a formative role in realms beyond international law and the development of modern just war theory), just war thinking bears a particular responsibility for those wicked problems.

I have argued here that the US Constitution is one place to find just war thinking outside international law and modern just war theory. The people that it shapes and the form of government that it defines therefore bear their own responsibility for these problems. That is, the US Constitution is a modern just war document and, as it enacts a particular way of understanding, shaping, and engaging political systems (the "state"), it reinforces the conditions that make dealing with climate-shaped conflicts so difficult. And it is not alone.

What, though, can be done about these wicked problems and the responsibilities borne by those who live within polities shaped by such constitutions? To start to get at an answer to that question, it is necessary to think about not only what makes the US Constitution a just war document but also what makes it a *Christian* just war document. That is the work of the next chapter.

4

State Time/Secular Time

The US Constitution as a Christian *Just War Document*

> The Congress, whenever two thirds of both Houses shall deem it necessary, shall propose Amendments to this Constitution, or, on the Application of the Legislatures of two thirds of the several States, shall call a Convention for proposing Amendments, which in either Case, shall be valid to all Intents and Purposes, as Part of this Constitution, when ratified by the Legislatures of three-fourths of the several States, or by Conventions in three fourths thereof, as the one or the other Mode of Ratification may be proposed by the Congress.
>
> The US Constitution, Article 5

> … the problem with constituting power is that it endlessly falls into theological modes of thinking and representation. This is because constituting power constitutes more than collective life. To succeed as an act of political making, constituting power must also constitute the overarching conceptions of justice, the conditions of social recognition and nonrecognition, that transcend political community and through which political community is achieved.
>
> Graham Hammill, *The Mosaic Constitution*

> Were it not for the possibility of the future, social and political change, revolutionary change, would forever remain a foregone conclusion. However, understanding the very richness of the possibility the future presents to us raises concerns about the relation of hope to contemporary life.
>
> Rochelle Green, "Nostalgia, Past and Future Hopes"

At first blush, treating the US Constitution as a Christian document is even more absurd than treating it as a just war document. Setting aside the First Amendment's famous and famously misunderstood statement that "Congress shall make no law respecting an establishment of religion, or prohibiting the free exercise thereof,"[1] its tone throughout is notably secular. Unlike, for instance, the Declaration of Independence, it makes no claims about the origins of the rights it establishes. Unalienable rights do not come from a Creator in the Constitution; they don't really come from anywhere. They just are. There are no references to God in its text, which isn't interested in locating any sources – however immanent or transcendent they may be – from which to derive its structuring laws (beyond "We, the people") so much as it is interested in describing those structures. And yet I will argue here that we should treat it as a kind of Christian document arising during the transition from the late medieval to the early modern period and that, as such, it not only carries the traces of that transition but also complicates the ways we will move forward into the Anthropocene and climate-shaped conflicts.

TWO WAYS THAT THE US CONSTITUTION IS NOT A "CHRISTIAN" DOCUMENT

Almost since its ratification, many Christians in the United States have found the absence of attention to any religious substructure for the nation's founding document to be profoundly disquieting. This disquiet reflects, in part, the complexity of the way religious and nonreligious perspectives have engaged each other since the early years of colonization. It also reflects confusions about how to understand "religion" as a category and Christianity as a particular expression of religion, about the relations of morality to religious perspectives, and about the relation of law to morality. And this disquiet is reflected in a range of narratives about the history of the interplay between culture and politics, some of which are quite recent.

One of those narratives is that the United States is a "Christian nation." The founder of the Moral Majority, Rev. Jerry Falwell, argued that "our founding fathers established America's laws and precepts on the principles recorded in the laws of God, including the Ten Commandments. God has blessed this nation because in its early days she sought to honor God and the Bible, the inerrant Word of the living God. Any diligent student of American history finds that our great nation was founded by godly men upon godly principles to be a Christian nation."[2] This

[1] The Constitution of the United States of America, Amend. I.

[2] Jerry Falwell, *Listen America!* quoted in James Calvin Davis, *In Defense of Civility: How Religion Can Unite America on Seven Moral Issues That Divide Us* (Louisville, KY: Westminster John Knox Press, 2010), 22.

narrative had nestled into the language and vision of the religious right by the late 1970s and expresses itself – especially during election seasons – within their writings and television programs, including those by Pat Robertson, James Kennedy, and others since them. Unsurprisingly, the primary purpose of the narrative is not so much to explicate history as it is to reinforce the contemporary tribal identity of conservative American Christians in order to motivate their political activities on issues associated with, among other things, homosexuality and abortion.

This narrative is also nonsense. The founding fathers were a diverse group who did not agree on theological matters and some of the most notable among them (including George Washington, Thomas Jefferson, and Benjamin Franklin) held beliefs that contemporary conservative Christians – or, indeed, most orthodox Christians – would find heretical. Influenced by Enlightenment thinkers who feared the impact of religion on public issues and its power to motivate division and violence, many of the framers of the Constitution, including those who would be more recognizably orthodox, worked hard to prevent those motivating qualities from infecting American public life.

For that matter, the religious right itself is a contemporary political invention and those who identify with it can only claim to follow in the footsteps of the founding fathers by dramatically misreading not only the history of the country's founding but also their own history as well. The term "evangelical" has changed meanings repeatedly over the course of several hundred years, and in its current form it looks little like what it meant 150 years ago. The idea of biblical inerrancy, for instance – an idea so closely linked to modern evangelicals – is a late nineteenth-century phenomenon. Ironically and in spite of being prohibited from doing so by the very same Ten Commandments that such Christians valorize, promoters of this narrative bear false witness against themselves, their neighbors, and the nation. Given its deleterious impact, the sooner this narrative disappears, the better. If the US Constitution is to be interpreted as a Christian just war document, it cannot be done via this narrative.[3] The polity of the United States simply was not structured on the basis of explicit Christian thought nor with the explicit intent of shaping a Christian nation.

[3] For more on the failures and falsities of this narrative, see Davis, *In Defense of Civility*, esp. ch. 2. See also William Lee Miller, *The First Liberty: America's Foundation in Religious Liberty* (Washington, DC: Georgetown University Press, 2003); Steven K. Green, *Inventing A Christian America: The Myth of the Religious Founding* (New York, NY: Oxford University Press, 2015); and Kevin M. Kruse, *One Nation Under God: How Corporate America Invented Christian America* (New York, NY: Basic Books, 2015).

One of the weaknesses of that narrative is its failure to deal adequately with the complexity of the interactions between politics, religion, and culture. For while it is not the case that the American political project is based in a particular religious tradition, it is nevertheless the case that the dominant religious force in American culture has long been Christianity (and, particularly, Protestantism). The founding fathers may not have all agreed with each other about what it meant to be Christian or how important Christianity should be in the American public scene, but they were generally shaped by the forms of Protestant Christianity that made their way from Europe during colonization (including Calvinism, Methodism, Quakerism, and the various forms manifest in the many free church approaches such as Baptist).

Attention to these varied forces motivates a second narrative about the relation between the laws of the United States and Christian thought: that Christianity, and particularly Protestantism, has strongly shaped American culture and American culture shapes America's laws. Protestant Christianity not only influenced the language and perspective of the founders, but it has also continued to shape American culture to the present day in a wide variety of ways from the relatively innocuous (e.g., printing "In God We Trust" on currency) to the beneficial (e.g., funding Abolitionism and the Civil Rights movements) to the wicked (e.g., defending slavery and the subjugation of women).

When Thomas Jefferson wrote to the Danbury Baptist Association about erecting a "wall of separation" between church and state, he was neither offering a definitive interpretation of the First Amendment nor suggesting that religion could or should be kept out of American public life. Not only has that wall proved to be quite porous but also its utility as a metaphor has funded a remarkable degree of voluntary religious engagement and affiliation in the United States. Sometimes it has done so by protecting and promoting the freedom of minority religious groups (broadly, first Roman Catholicism, then new flavors of Protestantism such as Seventh Day Adventism, then Judaism and new religions such as the Church of Jesus Christ of Latter-Day Saints, then the wider religions of the world) to give voice to their own religious perspectives on public matters. Sometimes it has done so by allowing dominant Protestant voices (and others) to offer their own forms of support to important social movements (including Abolitionism, Progressivism, Suffrage, and Civil Rights). And sometimes it has protected legislators from constituencies that would demand adherence from them on issues over which those constituencies burned hot but neither could nor should be legislated. The First Amendment keeps the institution of the state from controlling religious institutions and religious institutions from controlling the state (imperfectly, certainly, as any number of Supreme Court cases have demonstrated); it does not keep religious perspectives and religious language out of the public sphere.[4]

If anything, the American experiment of separating church and state, while permitting religion into American public life, has funded not only the state's resilience but also religion's vitality. Yet recognizing the roles of religion and religious language in culture – including addressing distinctly public issues – doesn't translate into a reason to think of the US Constitution as a Christian (let alone a theological) document. For one thing, the very difference I've named here between religion in public life and religion in the governing structures of the United States would need to be addressed. Culture shapes politics and vice versa, but they are not the same. For another, the Constitution has shaped a culture in which a plurality of religious voices can all speak; getting to a coherent theological perspective by beginning with the cultures of American public life is a fool's errand. To make the stronger claim I am making here – that the Constitution is a Christian just war document – will mean pursuing a track other than ones that have thus far wound through America's public life.

Toward making my larger argument, I turn both from a false political narrative about America being a "Christian nation" and from a true but inadequate narrative about the way the interactions of culture and politics have shaped US law and toward the insights of political theology.

SCHMITT, POLITICAL THEOLOGY, AND THE POWER TO DETERMINE EXCEPTIONS

The move to political theology complicates an already complicated task because the phrase "political theology" has collected a number of meanings over the past several decades. It can mean explorations of the political implications of various theological frameworks. This is the meaning that scholars such as Johann Baptist Metz, Dorothee Sölle, Jürgen

[4] For a more extended exploration of religious language in the public sphere, see my *Believing Aloud: Reflections on Being Religious in the Public Sphere* (Eugene, OR: Cascade, 2010).

Moltmann, and a range of liberation theologians have used for it.[5] It can mean explorations of the theological bases for a particular political perspective. This is the meaning applied to the term when it is used to refer to the ideas of thinkers such as John Courtney Murray or Reinhold Niebuhr, whose theologies reveal their political thought.[6] It can mean the attempt to shape a theology of politics, which is what Oliver O'Donovan and Charles Mathewes do.[7] Indeed, the range of writings and perspectives in an anthology such as *An Eerdmans Reader in Contemporary Political Theology* reveal just how varied the understandings of the term are.[8] Each of these meanings connects to the others at various points and all connect, at least tangentially, to the way I am using the term.

Here, though, attending to political theology means addressing the work of German political theorist Carl Schmitt and the implications that have followed from it. Schmitt's writings not only informed his defense of Nazism but also, over the past few decades, have captured the attention of both religious and secular scholars, including Leo Strauss, Jacques Derrida, Giorgio Agamben, and Slavoj Žižek, and his approach begins in the idea that political systems have theologies that sit behind and shape them. To stave off quick criticism, I hasten to add that such a meaning may play fast and loose with the term "theology" (and even, perhaps, "politics") as it does not necessarily link the idea of theology to a particular religious perspective. Indeed, Graham Hammill and Julia Reinhard Lupton begin their book *Political Theology and Early Modernity* with the provocative sentences: "Let's get this straight. Political theology is not the same as religion. Instead, we take it to name a form of questioning that arises precisely when religion is no longer a dominant explanatory

5 See, among others, Johann Baptist Metz, *Faith in History and Society: Toward a Practical Fundamental Theology* (New York, NY: Crossroad, 2007); Dorothee Sölle, *Political Theology* (Minneapolis, MN: Fortress, 1974); Jürgen Moltmann, *The Crucified God* (Minneapolis, MN: Fortress, 1993); Jon Sobrino, *Christology at the Crossroads: A Latin American Approach*, trans. John Drury (Maryknoll, NY: Orbis, 2002); and Jean-Marc Ela, *My Faith as an African*, trans. John Pairman Brown and Susan Perry (Maryknoll, NY: Orbis, 1988), as a start.

6 See John Courtney Murray, *Religious Liberty: Catholic Struggles with Pluralism* (Louisville, KY: Westminster John Knox Press, 1993); and Reinhold Niebuhr, *Moral Man and Immoral Society: A Study in Ethics and Politics* (Louisville, KY: Westminster John Knox Press, 2013).

7 See Oliver O'Donovan, *The Ways of Judgment* (Grand Rapids, MI: Eerdmans, 2005); and Charles Mathewes, *A Theology of Public Life* (Cambridge, UK: Cambridge University Press, 2007).

8 William T. Cavanaugh, Jeffrey W. Bailey, and Craig Hovey, eds., *An Eerdmans Reader in Contemporary Political Theology* (Grand Rapids, MI: Eerdmans, 2012).

or life mode ... The norms and forms of religious life are ... not exactly what concerns political theology" (though a mere paragraph later, they offer a definition that would seem to blur terms that they've been trying to distinguish: "We take the phrase 'political theology' to identify the exchanges, pacts, and contests that obtain between religious and political life, especially the use of sacred narratives, motifs, and liturgical forms to establish, legitimate, and reflect upon the sovereignty of monarchs, corporations, and parliaments").[9] Given the range of meanings and writers using the term after Schmitt, it sometimes seems that the only thing that binds all the thinkers and ideas associated with this understanding of political theology is that they all respond to Schmitt's work.

Schmitt was occupied with the question of what happens to religious ideas in the political sphere at the dawn of the secular age; to begin to make sense of Schmitt's work, it is therefore important to get a sense of what he means by "secular." Secular, for Schmitt, does not mean the opposite of "religious" but something more like the opposite of "eternal." To link it to Charles Taylor's use of the term, the secular realm is that realm that is immanently framed. All things can be adequately (and even exhaustively) described without reference to or connection with the transcendent. Prior to modernity, the argument goes, earthly powers in the West maintained themselves by linking their existence in history to a corresponding power in eternity: The church was composed of all the saints, living and dead, and the visible church should be understood as the worldly expression of the Kingdom of God; kings and queens were chosen by God to be God's earthly representatives and therefore ruled by divine right; acts of generosity on earth yielded blessings in heaven. Earth mirrors heaven.

That mirror shattered as it entered the secular age. Once things such as church, political rule, and charity can be understood without remainder in an immanent frame, they are full participants in the secular order: The church was that society of persons who shared a common belief system (and/or, perhaps, worked to make the world a better place); rulers were selected by the people to enact and enforce that people's will; acts of generosity were evidence of a good upbringing that helped stabilize the social order. Where the temporal mirrors the eternal, power comes from on high and authority exerts its power in the rituals that sustain connections with the divine. Where the temporal is the sum of all things, though, power moves through systems and authority exists wherever

[9] Both quotations are from Graham Hammill and Julia Reinhard Lupton, eds., *Political Theology and Early Modernity* (Chicago: University of Chicago Press, 2012), 1.

power is recognized to have pooled for a period and legitimized. As a result, the secular age is a vertiginous one – and people, Schmitt thought, aren't especially comfortable with vertigo.

Schmitt's most basic argument is that modern political theory responds to that vertigo by secretly importing the transcendent into wholly immanent political forms. When Schmitt argued that modern political theology is a covert form of secularized theology, he meant that modern political theory took overtly theological ideas such as divine sovereignty, the revelation of divine law, and the classical attributes of God (omniscience, omnipotence, omnipresence, etc.) and used them to frame the powers and purposes of the state. The state maintains its own sovereignty. The laws of the state not only express the state's power but also reinforce their own embeddedness in a nation's culture such that violating them is a violation of the proper order of things and an offense against the state. The range of the state's power, especially within its borders, extends into all components of citizens' daily lives through its bureaucratic regulatory functions and rights-extending provisions. As Schmitt famously wrote,

> [a]ll significant concepts of the modern theory of the state are secularized theological concepts not only because of their historical development—in which they were transferred from theology to the theory of the state, whereby, for example, the omnipotent God became the omnipotent lawgiver—but also because of their systematic structure, the recognition of which is necessary for a sociological consideration of these concepts. The exception in jurisprudence is analogous to the miracle in theology. Only by being aware of this analogy can we appreciate the manner in which the philosophical ideas of the state developed in the last centuries.[10]

Schmitt, then, wasn't doing political theory or theology in the first instance so much as phenomenology: He was trying to describe the visible expressions of a pattern he saw in the early modern thought of, for example, Thomas Hobbes or Jean-Jacques Rousseau, as they attempted to define the modern state and situate that state (and, especially, its sovereignty) at the center of political thought.[11] This new thing called the

10 Carl Schmitt, *Political Theology: Four Chapters on the Concept of Sovereignty*, trans. George Schwab (Chicago: University of Chicago Press, 1985), 36.

11 As Schmitt would note much later, "[m]y own book *Political Theology*, from 1922 ... does not deal with theological dogma, but with problems in epistemology and in the history of ideas: the structural identity of theological and juridical concepts, modes of argumentation and insights." See Carl Schmitt, *Political Theology II: The Myth of Closure of Any Political Theology*, trans. and introduced by Michael Hoelzl and Graham Ward (Malden, MA: Polity Press, 2008). First published in German in 1970.

modern nation-state maintains its integrity through the enforcement of its sovereignty via its monopoly on the legitimate use of violence. Much like God, Schmitt argued, the state is free to determine how and when to use violence even as it uses law to tell its citizens when they can and cannot resort to violence. When the state resorts to violence, its actions are beyond questioning and its laws do not apply to its own actions any more than God's laws prevent God from performing miracles by violating divine laws.

Justifiably, much of the focus on Schmitt's work since his writings has been on his analogy between the exception in jurisprudence and the miracle in theology.[12] One of the reasons for this is that Schmitt used his notion of exception to justify the right of Hitler and the Nazi state, as sovereign, to violate the standing conditions of just jurisprudence by eliminating Jews, the Roma, homosexuals, and those with various physical and mental disabilities from their midst. When the state reached a condition of exception in which the normal rules no longer applied, it was free to do as it wished much in the way that when God performs a miracle, God does so with the freedom not only to violate the laws of nature but also to exercise power arbitrarily.[13]

A second reason to focus on Schmitt's analogy between exception and miracle is that Schmitt's preoccupying questions ("Who makes decisions in political systems and on what basis?") are characteristic of modern political thought in which authority and agency are of paramount importance. This preoccupation is even more evident in some of Schmitt's later books, most notably *The* Nomos *of the Earth in the International Law of the* Jus Publicum Europaeum, in which Schmitt pursues questions about the history of international law in much greater historical detail.[14] In that text, Schmitt examines constitutions as a way of addressing those preoccupying questions and adding much needed context to the more grandiose claims of *Political Theology*, though not in a way that modifies either those claims or their implications for

[12] See, for example, Giorgio Agamben, *State of Exception*, trans. Kevin Attell (Chicago: University of Chicago Press, 2005).

[13] See, among many others responding to Schmitt's idea that the genuinely sovereign is free to determine both the point of exception and to respond as he/she/it wishes, Bonnie Honig, *Emergency Politics: Paradox, Law, Democracy* (Princeton, NJ: Princeton University Press, 2009); and Ted Smith, *Weird John Brown: Divine Violence and the Limits of Ethics* (Stanford, CA: Stanford University Press, 2015).

[14] See Carl Schmitt, *The* Nomos *of the Earth in the International Law of the* Jus Publicum Europaeum, trans. and introduced by G. L. Ulmen (New York, NY: Telos Press, 2003).

his thought. So, for instance, Ulmen notes in his introduction to *The* Nomos *of the Earth* that "[s]ince [the new order of the *jus publicum Europaeum*] was international, jurists had assumed a pivotal position in its creation and maintenance, until ... when state bureaucracies increasingly began to instrumentalize the legal establishment. Schmitt was seeking to reconstitute European jurisprudence in opposition to bureaucrats and technocrats, who systematically reduce it to regulations and procedures."[15] If anything, Schmitt's work in that book reiterated his focus on the freedom of the sovereign.[16]

Finally, a third reason to focus on Schmitt's analogy between exception and miracle is that his vision of political theology emerged – twice – at points in history when confidence about existing political structures in the West was growing increasingly tenuous. The failings of the Weimar Republic to address the crises growing out of World War I reinforced suspicions about the limits of social contract-based theories of the state and the presumption that increasing citizens' freedom resolved problems associated with the growing power of economic and technological powers on the world stage. Later, as the moral failings of neoliberal economic strategies became more evident and divisions between haves and have-nots became more pronounced during the postcolonial moment, crises at the beginning of the twentieth century (democracy in retreat, ideological retrenchment into utopian thought) recurred at the beginning of the twenty-first. As Jeffery Robbins has noted, "political theology first emerged and continues to thrive directly as a consequence of the despair over the perceived failure of modern liberalism. As such, political theology is originally conceived as the first and greatest rival to the dominant but increasingly ineffectual political order."[17] Change and the perceptions of political instability draw attention toward the popularity of illiberal solutions to political systems.

[15] G. L. Ulmen, "Introduction," in Schmitt, *The* Nomos *of the Earth*, 23.

[16] Schmitt's work in *The* Nomos *of the Earth*, incidentally, meshes well with the assumptions of temporal stability in the face of structuralist interpretations of history such as those made prominent by Fernand Braudel in *The Mediterranean and the Mediterranean World in the Age of Philip II*, 2 vols., trans. Siân Reynolds (New York, NY: Harper & Row, 1996). Braudel introduces the need for historians to attend to the natural world (or at least its geological features), but in a way that assumes their stability and ongoing power in ways that are both too simple and inattentive to change. That is, both Schmitt and Braudel are occupied with space, not time, in thinking through modern politics.

[17] Jeffery W. Robbins, *Radical Democracy and Political Theology* (New York, NY: Columbia University Press, 2011), 4.

SCHMITT'S POLITICAL THEOLOGY: JOINING WICKED POLITICS TO BAD THEOLOGY

The events of history (including the Shoah) justify the need to respond to Schmitt's notion of the freedom of the state within the context of the exception. The political implications of Schmitt's ideas about the power of the sovereign to declare exception have proven those ideas to be wicked, as many scholars across a range of disciplines have pointed out. Yet the wickedness revealed in those political implications is not the only failure of Schmitt's thought. His conceptions of law and theology also fail. Since Schmitt's work has continued to attract interlocutors (including those who have accepted his basic thesis that modern political theory is covertly theological while rejecting his arguments analogizing the exception in jurisprudence to miracle), damning his arguments about the exception isn't sufficient. A closer look at the way Schmitt thinks about law and theology helps to locate the failings within Schmitt's thesis while also clearing space within which to take up a chastened form of that thesis and use it to diagnose failings that have helped to shape modern environmental crises and inadequate contemporary political responses to them.

Let us begin with Schmitt's conception of law. Setting aside the fact that Schmitt was more comfortable discussing early modern political philosophy than early modern jurisprudence (one searches *Political Theology* in vain for, e.g., references to the Salamanca School, Alberico Gentili, or Hugo Grotius, even though some of them might be helpful to his argument; Samuel Pufendorf and Emmerich de Vattel get single mentions; and the two jurists that most occupied him in the book – Krabbe and Kelsen – were both his contemporaries), his relative neglect of attention to constitutions is telling. Constitutions are more than the products of thought; they shape thought as they shape law and direct policy. They are, moreover, the paradigmatic expression of modern jurisprudential thought and the states such thought creates. They are artifacts, tools, forces, and signs all at once. Attention to the politics of creating and maintaining constitutions; to the growth, destruction, and transformation of constitutions over time; and to contending schools of constitutional jurisprudence should be a requirement of any theory that seeks to describe the sources and powers of modern political forms. Yet Schmitt barely glances at constitutions.

A paragraph after his claim that "[a]ll significant concepts of the modern theory of the state are secularized theological conceptions … ," he offers a loose connection between modern constitutions and theology: "The idea of the modern constitutional state triumphed together with deism, a theology and metaphysics that banished the miracle from the world."[18] The sentence not only fails to describe the type of connection between constitutions and deism (causal? correlative? accidental? illusory?), it blurs a kind of state with a form of law that some such states maintain. Even if his description is correct, is he implying that constitutional states can evade the dangers that come with sovereigns declaring jurisprudential exceptions? And if so, has he offered a solution to the very problem of the exception that has preoccupied his critics? How confident should we be that constitutional states don't run the same risks that other modern states do? There are, after all, resources in constitutional law that reinforce the quasi-sacred character of state sovereignty. One thinks of presidential immunity from civil liability for official presidential acts or the right of the president to pardon offences against the United States provided in US law, for instance.

The fuzziness does not stop with that sentence. In one of his few other references to constitutions, he notes that "[t]he tendency of liberal constitutionalism to regulate the exception as precisely as possible means, after all, the attempt to spell out in detail the case in which law suspends itself."[19] The sentence is so woolly it practically invites shearing. What is the force of the adjective "liberal?" Does Schmitt distinguish liberal and illiberal constitutionalism? Are there nonconstitutional forms of liberal jurisprudence and, if so, what are they? When Schmitt suggests that liberal constitutionalism tries to regulate exceptions, what actual constitutions, if any, is he referring to? Does the law suspend itself or does the sovereign suspend the law – and what is the relation between the law and the sovereign, given what he will say in the next essay, anyway? What could it possibly mean for Schmitt to say that the exception can be regulated, let alone regulated precisely, if the exception is tantamount to a miracle?

There are a few other fleeting references to constitutions in *Political Theology*, but they offer little further clarity to those fuzzy sentences. Moreover, the single reference to an actual constitution is an extended paragraph on the German constitution of 1919. As much as that constitution expressed the liberal and democratic confidence of the Weimar Republic, it was notably deficient – even compared with many other

[18] Schmitt, *Political Theology*, 36.

[19] Ibid., 14.

constitutions – in allocating presidential powers, inhibiting splinter parties from disrupting constitutional processes, and maintaining checks and balances.[20] It is hardly representative of all constitutions in this regard. And when Schmitt does reference article 48 of that constitution (which addresses presidential power to act without the consent of the Reichstag in cases of emergency), he is asking one article of one constitution to carry the weight of representing the tensions between all constitutional forms and absolute sovereignty. Implicit in his argument is the conviction that the power of the right leader to declare exception allows that leader to either legally manipulate or legitimately trump the checks on power developed in constitutional law. Here he seemingly offers a model of the "great man" myth of history discussed earlier in this book.

There are, moreover, any number of ways to link constitutions and states shaped by constitutions to theology (via, e.g., comparisons with covenants and covenantal or federal theology, comparisons with other forms of law, or theological perspectives on concepts such as human rights that may fund constitutions). Indeed, I want to explore one way of linking constitutions to theology momentarily. For the moment, though, note that Schmitt's concept of the exception in jurisprudence as analogy to miracle loses considerable force when he offers the concept disconnected from the prototypical expression of jurisprudence for the time.[21]

None of this is to say that closer attention to and a different perspective on the shape and significance of constitutions in modern political thought will resolve the problems of the exception or, more broadly, the dangers of covertly importing theology into modern political thought. It is, though, to say that Schmitt's general disregard of constitutions, on the one side, and his limited if peculiar attention to them as a way of rejecting the conditions of the exception, on the other, undermine the credibility of his larger political vision. Moreover, interpreting that general

[20] See William Shirer, *The Rise and Fall of the Third Reich: A History of Nazi German*, reissued ed. (New York, NY: Simon and Schuster, 2011), ch. 3, especially the footnote on p. 53.

[21] Perhaps one reason for Schmitt's failure (or unwillingness) to focus on constitutions is his suspicion of the idea that a people can actualize their own will apart from submitting that will to a supreme leader. On their face, constitutions suggest that the people can do this; Schmitt's analogy of sovereign to God suggests he does not. Graham Hammill suggests that Schmitt's argument is premised on his rejection of Emmanuel-Joseph Sieyès' concept of a "constituting power" that can bring about a new political order and that arises out of the nation and its myths. See Graham Hammill, *The Mosaic Constitution: Political Theology and the Imagination from Machiavelli to Milton* (Chicago: University of Chicago Press, 2012), 4–5.

disregard and peculiar attention as signs in need of interpretation may offer insights into the ways modern political theory presents itself and the constraints modern political theorists place themselves under when they accept modernity's self-descriptions as adequate descriptions.[22] Schmitt simply got early modern political theory wrong when he under-regarded constitutions. And though later in his life he would give much closer attention to constitutions, it is hard to see how his later focus led him to revise his earlier claims.

Likewise, Schmitt's vision of theology goes wanting. Given the legitimate preoccupation with sovereignty – of both persons and nations – in early modern political thought, Schmitt's focus on the analogy between the sovereignty of the state and the sovereignty of God is understandable. Yet how (and how much) was divine sovereignty a preoccupation of premodern Christian theology? Schmitt's argument is that when the Western world entered modernity, its attempt to do a new thing was limited by the language available to describe that new thing. It therefore used existing language (which was both theological and also fit within a world in which earth mirrors heaven) to describe a new thing in which that language should play no role and so must be covertly deployed.[23] For Schmitt's arguments about sovereignty to hold sway, the landscape of premodern Europe should be littered with Christian language about the sovereignty of God. Was it?

The answer to that question is complicated. On the one hand, premodern Catholicisms stretching back to the very early church, although attentive to theological doctrines such as providence and political language for God such as "Lord," don't give nearly as much attention to such matters as they do to other complicated theological questions. Likewise, various Protestant communions (notably those of the Radical Reformation, but also Lutherans and, to a lesser degree, Anglicans) actively resisted including or prioritizing such doctrines and language

22 This is where other critical readings of modern political theory gain traction. See, among scholars offering such critical readings, Hans Blumenberg, *The Legitimacy of the Modern Age*, trans. Robert M. Wallace (Cambridge, MA: MIT Press, 1983); Bernard Bourdin, *The Theological-Political Origins of the Modern State: The Controversy between James I of England and Cardinal Bellarmine*, trans. Susan Pickford (Washington, DC: Catholic University of America Press, 2004); and Péter Losonczi, Mika Luoma-Aho, and Aakash Singh, eds., *The Future of Political Theology: Religious and Theological Perspectives* (Surrey, UK: Ashgate, 2011), among others.

23 This is, roughly, Hannah Arendt's reading of Schmitt. See Hannah Arendt, *On Revolution* (New York, NY: Penguin, 1990). Arendt uses this reading to make sense of the distinctiveness of the American Revolution.

in their theological constructions. Moreover, even when many of these communions did approach such doctrines and language, they subverted them by locating them in larger theological contexts having to do with questions about how to follow a "servant leader" as servant leaders, how Jesus of Nazareth could be identified with the first person of the Trinity, how this Jesus mediated relationships between creator and creature, how the visible church both participated in and failed to fully express itself in the life of God, and how to understand the growing delay between Jesus' departure and his return. That is, whatever else the sovereignty of God meant to these communions, it didn't mean the things that Schmitt meant when he talked about sovereignty. Even a quick glance at the confessions that guided these various communions (beginning with the Nicene and Apostles' Creeds but including, e.g., the Athanasian Creed, the Augsburg Confession, and the Schleitheim Confession) shows that their theological concerns tended to lay elsewhere. While one finds sovereignty language (or at least language about the "omni-" attributes of God) in many of those confessions, it is consistently either submerged below more pressing and significant theological concerns having to do with, for example, Christology, or restricted from particular realms, for example, those having to do with human free will.

On the other hand, not only do the language and themes of sovereignty manifest themselves throughout the history of Christianity (though the sheer diversity of ways that language and those themes express themselves ought to give us pause on any 1:1 translation of theological sovereignty to political sovereignty) but also some Christian communions have placed divine sovereignty at the center of their theological visions of the world and its political forms.[24] Notable among such communions are those arising out of the Reformed tradition, including the various branches of Calvinism that began in Geneva in the sixteenth century. Within those communions, convictions about the providence

[24] One possible connection here arises out of the recent revival of realist/nominalist debates (or, rather, debates about the realist/nominalist debates) that began with William of Ockham in the fourteenth century and drove forward a voluntarist perspective that would deeply impact not only Protestantism but also Enlightenment thought. Ockham's nominalist God is sometimes portrayed as a supreme being whose freedom is all but defined by a willful rejection of obligations toward natural law: the very definition of the sovereign who can declare exception. The problem with such a reading is not only that it ignores Ockham's more complex understanding of natural law but also that it renders to Ockham more political influence in early modern thought than seems warranted, given the other sources of authority at work at the time. He was influential, to be sure, but not nearly so giant a figure as recent debates suggest.

of God manifest in history, the predestining power of God to save and damn as God wills, the idea of human vocation as participation in the work of God, divine command ethics, and the priority of recognizing the mystery and glory of God in worship all at least hint at a willingness to accept the freedom of the sovereign to decide the exception. And while they all grow out of only one theological branch of a larger tree of Christian communions, the Reformed traditions have played an outsized role in shaping modern political theory.[25]

Yet the Reformed tradition has at least implicitly recognized the dangers of imputing the freedom of God onto human projects (including political ones) with a corresponding emphasis on the dangers of idolatry. God may be free to set the rules and break them; human beings are not. Moreover, the Reformed preoccupation with the authority of Scripture as the rule of faith and life offers at least some constraint on human action. Whether the Reformed emphasis on divine sovereignty expresses itself in traditional Calvinism, Edwardsian puritanism, Kuyperian neo-Calvinism, Barthian neoorthodoxy, or some other variant of Reformed theology, the Reformed theological tradition includes brakes on the abuses of sovereignty within its theological structure.

As Schmitt's phenomenology about the covert importation of theological concepts into modern political theory certainly allows for (and even favors) a willy-nilly picking and choosing of concepts, he and those who would come after him at least have the responsibility of explaining how modern political theorists could pick up one theological concept without dragging along any number of closely associated theological concepts that might come trailing after it. Treating theological doctrines as isolable from each other is itself more likely to be the product of modern assumptions about the atomistic qualities of ideas than the weaknesses of the relations between those doctrines. Picking and choosing theological concepts as if they are not part of a larger theological schema through which they become coherent in the first place suggests that Schmitt ignored the complex interplay of theological concepts in order to make a political point. That is, he treated the sovereign's power

[25] See, again, Michael Walzer, *The Revolution of the Saints: A Study in the Origins of Radical Politics* (Cambridge, MA: Harvard University Press, 1982). There has been some interest, parenthetically, in comparing the way Carl Schmitt thinks about sovereignty, the exception, and the "borderline case" (*Grenzbegriff*) and the way Reformed theologian Karl Barth shapes his divine command ethic to include "limit cases" (*Grenzfall*), though that is beyond the range of this project. See Karl Barth, *Church Dogmatics II.2*, ed. G. W. Bromiley and T. F. Torrance (Edinburgh: T&T Clark, 1957).

manifest in the state of exception not as an actual and previously existing theological concern but as a contemporary political concern for which he had found a useful metaphor.[26]

It follows, moreover, that Schmitt's claim that modern political thought draws from premodern theological views has the effect of undermining his own understanding of the secularity of the modern age. That is, he treats the modern age as bereft of religious sensibilities rather than as an age in which the immanent frame has made understanding the religious sensibilities operative in modernity difficult to understand and describe. Yet as complicated as it has been for modern thought to imagine religious perspectives shaped by the mirroring of the transcendent in the immanent, it is nevertheless the case that religion and the religious perspectives that existed prior to modernity didn't disappear. Instead, they continued to express themselves, albeit in more complicated ways, in modernity. Schmitt is trying to have it both ways: to treat the secular as lacking religion (and so needing to draw from premodern sources) and as manifesting religion (albeit in ways that it lacks the language to adequately describe).

So Schmitt's vision of theological concepts is appallingly emaciated. Even setting aside the oxymoronic quality of the term "secularized theological concepts," it's unclear what Schmitt could actually mean by "theological" beyond using the term as a placeholder for a particular view of sovereignty that is more likely to come from modern theory than to precede it. When he developed arguments about how modern theories of the state are secularized theological concepts and chose to focus on sovereignty and exception, his arguments sounded like what someone who doesn't talk about theology thinks theologians talk like. Indeed, that there are religious traditions in which divine sovereignty plays a more central theological role than it does in Christianity suggests that either Schmitt is thinking about Christianity as one "religion" in a world of religions (which is, itself, a modern perspective)[27] or he is far less interested in the theological building blocks of the modern state than his phenomenological theory suggests. Again, none of this is to deny the possibility that Schmitt's basic argument is right: The modern theory of the state may well be founded in secularized theological

[26] This is, roughly, Hans Blumenberg's point about Schmitt's literalizing of metaphor in *The Legitimacy of the Modern Age*.

[27] See, for example, Tomoko Masuzawa, *The Invention of World Religions: Or How European Universalism Was Preserved in the Language of Pluralism* (Chicago: University of Chicago Press, 2005).

concepts. It is, though, to suggest that if that basic argument is right, Schmitt's focus on sovereignty and exception isn't an especially good starting point.

By way of illustrative contrast, imagine taking a theological concept such as the Trinity as the starting point for thinking about the conceptual origins of the modern state. That concept at least has the advantage of being central to almost all Christian understandings of God and how God relates to the world. Rather than focusing on the power of the sovereign to decide the exception, a Trinitarian starting point for a Schmittian political theology might focus on the way the perichoretic qualities of three-in-one both reinscribe and blur identity. By way of secularized analogy, the modern state demands that citizens honor distinctness while blurring their other distinct identities (national, tribal, familial, vocational, etc.) in order to find common purpose and pursue common good. This is the "melting pot" metaphor for national identity as secularized theological concept. Or a Trinitarian starting point might focus on the way the economic Trinity diffuses individual responsibility: Who is responsible for what actions becomes less clear even as those actions remain vital for human flourishing. Secularized, this starting point could offer insight into the bureaucratic features so distinctly characteristic of the modern state. This is the Department of Motor Vehicles as expression of secularized theological concept.[28]

Setting aside snarky hyperbole and facile equivalencies, this illustration gets at another of the limits of Schmitt's arguments as he offered them in *Political Theology*. Just as sovereignty is more an acute modern political concern than a central and abiding theological one, so Trinity is a clearer theological concern than it is an acute modern political one. If this is the case, then not only might it be important to think more carefully about which theological-political metaphors are best for making the Schmittian case that "[a]ll significant concepts of the modern theory of the state are secularized theological concepts" but also that we should recognize that a Schmittian approach should include attention to the convenience, productivity, and complexity of those metaphors.

[28] Much of Schmitt's book *Political Theology II* is a refutation of Erik Peterson's claim that the doctrine of the Trinity subverts all forms of political theology. Clearly that's not the direction – even hyperbolically – that I'm going here. It seems to me that Peterson's argument, once shorn of its anti-Semitic qualities and iffy theological vision, has more to do with rejections of Schmitt's vision of sovereignty than anything else.

I've offered this extended series of criticisms of Schmitt's argument not to wholly undermine it but because I want to learn from it. I find his overall argument – that the concepts of the modern theory of the state are secularized theological concepts – compelling, but to go forward with him, I'd like to avoid not only his more egregious and commented-upon conclusions (including those that led him to become an apologist for the horrors of Nazism) but also the errors that arose out of his methodological blind spots. As I make the case for the US Constitution as a Christian theological document, I obviously need to avoid Schmitt's blind spot vis-à-vis the power of literature and literary forms to reveal and shape a modern theory of the state. Equally obviously, as I make the case for the US Constitution as a *Christian* document, I need to avoid the blind spot of denying theological reflection its own integrity in shaping the conditions that lead to the modern state. And, finally, as I make my case, I need to avoid the blind spots of demetaphorizing a genuine theological concept or treating the secular as an areligious realm.

CHANGING METAPHORS: THE US CONSTITUTION AND A THEOLOGY OF LAW

One possible starting point for pursuing better metaphors – and one much more amenable to the case I'm making here – would be to attend to the Exodus and the giving of the law on Sinai. Not only does this doubled event (i.e., leaving Egypt and receiving the law) get at some of the complexity that comes with a law that forms a people into a state but also the events of Sinai, especially, are significant in the literature of early modern political thought. As Graham Hammill argues,

> the Mosaic constitution had an integral role to play in the emergence of early modern concepts of the state. Like the story of Oedipus or Antigone, the Mosaic constitution was one of the founding fictions of early modern life ... Through the revision and expansion of Moses's constitution of Israel as a people unified by law and collected under the rule of a single sovereign, early modern literary and political writers attempted to grasp and probe key problems, dynamics, and relationships that defined the early modern state, including toleration and the conflictual character of politics; the fraught relations among revelation, political authority, and literary authorship; and the role of fiction in defining and delimiting obedience, liberty, and the right to resist coercive authority.[29]

[29] Hammill, *The Mosaic Constitution*, 2. Hammill also points to Barbara Johnson, *Moses and Multiculturalism* (Berkeley: University of California Press, 2010), as making a common argument.

Hammill's arguments build on the recognition that political ideas are nomadic, not monadic. They travel through time and find new modes of expression, but they don't do so in isolation from each other. Nor do these new modes of expression arise simply and clearly. They provide new insights but they also shape new problems and constrain other approaches. Their use requires imagination because they manifest as images, metaphors, and narratives. They function because they are flexible and open to multiple interpretations. They reveal their interpreters as believers (that is, as having convictions about the way things are and should be that cannot be fully deduced from empirical observation or logical conclusion) and they reveal what their interpreters believe. To quote Hammill again, "[t]he Mosaic constitution had such powerful appeal because it was a narrative that both relied on belief and gave a flexible set of terms for analyzing relations between and among obedience, authority, community, and constitution as problems concerning representation, communication, rhetoric, and fiction."[30] Over the course of chapters on Machiavelli, Spinoza, Marlowe, Harrington, and others, Hammill unites literary analysis, political theory, and a history of the uses of the Mosaic covenant to offer a thoughtful alternative to Schmitt's version of political theology.

Hammill's attention to the Mosaic constitution is, for obvious reasons, helpful to my argument. He's paying attention to constitutions, literature, and the complex ways that theological (or at least religious) ideas moved from one social imaginary into another: "Put most broadly, this book's central thesis is that the Mosaic constitution figures a productive intersection – or what philosopher Charles Taylor calls a point of cross-pressure – between religious belief and paradigms of human creation that is unique to an early modern sense of political history."[31] He resists extricating one religious idea from the larger conceptual contexts that frame it. He's attentive to the way metaphors create some problems even as they resolve others. He thinks of modern political theology as shaped by the entanglements between the secular and the eternal during a time that the secular is ascendant. His focus is on early modern political thought and the way that the Mosaic constitution was an actual topic for reflection in that thought. He evades many of the blind spots that undermined Schmitt's arguments and haunt those of us who have engaged Schmitt.

[30] Hammill, *The Mosaic Constitution*, 16.

[31] Ibid., 3. The reference to Taylor is from Charles Taylor, *A Secular Age* (Cambridge, MA: Harvard University Press, 2007), 595.

There are, though, limits to Hammill's project. His equation of "constitution" and "covenant" betrays a modernist impulse that disregards real distinctions between the terms (it is telling that the term "covenant" does not appear in his index, nor do notable modern theologians who wrote on covenantal theology). He bypasses differences in the ways Jews and Christians have interpreted the Exodus and Sinai and how those differences may have shaped the early modern thinkers he discusses. As a result, as helpful as he is in offering a rich literary analysis of the role of the Mosaic constitution in early modern thought, he still treats it as a metaphor that can be brought into modern thought without theological remainder. While he doesn't abuse the metaphor the way Schmitt abuses "sovereignty," neither does he pursue the metaphor with much attention to the religio-historical reasons that it could function so successfully in such a variety of early modern ideas. As a result, he misses something of the way the metaphor of the Mosaic constitution can be a "productive intersection … between religious belief and paradigms of human creation" because he's so elevated its uniqueness for "an early modern sense of political history."[32] The variety of its uses at the time, undoubtedly shaped by the many ways that religious groups thought about covenant and left their ideas strewn about the early modern West, would suggest reasons for a range of entanglements of the secular and the sacred in political theology. Theology is as epiphenomenal for Hammill's project as it is for any project that presumes to fully engage religious thought from within an immanent frame – which is to say, Hammill's project repeats one of the tensions of modern thought without doing much to name and address that tension.

This is not necessarily a problem with Hammill's thought. He is, after all, one of the editors that wrote: "Let's get this straight. Political theology is not the same as religion,"[33] and in the secular age there is no obligation to talk about religion or speak from theological perspectives. Hammill does a great deal of space-clearing with his work that is valuable for political theology after Schmitt; among the most valuable tracts of cleared space is a way of addressing the problem of how the law can shape and maintain order in society when it doesn't have the power to compel obedience from citizens. Where Schmitt argues that the sovereign can compel obedience by taking on divine authority, Hammill argues that obedience need not be compelled by force if citizens are seduced

[32] Ibid.
[33] Hammill and Lupton, *Political Theology and Early Modernity*, 1.

into agreement through their acceptance of powerful narratives (and the images and metaphors that constitute those narratives) with which they identify. It is those modern narratives, Hammill claims, that are infused with secular theological concepts; it is the power of political theology to investigate these narratives.

Treat the US Constitution, then, as a metaphor for a larger narrative about a particular modern state in which "We the people" structure their political relations toward the formation of a union in which justice, tranquility, security, welfare, and freedom will reign. This is social contract theory writ large and infused with teleological perspective. Why should citizens be obedient in such a system? Because the promises offered to them and their children in return for their obedience are so enticing that choosing disobedience carries the threat of being perceived as (or thinking of oneself as) irrational, unreasonable, and/or unhopeful. The US Constitution is the American Dream sublimated into law. Although few US citizens know much if anything about their constitution,[34] they honor (and even venerate) it as an abstraction because it occupies such prominent roles in their imaginations about what makes their country great and in their narratives about how to keep their country great. Its mythic qualities exceed its practical ones – as, indeed, they must if its practical qualities are to continue to be sustained by Americans' willingness to obey the laws it lays out. To paraphrase Hammill, the US Constitution has such powerful appeal because it offers a narrative that both relies on belief and gives a flexible set of terms for analyzing relations between and among obedience, authority, and community.[35]

THE US CONSTITUTION, POWER, AND TEMPORAL DEFERRAL

Note, though, the way that the US Constitution instantiates this power. Unlike the covenant at Sinai, it doesn't give a detailed list of rules – a series of dos and don'ts – to which the people promise to be beholden so that blessings will flow their way. Unlike monarchies, it doesn't offer

[34] See Annenberg Public Policy Center, "Americans Are Poorly Informed about Basic Constitutional Provisions," www.annenbergpublicpolicycenter.org/americans-are-poorly-informed-about-basic-constitutional-provisions (accessed November 10, 2017).

[35] Hammill's original statement is: "The Mosaic constitution had such powerful appeal because it was a narrative that both relied on belief and gave a flexible set of terms for analyzing relations between and among obedience, authority, community, and constitution as problems concerning representation, communication, rhetoric, and fiction." See Hammill, *The Mosaic Constitution*, 3.

a particular figure endowed with the power to bless or withhold blessings based on the obedience of the people. Unlike collectives, it doesn't revolve around an explicit social contract in which costs and benefits are distributed according to predetermined agreements.

Instead, the US Constitution presents itself through a series of temporal deferrals that maximize the space within which citizens implicitly engage questions of self and community, authority and obedience, promise and threat, agency and impotence, identity and otherness, self-regard and concern for others, and the many other concerns central to the creation and maintenance of a coherent society.

The most obvious temporal deferral is present in its opening articles, which don't set the parameters for decisions but instead create a variety of decision-makers. Rather than assume its ability to answer all the questions that citizens will ask of law and government, it empowers those decision-makers by enumerating the basic functions of their offices and then, derivatively, provides those decision-makers with resources to guide their decisions (not only amendments, case law, philosophies of jurisprudence described in court decisions, and access to majority and minority opinions but also the voting will of constituents, the wisdom of cabinets of advisors, etc.). In a paradoxically self-effacing project, the Constitution, as text, favors the wisdom of interpreters to the guidance of a text. "Texts, themselves," it seems to say, "cannot resolve your questions, so do not idolize them and do not become contented with them. Instead, let those questions guide you in your quest not only for answers but also for better understandings of the motivating power of those questions." Such quests take time for deliberation and the profoundest of them may never achieve conclusions.[36] The Constitution seemingly accepts the possibility of the indefinite deferral of some solutions even as it offers processes for resolving most of them.

Such wisdom is gained, moreover, not only through the deep study of texts but via the lived experiences of decision-makers arising out of and in conversation with the wider populace. This is obviously the case in the legislative and executive branches, but it is no less the case in the judicial branch, which relies for its appointments on the other branches of the government and is guided but never entirely constrained by the

[36] See the work of Elizabeth F. Cohen on questions about the political significance of time (and taking time) in democracies, particularly in her book *The Political Value of Time: Citizenship, Duration, and Democratic Justice* (Cambridge, UK: Cambridge University Press, 2018).

interpretations of earlier generations. Indeed, the opening articles not only shift away from answers and toward decision-makers; they also shift by putting those decision-makers in tension with each other through a system of checks and balances.

This shapes a second temporal deferral: forcing time for reflection by pitting answerers against each other. The Constitution rejects the idea that power should pool in one place by diffusing authority through the political system: The three branches check and balance each other, and each – either directly or indirectly – is answerable to the wider populace. Built into the system, then, are assumptions about the wisdom of citizens not only exercised in diverse ways (voting, running for and holding office) but also tested against the wisdom of other citizens. "The wisdom of leaders," the text seems to say, "should not be trusted more than the wisdom of citizens, so do not deify leaders and do not allow them to remain in place in perpetuity. Prefer, instead, the wisdom of the wider community, full of differences of opinion and diverse passions though they be, to constrain power and to legitimate authority." Since such wisdom is contextual and practical (as is all wisdom), it remains fundamentally open-ended in the face of changing contexts, mores, and technologies. The Constitution favors a kind of temporal modesty, trusting that new contexts will surface new approaches to questions that may connect to the insights of established answers but will not be bound to them.

Nowhere is the legitimacy of new approaches to questions and new insights (which is to say temporal modesty) more clearly built into Constitutional processes than in the process for amendment. That parts of the very law that orders basic political structures could, themselves, be changed not only by the addition of new amendments that can modify or specify existing articles but also by the actual overturning of existing amendments reveals not only temporal modesty but also epistemic humility. This shapes a third temporal deferral: openness to amendment. The Constitution rejects the idea that any group of citizens, including its framers, can attain and display wisdom for the ages. Built into the system, then, is a recognition of the limits to the wisdom of any particular collection of citizens based on assumptions about the inadequacies of their understanding of the world around them. "The wisdom of any group of people," the text seems to say, "is necessarily open to review and modification, so venerate neither it nor your own insights. Instead, approach the future with openness to new perspectives and humility about your current understandings." Building the capacity for amendment into basic political structures makes them resilient and adaptive rather than brittle and unyielding. The Constitution offers itself as a living document, trusting that change is an expression of growth rather than random movement.

Yet such growth is moderated rather than chaotic. The Constitution favors reform over revolution not only by making the process of amendment slow and difficult but also by undergirding the processes for change with a set of restrictions about where and how the state can act, which are most visible in its opening amendments. Enumerating a series of individual rights, these first amendments set boundaries not so much on how dramatically it can change but on how intrusive those changes can be. That its political structures can reach only so far into the lives of citizens guarantees them safe spaces into which they can retreat not only to protect themselves from the state but also to organize their thoughts away from the cacophony of a many-voiced public.

This shapes a fourth temporal deferral. The US Constitution not only tacitly rejects the power of any governing body to unduly intrude into the lives of citizens (therein rejecting the idea that one group can have and exercise a kind of gnostic wisdom about the underlying structures of the age) but also mandates the shape of its citizens' public engagement. While they can organize their private lives in a wide variety of ways, when they enter into public spaces they are obligated to give publicly accessible reasons to those with whom they share citizenship. That is, the Constitution is shaped to give preference to conversation rather than fiat as the fundamental practice of political change and to give preference to engagement and persuasion rather than threat or coercion as the basic rhetorical structure of conversation. "Revolutionary and idiosyncratic brilliance may be awe-inspiring," the text seems to say, "but it is also dangerous and not to be trusted above the time-consuming accrual of shared wisdom that follows from the processes of giving and receiving publicly accessible reasons. Approach visions of the future tentatively, testing the ground in front of you before you step into the unknown." Demarcating a boundary between public and private not only suffocates totalitarian visions of the world and rejects their implicit claim that we have reached the end of history, but it also animates the time-consuming practices that come with learning to talk with strangers. The Constitution offers itself as a constrained and fundamentally conservative document, trusting that conversation is better than command at funding the continuing process of turning *pluribus* into *unum*.

All of which shapes a final temporal deferral that I'll name here: the project of turning a collection of persons into a people. I have earlier described the rhetorical sleight of hand involved in beginning the Constitution with a "We the people" Preamble that creates the very body it would then govern. This act of creation is neither an all at once nor a once for all project: It is a continuing task as new citizens enter into the people it creates and current citizens move deeper into or away from their citizenly obligations. Lacking any formal ruling power (unlike its articles, the Constitution's Preamble offers neither precedent nor legal guidance), the Preamble nonetheless exerts a kind of persuasive power in that it invites persons to accept its basic governing structures by being charmed into accepting the privileges and responsibilities that come with being involved in forming a more perfect union. Invitational rather than commanding, the Preamble makes no assumptions about the kinds of persons that can be included and erects no barriers to prevent any categories of people from entering into the project of governance (though the Constitution's articles have done so). That is, it seemingly imagines an unending and continual process of nation-creating as if it has all the time in the world for projects of national improvement. "Join this people," the text whispers, "and claim your own role in shaping something that is bigger than you are but that never asks you to surrender your own identity in that project. In so doing, not only you but your children and all who come after you will benefit." Without enumerating them or describing the processes for their growth, the Constitution suggests that there are virtues of democratic citizenship and, in linking the Preamble to the articles, offers governing structures for making the space in which such virtues can be pursued. More than any other portion of the Constitution, the Preamble signals the entire document's fundamentally aspirational character, treating the future as both focus and promise.[37]

None of this is to say that this series of temporal deferrals has worked out the way the Constitution's writers envisioned (or even that they were preoccupied with attempts to shape such a series of temporal deferrals as they negotiated its passage). Nor is it to say that such deferrals are necessarily a good thing or that they avoid creating lacunae and problems in

37 The significance of the Preamble to the Constitution obviates the debate about how to interpret the king's political body (as described by Ernst Kantorowicz in *The King's Two Bodies*) in at least the US republican context: either the people or the law. On my interpretation, the law and the people make each other. See Ernst Kantorowicz, *The King's Two Bodies: A Study in Mediaeval Political Theology* (Princeton, NJ: Princeton University Press, 1957). For an efficient summary of the debate between people and law, see Smith, *Weird John Brown*, 17–20.

shaping governance structures. These deferrals involved compromises on the way to the Constitution's ratification and they have regularly failed to achieve whatever goods they may have intended; indeed, one of the dangers of such a future-oriented document is that the goods it promises can be infinitely deferred: justice, tranquility, general welfare, and the other goods it promises in the Preamble not only remain future hopes, but they are also eschatological ones.[38]

Momentarily, I will raise significant concerns about the consequences of building temporal deferral into governing structures. First, though, I want to note that this collection of temporal deferrals was one of the reasons that Schmitt was antiliberal and was likely one of the reasons that he spoke so poorly of constitutions. Noting the acceleration of social and technological change, Schmitt described a concomitant acceleration in politics, which he referred to as the motorization of lawmaking.[39] As the rate of acceleration in Western societies increased, Schmitt found the processes of deliberative inquiry increasingly burdensome and unable to keep up with the rate of change, with twinned results that jurists find themselves dealing with increasingly incoherent laws (and often functioning either as a brake on legislation in an essentially conservative mode or legislating from the bench) and the power of the executive must grow.

William E. Scheuerman has done important work in challenging Schmitt's approach while also recognizing the impact of the social acceleration of time on liberal democracies. In response, he has argued that there are possible transformations within liberal democracies that can not only address but also benefit from the social acceleration of time:

> Might we successfully reconfigure liberal democracy for a temporal context distinct from that in which its eighteenth- and early-nineteenth-century defenders first hinted at the possibility of its realization? ... A substantial dose of institutional imagination is called for if we are to update liberal democracy to make it mesh with the dictates of speed. Although my own contribution here to a revitalization of liberal democracy [via advocacy for reflexive law and transnational democracy] has undoubtedly been a modest one, I hope at the very least to encourage others to

[38] This is one of the reasons, incidentally, that I imagine the United States has so struggled to deal with issues of race: not only have the laws of the land regularly instituted, shaped, justified, and reinforced racism, but also the very constitutional tool that many have used to undermine or thwart that racism (its aspirational character) offers neither deadlines nor benchmarks for correcting the nation's racism. Even when it can mitigate some of the consequences of racism (in, e.g., its Thirteenth, Fourteenth, and Fifteenth Amendments), it does so while making inexecutable promises for an indefinite future.

[39] Carl Schmitt, *Die Lage der europäischen Rechtswissenschaft* (Tübingen: Universitäts-Verlag, 1950).

confront the difficult unanswered questions posed by changes in the temporality of contemporary society for political life.[40]

While I am attracted to Scheuerman's vision of the possibilities for a robust liberal democratic response to the social acceleration of time, I worry that Scheuerman's approach fails to adequately note the way technology, society, politics, and – given, especially, the nature of this project – climate change are all accelerating at different rates and in different ways. Perhaps Scheuerman and other wise political theorists can have some success in modifying liberal democracies to address their concerns about the social acceleration of time in political systems. The accelerations that preoccupy them, though, are occurring in a human-sized temporal frame and a human-sensory-oriented temporal shape: We can see the technological and social changes around us and imagine shaping responses to those changes in ways that are meaningful in our lifetimes.

While the climate is changing at a rate we can now see in our lifetimes – albeit in ways that have been, up until fairly recently, visible only to those who were not only paying attention to the myriad effects of climate change but also linking those effects to a changing climate – the way its change has accelerated has meant that much of that acceleration began before even most scientists began to take note of it, and the momentum it has built up during its acceleration is unlikely to be arrested in our lifetimes, by which point it may be too late to arrest. Scheuerman's responses to the social acceleration of time may help escape the political dangers of authoritarian regimes; it is an open question, though, whether they are adequate to address the political dangers of climate-shaped chaos. To make sense of why – and to see why both Scheuerman and Schmitt operate within a shared vision of time and politics even if they disagree about the best political response to that vision of time – I return to the temporal deferrals I described above.

TEMPORAL DEFERRALS AND THE CHRISTIAN POSTURE OF POLITICAL THEOLOGY

My purpose in naming these temporal deferrals (and linking them to Scheuerman and Schmitt) is to drive forward an inquiry about where they came from. Was there something about the social imaginary of

40 William E. Scheuerman, *Liberal Democracy and the Social Acceleration of Time* (Baltimore, MD: Johns Hopkins University Press, 2004), 225–226.

early modern political thought that promoted temporal deferral rather than, for example, treating time as either immaterial (because political systems mirror the politics of heaven) or preoccupying (because political systems must address the urgent)? Where, that is, do the assumptions that time matters and that it is plentiful come from? They come, I argue, from early modern Protestantism.

Any discussion of the concept of time risks plunging into philosophical, theological, and sociological depths far beyond what current space allows. And any discussion of early modern Protestantism risks blending and simplifying diverse social, political, economic, and religious patterns and changes in ways that violate the complexity of their interactions, especially during the transition from one social imaginary into another. Nevertheless, it is undeniably the case that Protestantism was simultaneously influenced by and an influencer of changing understandings of time in early modern thought even as Christianity itself has profoundly reshaped Western senses of time,[41] by bringing forward a linear vision of time in which change, growth, and transformation were not only possible but also meaningful.

Though the earliest church thought of itself as coming at the end of history and the beginning of the eschaton, its sense of an imminent end receded as Jesus' immediate return seemed more and more unlikely and as the early church made peace with its surrounding cultures. Replacing its sense that it sat on the cusp of the end, the church after Constantine and Augustine increasingly thought of time as subjectively experienced through the transitions of anticipation into memory and objectively experienced through the rhythms of seasons that could be liturgically marked through the repetitions of feast days over the course of the year and designated times of prayer over the course of the day.[42] While Christian

[41] Evidenced, for instance, by the fact that it has traditionally measured years according to "Before Christ" and "*Anno Domini*": BC and AD.

[42] "The Constantinian revolution of the 4th century marked not only the beginning of a new social order, but of a new temporality. Henceforth, time passed differently, imprinting Christ's life on the temporality of the social body through a recurring series of holidays: Christmas, Easter, and the more frequent, seven-day return of the Sun's day (*dies Solis*), when all but farmers were to cease from labour. Through repetition the passage of time is overcome and its creator glorified, in recollection of his rest on the seventh day, after his labours (Gen. 2:2). More intense patternings of time – chronotypes – would be constructed in churches, and above all monasteries, where the monastic body, individual and communal, would repeatedly invoke the eternal God ..." Gerald Loughlin, "Time" in *The Oxford Companion to Christian Thought*, ed. Adrian Hastings et al. (Oxford: Oxford University Press, 2000), 708.

visions of time have always joined together linear conceptions of time (moving from creation through fall to Exodus and exile, then crucifixion and resurrection and then on to consummation) with cyclical ones (e.g., Christians repeatedly loop back in time as they participate in Jesus' life, death, and resurrection within rituals such as the Eucharist and the movement from Advent through Pentecost as expressed in the liturgical calendar), medieval Christianity forwarded cyclical conceptions as its way of mirroring the divine liturgies of eternity in daily life. Primarily agrarian and nonliterate cultures reinforced this cyclical perspective as they attended to changing seasons and made sense of the narratives of the faith through images more than texts.[43]

The Reformation was both the product and the cause of changing understandings of time as Protestantism arose while cyclical perspectives of time were becoming less pronounced and linear perspectives were growing in prominence. Cities were rising and with them a merchant class whose work was not ordered by seasons so much as business hours. The beginnings of the secular age drove a preoccupation among some with the possibility that history was coming to an end and Jesus would finally be coming back. As reading and writing became more common after the invention of the printing press, they reshaped the patterns and foci of liturgical sensibilities, including Bible reading. As nation-states came into being, suspicions about the authority of the Roman Catholic Church – including deep suspicions in some parts of Europe about Pope Gregory XIII's replacement of the Julian calendar with the Gregorian one – were expressed more stridently.[44] Each of these social forces both impacted and was impacted by the Protestant Reformation.

Among the most radical Protestant attempts to revise the concept of time were those coming from reformers who saw signs of Jesus' impending return and, like the earliest Christians, believed they lived at the end of time. Sixteenth- and seventeenth-century Protestants such as Thomas Müntzer, John of Leiden, and the Fifth Monarchists all preached the coming of the end times as described in the Book of Revelation and

43 One exception here being the development of clocks within monastic communities for keeping ordered times of prayer through the night. Recognize, though, that clocks arose in the monasteries (which were among the few institutions wealthy and public enough to house them) in order to maintain the cyclic patterns of prayer.

44 For a fascinating review of the politics of time (past and present) with special attention given to the shift from the Julian to the Gregorian calendar, see Michelle Stacy, "Clash of the Time Lords," *Harper's Magazine* (December 2006).

resorted to armed resistance against the pope, the established rulers of the land, and even other, more moderate, Protestants. Rules and mores that guided and stabilized social interactions were set aside: Why maintain the niceties of politeness when the world is coming to an end and most of the people around you are doomed to hell? For that matter, why wait out time until the revolution of the eschaton when you could help to bring it about through robust evangelism and/or physical force? Yet while their eschatologically fired rebellions led to the deaths of thousands, their brands of rebellion did not represent the centers of Reformed thought about time.

Most reformers paid less attention to a changing conception of time than they did to ideas such as the priesthood of all believers, the significance of Scripture, corruption within the church, egalitarian ecclesial structures, and the marks of true worship. Yet these ideas had implications for how they understood time so that even where a vision of time was not central to some Protestants' thoughts, their preoccupations did contribute to the reshaping of understandings of time. Martin Luther's emphasis on the priesthood of all believers removed any need to pray via Mary or the other saints and so celebrating particular saints on feast days became not only unnecessary but also cause for concern: Focusing on the saints signaled a failure to understand one's unmediated relationship to God. Reading the Bible methodically, working through individual books over time, supplanted cyclical lectionary readings and fed a linear sensibility about how narratives move through time. The Radical Reformers' treatment of baptism as a sign that one accepted Jesus Christ as savior and was ready to enter the community of the faithful led them to downplay the mimetic qualities of dying and rising with Jesus that animated Roman Catholic understandings of baptism. Huldrych Zwingli's insistence that the Lord's Supper was a memorial meal implied that celebrating the sacrament meant remembering something that happened in the past rather than participating in an event that was recurring in the present. The sacraments were reduced and linked to the two actions that Jesus had ordered believers to take (baptism and communion) rather than to the significant events in the life cycle of human beings; no longer did sacraments connect so obviously to the cycle of life. Churches became increasingly urban and, with their urbanization, adapted to the patterns of urban life, including the work week and the insistence that believers gather together each Sunday rather than occasionally attending Mass, which was offered every day. Learning to live with others across differences – whether those differences were religious, ethnic, or national – after conflict meant promoting tolerance as a way of keeping past offenses in the past. And the God of deism became, of all things, a watchmaker.[45]

Perhaps the reformer whose thoughts most prominently changed Europe's sense of time was John Calvin. He argued that Christians should treat their lives as a kind of pilgrimage toward God, making use of each moment to pursue sanctification. For Calvin, life was a project of disciplining one's desires and learning to treat all of one's tasks as opportunities for practicing at glorifying God. Time, as a gift from God, should be used responsibly. Indeed, while he rejected jewelry and ornamentation, Calvin approved of watches as a way of keeping track of and behaving responsibly in time.

Rather than oscillating back and forth between periods of celebration and periods of penance, Calvinists trod temperately through life, abstaining from the excesses permitted in cyclic visions of time and attentive to the continuous obligation to behave faithfully and responsibly. Through their preoccupation with the orderly, step by step, and encompassing progress of one's life through the time God had given them to grow in holiness, Calvinists – including the Puritans – would exercise an outsized influence over Western culture, shaping both modern conceptions of capitalism and modern democratic systems,[46] each of which relies on a particular vision of the disciplined life for its success. Modes of thinking about time that Calvin started for theological reasons quickly took hold throughout the continent and beyond. In the process, those reasons were forgotten, replaced by different reasons that provided for the same modes of thought, including making money, maintaining bureaucracies, using

[45] The literature on changing conceptions of time during the early modern period and its connection to the Protestant Reformation is voluminous. Among such resources, one might note David S. Landes, *Revolution in Time: Clocks and the Making of the Modern World*, revised and enlarged ed. (Cambridge, MA: Harvard University Press, 2000); Jacques LeGoff, ed., *Time, Work and Culture in the Middle Ages* (Chicago: University of Chicago Press, 1982); Natalie Zemon Davis, "The Sacred and the Body Social in Sixteenth-Century Lyon," *Past and Present* 90 (February 1981): 40–70; Keith Thomas, *Religion and the Decline of Magic: Studies in Popular Beliefs in Sixteenth- and Seventeenth Century London* (New York, NY: Penguin, 2003); and Catherine Pickstock, *After Writing: On the Liturgical Consummation of Philosophy* (Hoboken, NJ: Wiley-Blackwell, 1998).

[46] Recognizing the limits of his vision, it is nevertheless the case that Max Weber makes important points about the origins of modern capitalism in Calvinist thought. See Max Weber, *The Protestant Ethic and the Spirit of Capitalism* (New York, NY: Oxford University Press, 2010). See, again, Walzer, *The Revolution of the Saints*.

precise measurements in scientific research, and adhering to the social obligations and courtesies of timeliness and keeping well-ordered lives.

This is the conception of time that animates the temporal deferrals of the Constitution: maintain orderliness, trust in future-facing projects, imagine the possibilities for growth and progress, and honor the aspirational vision of a world in which things can get better over time. This particular way of living between "the already" of work God has done and "the not yet" of that work being completed funds a kind of confidence about political engagement: As political systems participate in work God has already done and is doing on the way to eschatological completion, they participate in a world drenched with meaning even as awareness of their imperfections funds hope about how they might change over time. At the same time, this particular way of living between the already and the not yet mandates a kind of modesty about the politics of the present moment: God's completed work includes judgment against political systems and God's yet-to-be completed work mandates their transformation. Time for politics is a gift but an ambivalently good one.[47]

47 Though I will not pursue them here, this idea that temporal deferral is a secular expression of a major early Protestant vision of eschatology has far-reaching consequences, including for how we think about the origins of human rights. After all, the vision of time and political engagement I'm offering here also funds a particular way of thinking about human beings that has to do with them not only as actors in the world but also as objects upon whom actions have effects. In a world drenched in meaning, some of that meaning inheres in human beings and, therein, shapes obligations about how they are treated. Nobly, those obligations take expression in the notion of rights that inhere in persons and set boundaries about how they will be treated. Maliciously, those same obligations shift debates about the treatment of women and persons of color, for instance, away from how to treat such people as human beings toward whether to think of them as fully human. Also unsurprisingly, this vision of human beings reshapes just war thinking, most notably by driving a shift in attention from *jus ad bellum* to *jus in bello* concerns. How to understand and treat the enemy is a far more pressing question when one recognizes that enemy not simply as a fellow human being but also as a human being with rights. This, in turn, has implications about how one understands the movement of the Christian just war tradition through time. So, for example, it suggests that James Turner Johnson's argument about the way the Christian just war tradition transformed itself into international law in early modern thought is right but incomplete. It isn't simply international law but modern law itself that was shaped by the Christian just war tradition. Rights-based answers to questions about how to think about and treat human beings are secular expressions of long-standing theological claims about the value of human beings. Moreover, on my reading, international law doesn't replace Christian just war thinking; it comes alongside of it. Johnson tacitly advances a myth of progress in that things replace each other; I, on the other hand, am arguing that the way traditions change as they move through time is considerably more complex than this – and one piece of evidence for this is the way that the Christian just war tradition changes into a secular form as it enters modernity.

The difference between this vision of time and Schmitt's vision of time is evident in the degree to which Schmitt relies on the work of Thomas Hobbes. Hobbes' vision of time differs markedly from the one I describe above.[48] Hobbes' natural man is animated by fear of death and hope for glory to enact a social contract that will steady him in turbulent political systems. The past is pain and the future is a mix of threat and vainglory. Hobbes' political man seeks neither a return to the past nor entrance into a different future; instead, he pursues politics as a way of stabilizing himself against threats of chaos. Even if this individual pursues politics calmly and with a sense of detachment, he does so with natural man still whispering to him from within his soul.[49] Hobbes has a sense of history to the extent that, for him, political order made it possible for history to exist rather than leaving human beings in a state of nature where the "links between past, present, and future had been snapped, and [human beings] were left alone in an awful stillness where the past was silent and the future unbeckoning."[50]

A future that only becomes meaningful when politics escapes nature is tantamount to a future lacking either foundation or direction. In such a realm, politics moves into the future passively by obsessing on the present. Hobbes didn't believe he lived during the transition from the already into the not yet so much as he believed he lived between the past and the future. Debates about whether Hobbes believed in God to the side, his political vision finds its origins and completion within an immanent frame that offers neither the confidence of

48 To be clear, I am not addressing Hobbes' conception of time (i.e., his idea of time as "the phantasm of before and after in motion" from *De Corpore*) so much as the assumptions about time that manifest themselves in his political thought. Nor am I addressing Rousseau's or Locke's or even Kant's conception of time (though Kant's understanding of the relation between time and eternity in his later work fits better with my project than do many). That is, I'm trying to answer the question, "What must one assume about time and one's movement through it in order for this political vision to be operative?" So, for example, one bit of evidence that Machiavellian thought played very little role in shaping the Constitution comes from the fact that Machiavelli defends the superiority of republican Rome and a glorious past rather than the possibilities of an as yet unwritten future.

49 I am indebted to Gabriella Slomp for highlighting the emphasis on glory in Hobbes' work. While I think she may overstate her case (her thesis has been challenged for its willingness to read all of Hobbes' work through the lens of *Leviathan*), her attention to glory is a worthwhile corrective to the "nasty, brutish, and short" life-as-fear-driven conventional read of Hobbes. See Gabriella Slomp, *Thomas Hobbes and the Political Philosophy of Glory* (New York, NY: St. Martin's Press, 2000).

50 Sheldon Wolin, *Politics and Vision: Continuity and Innovation in Western Political Thought* (Princeton, NJ: Princeton University Press, 2004), 237.

created orderliness, nor the resources of eschatological hope, nor the condolences of eternity.[51]

Nor did other contending early modern religious visions of time find much purchase in the assumptions about time manifest in the Constitution. Post-Tridentine Roman Catholicism implicitly rejected this immanent frame partly as a way of maintaining faith in its traditions as they had traversed medieval thought.[52] That church, as mirror of the Kingdom of God, was caught up in eternal things and so had no need to either anticipate the immediate return of Jesus or be driven by fears of its own end. If anything, its political thought reinforced its claims about its own permanence. Its theological understanding of its existence between the already and the not yet, teleological though it was, wasn't conditioned by concerns about temporality so much as it was by attention to spiritual maturity within a realm where eternity was already present.[53] And millennialist preoccupations about an immanent end, although expressed among a few of the Reformed theologians who would shape pietism (e.g., Cocceius and Witsius), neither funded a coherent political theology nor exercised much political influence on the framers of the Constitution. All of which is to say that among the various visions of time that floated through early modern thought, that vision associated with what would later come to be understood as mainline Protestantism was the one that would most seem to have influenced the Constitution.

Nor, as a final argument about the alignment of temporal deferral in the Constitution and a Reformed sense of time, does this posture of temporal deferral express itself in nearly so clear a way in other

[51] Likewise, Schmitt's (more meager) attention to Rousseau. Unsurprisingly, among social contract theorists, John Locke's conception of time might best align with the vision I'm describing here. Not only are Locke's fingerprints the most visible in the US Constitution, but also Locke's thought aligns better with the major streams of early modern Protestantism than do those of Hobbes and Rousseau (which probably only reveals that Locke was shaped by the same streams of thought as the writers of the Constitution were).

[52] Schmitt repeatedly argues that the post-Trent Roman Catholic church spent its energy resisting the notion that times were changing, in part as a way of rejecting the millennialist movements of the Reformation. While this is, I think, overstating his case, his suspicions about Roman Catholic resistance to change (and, therein, to history) are congruent with my claims about early modern Roman Catholicism's comparative lack of attention to temporal deferral.

[53] This suggests, parenthetically, how far removed post-Trent Roman Catholicism is from the influence of Augustine (and how significant Augustine was for reformers such as Luther and Calvin). For more on Augustine, politics, and time, see Wolin, *Politics and Vision*, 111ff.

religious traditions as it does in (Protestant) Christianity. While Eastern religions such as Hinduism and Buddhism offer cyclical descriptions of time that might make one think in terms of deferral, they don't so much defer judgment as they do sentencing: Karmic debts are being constantly accrued. Abrahamic traditions are more eschatologically oriented, but among these traditions, eschatology is far more prominent as a doctrine for making sense of Christian theology than it is for making sense of other Western religious traditions (i.e., Judaism and Islam). Waiting for Jesus to return all but mandates a future-looking posture even if that return isn't imminent: Ultimate judgment and understanding, themselves, get pushed into the future, leaving Christians with a time between times that is still freighted with meaning and meaningful activity. And eschatology, shorn of theological claims about Jesus' return and the way that return completes the work of God already begun, looks like a myth of progress.[54]

The Constitution is, in short, the juridical expression of the myth of progress and this myth is vivified by a Protestant vision of time, albeit one shorn of the theological categories that lent it meaning in the first place. It is an immanentized Protestant eschatology.[55] Contra Schmitt, there is no obvious metaphor or set of metaphors taken from theology and then secularized that can apply to its vision of the world. There is, though, a basic posture in the Constitution vis-à-vis its sense of time that comes from Protestant theology and, in its secular form, animates it.[56]

Perhaps "We the people" can be interpreted as "ecclesia." Perhaps "here is the person to answer your question" can be interpreted as "prayer." Perhaps divided government can be interpreted as Trinity, openness to amendment as confession, or gradual reform as sanctification. These metaphors have at least as much carrying capacity as "exception = miracle" (especially since it isn't the miracle but the absence of the miracle that directs the Constitution's approach of what to do with exceptional cases).

[54] To repeat something I noted in the Introduction, there are many different sources and expressions of a myth of progress in modern thought. What I name here as a myth of progress need not align itself with any one of those sources or expressions. My more generic use of the term only intends to highlight the ideas that time matters, that change over time is gradual, and that change manifests as improvement over the past.

[55] This is, ironically, the very thing that Schmitt later accuses Metz and other theologians of the politics of Christianity of pursuing. See Schmitt, *Political Theology II*, 54.

[56] Schmitt misses this posture, in part, because of his understanding of a Christian vision of time, which he treats as "not a long march but a single long waiting, a long interim between two simultaneities ..." Schmitt, *Political Theology II*, 86.

But these metaphors only carry weight – to the extent that they do – because they are being sustained by a posture toward the movement of events through time that is the product of secularized Protestantism. Which is to say, a modified political theology reading of the Constitution allows us to see it as a document of (Christian) political theology because it finds its footing in the secularized expression of a particular Protestant theological vision of time. It is a Christian document not because it comes from or is for the church, not because it references God or makes theological arguments, not because it has borrowed a particular metaphor or set of metaphors from systematic theology, and not because it was designed to favor a particular set of cultural norms and values associated with a particular religious tradition, but because its assumptions are structured by an early modern Protestant theological posture toward time. It is posture, not metaphor, that defines the US Constitution as a work of Christian political theology.

A FIRST IMPLICATION OF THIS (THEOLOGICAL) TEMPORAL POSTURE: DEFERRAL AS DELAY

Chapter 3 argued that the US Constitution is a just war document. This chapter argues that it is a Christian theological one, albeit in a secular form. Together, they make the case for treating the US Constitution as a Christian just war document. What links these two chapters and why might this matter at the entrance into the environmental age?

The link, I think, is that the assumptions about time that animate the political theology of the US Constitution are congruent with those that animate modern Christian (and especially Protestant) just war thinking. That is, they share similar postures. Just war thinking is resistant to both the apocalypticism of millennialist movements and the nihilism of a world in which change lacks meaning. Even as it faces the crises of violence, just war thinking favors processes for taking time to discern the justness of causes, the sources of legitimate authority, and the proportions of overall good and bad that are produced. It prioritizes last (reasonable) resort. And it does so because it believes that wars, foul as they are, are still events in which ultimate judgments are deferred in order to justify meaningful if ambiguously good action in the near future. Moreover, just war thinking of the sort expressed in modern just war theory is, as noted in Chapter 1, shaped around a myth of progress similar to the one that funds the US Constitution. It is not simply for reasons having to do with political power or ideological commitments that mainline Protestants have been more likely to affirm just war than have their sisters and brothers in the peace churches, less inclined to promote war than many contemporary evangelical churches, and more willing to think about war than many Orthodox churches. It is primarily with their Roman Catholic brothers and sisters that mainline Protestant churches have been willing to talk war – and this willingness has only grown over the past half-century since Vatican II. Unsurprisingly, for mainline Protestants, just war thinking feels familiar and, disconcertingly, comfortable. The posture of constitutional and just war thinking fits mainline Protestants because it is the posture of politically engaged but temporally deferred behavior.

Why, though, does this matter at the entrance to the Anthropocene? There are at least three costly implications to this overlap of temporal postures when facing wars in a warming world. The first is the danger that comes with blurring deferral and delay.

Temporal deferral, after all, can invite not only deliberation but also delay. In principle, deferral and delay are quite different: Deferral shapes an opportunity for reflection, conversation, and second thoughts, whereas delay is a way of putting off these very things. However, the postures of deferral and delay can look very much the same in political contexts even if their effects on those contexts are quite different. The invitation to delay offers a kind of cathartic relief valve from the pressures of complexity by inviting citizens to put hard questions off until they decide they're ready to deal with them. As valuable as catharsis might be in easing the psychic tensions that come with considering imperfect solutions to complex questions, it won't resolve those questions (whether they are of politics, of violence, of environmental degradation, or of some combination of the three). Whereas temporal deferral ideally offers the space to engage such questions thoughtfully, delay offers the space to avoid them. Yet feeling better about a situation by ignoring the questions that it raises is unlikely to make the situation better.

Importantly, while the posture of temporal deferral may have its benefits (e.g., its mix of malleability and resilience as it favors gradual change over time, its recognition of its own incompleteness, even its theological recognition of living between the already and the not yet) and its political expression within liberal constitutional thought has manifested those benefits, such a posture is better oriented toward addressing some types of political problems than others. Problems that need to be addressed gradually – such as those that turn on changes in culture – can be addressed through temporal deferral. So, for example, US culture needed to discover the range and depth of the problems with legislating morality through banning alcohol before overturning the Eighteenth Amendment with the Twenty-First. Or complex problems that require deliberation such as those having to do with the impact of allowing presidents to serve more than two terms benefit from deferral and the slowly gained wisdom of casuistry.

Paradoxically, problems that arise rapidly, manifesting as sudden crises, can also be addressed via a posture of temporal deferral. Knowing that an initial attempt to deal with a crisis need not be the final attempt to do so frees authorities to act rapidly and learn slowly. Knowing who is responsible for making decisions in such moments (and, therein, who is accountable for dealing with the results of those decisions) speeds decision-making, especially when existing processes and policies don't seem to apply or offer ambiguous wisdom. The slow process of shaping an engaged citizenry, thoughtfully and transparently pursued if never finally achieved, at least makes popular buy-in more likely when rapid national action seems called for. The US Congress has not declared war since World War II, but that has not prevented the United States from pursuing military actions when it believes it must, as, for example, in the case of the actions it took in Afghanistan after September 11, 2001. A posture of temporal deferral makes provisional and imperfect decisions defensible even if those decisions shape new questions and create new problems.

The posture of temporal deferral is not, though, well oriented to address the types of problems that are arising in the Anthropocene. Climate change has been a slowly developing and poorly understood phenomenon in which revelations about its impact have occurred after a point when most of the things that could be done to prevent it have passed and during a time when most of the things that can be done to mitigate it are both limited and politically difficult. The catastrophic loss of biodiversity finds its causes in actions (e.g., the global movement of goods and persons, the building of roads, new farming technologies, the introduction of chemicals into ecosystems) that carried unanticipated consequences and are all but impossible to undo. The staggering growth of pollutants suffers from a global free-rider problem: Individuals are aware that they produce miniscule quantities of waste in comparison with the total amount produced and so lack much incentive to reduce/reuse/recycle, with the result being the continued growth of pollutants. By the time these problems have reached broadly recognized crisis points, the systems that have created them are too embedded into society to change rapidly and further deferral exponentially exacerbates them.

One way to place this into the context of political philosophy would be to suggest that, on the one hand, a posture of temporal deferral is a way to avoid the dangers of concentrated power that come with a strong sovereign and, on the other hand, the problems of the Anthropocene would seem to require concentrated power in order to bring about the types of large-scale systemic changes necessary to deal with them. Centralizing power to address climate problems invites abuse, and diffusing power to avoid abuse leaves systems ill-equipped to respond. Caught between the Scylla of political abuse and the Charybdis of political effectiveness, those who would address the environmental problems of the Anthropocene find themselves with underpowered political mechanisms (e.g., carbon taxes, cap-and-trade mechanisms, the Paris Climate Treaty) to shape adequate large-scale policy changes and so are forced to rely on inadequate moral suasion mechanisms ("reduce, reuse, recycle," "eat less meat," "buy local") instead. Irony rules and it has shaped the nihilism and handwringing that one often finds within the writings of those thinking about the implications of climate change.[57]

A SECOND IMPLICATION OF THIS (THEOLOGICAL) TEMPORAL POSTURE: CONFUSING OPTIMISM AND HOPE

A second implication of the posture of temporal deferral for the problems of the Anthropocene is the way that posture confuses optimism and hope. Optimism – a motivating sense of confidence about the future – promises the energy to imagine a better future and the freedom to experiment with varied means of getting there. It is also immanently framed: The future it imagines is continuous with the present and fed by a peculiarly egocentric vision of the past in which what has happened in the past hasn't only led to now but also has led to us – and we have (and are) the resources we need to make life better. It is a secular disposition.

The politics of optimism feed an expansive sense of human agency and moral clarity. At the beginning of the Anthropocene, they find expression in suggestions about geoengineering ourselves out of the catastrophes of climate change, the power of free markets to solve the environmental problems of energy production, and/or the benefits of global warming for climates

[57] See, for example, Howard Kunstler, *The Long Emergency: Surviving the End of Oil, Climate Change, and Other Converging Catastrophes of the Twenty-First Century* (New York, NY: Grove Press, 2006); and Harald Welzer, *Climate Wars: Why People Will Be Killed in the 21st Century* (Malden, MA: Polity Press, 2017).

currently too cold to be used for food production. Whether such an expansive sense of human agency is warranted, however, turns on one's willingness to downplay the deleterious consequences of earlier actions, ignore the moral ambiguities of human desires, and reject claims that human beings are more likely to be acted upon than to be actors in the dramas of their lives. Contemporary defenses of geoengineering, free markets, and expanding temperate zones, for example, are more likely to reveal the biases of those who voice them than to be driven by either empirical or historical evidence. The sheer scale of the problems of the Anthropocene strongly indicate that such suggestions cannot carry the salvific weight placed upon them and, therein, that the optimism that underwrites them is misplaced.

Hope, on the other hand, starts in neither the capacities of human agency nor the continuities of history. It questions the sufficiency of autonomous reasoning, defies the closures of an immanent frame, and suspects that both optimists and pessimists cherry-pick the data of the past in order to avoid facing the complexities – the ironies, tragedies, possibilities, and paradoxes – of history. "Hope" is a religious term, at least in the sense that it invites us to imagine something bigger than history at work, foregrounding a peculiar way of engaging the past that is willing to attend to complexity without being swamped by it. It uses a long view of history – looking both backward and forward – to argue for the possibility of discovering meaning that transcends history. It enlivens confidence with infusions of imagination and conviction. In ways that are both captured by and extend beyond traditional Roman Catholic understandings of the phrase, hope is a theological virtue.

The politics of hope are prophetic in a Hebrew Scripture sense of that term, mixing proclamations of judgment about corruptions in the present with promises of something better in the future. At the beginning of the Anthropocene, hope-driven politics finds expression in calls for everyone to step back and take a deep breath because this isn't the first time that the world has endured calamities (albeit at great cost) and it is unlikely to be the last. Its politics begins by offering accounts that replace both narratives of denial about the reality and costs of climate change and narratives of an apocalypse in which climate change brings about the end of human life (or life itself) on earth. It suggests that we don't yet have clarity about the shape or nature of the social imaginary of the environmental age and so modesty is called for. It doesn't so much fund the pursuit of an unclear future as suggest that our energies are better spent dealing with current symptoms of climate change than dreaming of how we can solve the future. It worries about the costs of environmental changes (to human life and health, to nature's resilience, to the continuation of civilizations) but also, perhaps even more importantly, about the disproportionate and unjust ways those costs are being and will be borne. Whether it offers the moral resources and motivating energies necessary to meaningfully engage in the pursuit of environmental justice and human welfare, though, is an open question. Hope is hard and its comforts usually involve grief.

This is, perhaps, one reason that modern thought of the type manifested in the US Constitution has blended hope with optimism. In contemporary Western societies, this blurring of optimism and hope manifests as suspicions about contemporary political structures, on the one hand, and romantic visions of what might come to be through determination and luck, on the other. It looks like the convictions that the Paris Climate Treaty, while insufficient, shapes a trajectory for global change that can at least mitigate the worst parts of climate change; that when individuals decide to drive less or shift to a plant-based diet, they can actually make a difference; that the vacuum of political leadership at national or international levels can be filled through the actions of civil society and nongovernmental organizations. It recites Margaret Mead's statement: "Never doubt that a small group of thoughtful, committed citizens can change the world; indeed, it's the only thing that ever has",[58] and June Jordan's: "We are the ones we have been waiting for."[59]

One of the dangers evident here is that the blurring of optimism and hope manifests in the rhetorically awkward and strategically disjointed linking of evidence-based statements about current problems and vague aspirational statements about the future. Temporal deferral does more than move answers into the future; when the questions driven by acute crises are answered by platitudes, temporal deferral conceptually separates answers from the energies of the questions that motivate the search for those answers. By sundering this connection between our sense of the present and our vision of the future, this blending of optimism and hope reinforces some of the worst parts of each. It draws from optimism

58 Although the quote has always been attributed to Mead, its source remains so uncertain that the Institute for Intercultural Studies, which Mead established, maintained a website for addressing it and those who would use it. See www.intercultusralstudies.org/faq.html (accessed January 12, 2018).

59 The origin of this quote, too, is uncertain. It is often associated with author Alice Walker, who attributes it to Jordan – though Jordan's epigraphic use of the sentence may itself signal an earlier existence. See Peter Levine, "Who First Said, 'We Are the Ones We Have Been Waiting For'?" at http://peterlevine.ws/?p=6105 (accessed January 12, 2018).

an unjustified confidence in human agency without also claiming optimism's wisdom that past, present, and future are linked. It draws from hope a vision of a future that might be different in kind from the present without accepting the hard truth of hope that big changes always threaten harm and carry grief within.

This blurring of optimism and hope sustains the myth of progress. That myth, which is a particular way of narrating our movement through time, not only points to actual times when things have gotten better but also believes that things will continue to get better. It ignores the way that things get better in one place (say, the countries of Western Europe) because things have gotten worse in other places (say, those places that the countries of Western Europe colonized and then extracted resources from, leaving them not only economically, environmentally, and politically poorer but also with economies that are more dependent on resource extraction). It ignores the way that things getting better for human beings generally (through, say, the creation and use of antibiotics) can lead to deleterious consequences not only for nonhumans (as when the proliferation of antibiotics harms beneficial bacteria in ecosystems) but also for humans (as when the misuse of antibiotics lead to strains of multi-drug-resistant bacteria). It even ignores the way catastrophes such as mass extinction events make new things (species, ecosystems, etc.) possible. The myth of progress doesn't leave adequate room for malice, tragedy, irony, and paradox.

The problem with the myth of progress isn't whether progress is possible; it clearly is. Things do change over time and that change regularly brings improvement: better to live in a world with antibiotics than one without them. Instead, the problem with the myth of progress is that it suggests that progress can be assumed as an act of faith. The affirmation of that assumption is far more dubious. A moment of reflection on history reveals that it is replete with too much carnage to justify such a faith. A moment of theological reflection reveals how paltry and infertile such a faith would be. The myth of progress – which is the problematic blurring of optimism and hope – not only cannot carry the weight being heaped upon it, but also its power constricts the range of resources that we need to move forward in time by making it harder to learn from the atrocities, tragedies, ironies, and paradoxes of human history.

The modern story of the history of the Christian just war tradition is premised on a myth of progress; this is one way to tell that it is a modern story. And, as noted earlier, the modern story of the history of the Christian just war tradition is oriented around the development of a theory (which, in itself, is evidence of its beholdenness to a myth of progress, since theories provide mechanical solutions to problems). That's another way to tell that it is a modern story. If, though, the US Constitution is a Christian just war document – as I have been arguing over these last two chapters – and its peculiar form of Christian political theology blurs optimism and hope, the modern Christian just war tradition is more than premised on a myth of progress. It is a basis for that myth and it continues to fund that myth. It feeds its own errors.

One bit of evidence for this failing is the fact that there doesn't seem to be any place for climate-shaped conflict in the modern narrative about the Christian just war tradition. In spite of the fact that scientists are offering increasingly compelling evidence that changes to the climate have triggered violence; in spite of the fact that a range of contemporary conflicts in, for example, Syria, Sudan, Somalia, and elsewhere are linked to climatic change; in spite of the existence of established programs in environmental security studies within academic disciplines such as political science and international studies; in spite of the fact that leading military figures are calling the effects of climate change among the greatest threats that nations face in the twenty-first century; in spite of the fact that many of the current low-level, climate-shaped conflicts (e.g., those between contending tribes of herdsmen in the Sahel) don't fit within modern just war categories; in spite of the fact that both pacifists and those who traffic in *realpolitik* are increasingly making connections between climate and conflict – in spite of all this, the modern Christian just war tradition has been all but silent on climate-shaped conflict. In the myth of progress, things shouldn't be getting worse and there definitely shouldn't be new kinds of and understandings of conflict arising. Modern just war theory's failings to address climate-shaped conflict start with its unwillingness to tell a problematic story about its own history.

A better telling of the history of Christian just war thinking will need to account for its own complicity in perpetuating problematic stories about itself. It will need to recover resources from within itself that its participation in the myth of progress has concealed (including, as the following chapters will suggest, better ways of thinking about its relationship to the natural world). It will need to imagine what it would mean to think more in terms of hope than of optimism and to be more cognizant of its limits and its non-just war thinking conversation partners than it thus far has done. And, especially at the beginning of the Anthropocene, Christian just war thinking will need to come to terms with the disjointed, disjunctive, disorienting, and regularly dysfunctional qualities of its own thought, especially as that thought moves into a new social imaginary.

A THIRD IMPLICATION OF THIS (CHRISTIAN) TEMPORAL POSTURE: FREEDOM AS FETISH

Finally, a third implication of the posture of temporal deferral is that it has contributed to and undergirded the fetishization of freedom. In societies that prioritize autonomy, freedom is a great and necessary good for coherent and resilient polities and temporal deferral sustains those polities. People need freedom in order to reflect, to choose, and to make those choices meaningful by allowing their decisions to bear fruit; temporal deferral creates the space for such freedom to function. The deferrals within the Constitution as I describe them above join with that document's emphasis on freedom in, for example, the Bill of Rights, to shape a more compelling and coherent argument for the type of liberal polity that the document defends. Without ignoring the way that the same document restricted and/or undermined freedom in wholly unjust ways (reifying a political defense of slavery through the Three-Fifths Compromise, restricting voting rights, etc.), its promotion of freedom and its posture of temporal deferral not only align but also are linked.

Neither a posture of temporal deferral nor the good of freedom, in themselves, are sufficient for the task of fully shaping a coherent and resilient polity. Other goods (e.g., order, concern for the common good, resources for adjudicating conflicts between goods) are also vital and also expressed in the US Constitution. In a liberal polity, though, freedom gets pride of place as the good around which other goods must revolve. They can shape, order, and at times even override it, but freedom leaves traces of itself even when other goods are given priority: Order mandates a polity in which theft is illegal in order to shape an economic system in which one's freedom to buy and sell go (relatively) unrestricted; the pursuit of the common good shapes a polity in which politicians represent their constituencies rather than their own interests, but those politicians can legitimately hold their offices only so long as people can freely vote for or against them; for the sake of the common good, access to and use of resources may be restricted during periods of scarcity but once such periods are over, fuller access should follow.

The priority of freedom shapes a necessary imbalance among goods in liberal polities. Other goods are justified by the way they can promote freedom whereas the value of freedom/liberty is assumed. As a result, freedom/liberty escapes any tight operative definition in liberal polities because the assumption of its value means that it will not be sufficiently described by comparing it to other goods. This absence of tight operational definition reveals a cipher-like quality to freedom: Its meaning not only gets determined in context but can also vary within contexts, depending on the concerns, priorities, and perspectives of those in that context. This is its fetishized quality: Its value is greater than any shared meaning available to it.

The fetishization of freedom finds expression not only in the way it is valued in many spheres of human activity (politics, economics, family, religion, psychology, etc.) but also in the way it functions as a kind of fiat currency within the exchanges of meaning-making. Not only is liberty a political priority; not only are free markets an economic necessity; not only is independence a sign of maturity; not only is free will a precondition for faith; and not only is self-differentiation a sign of emotional health, but also freedom traverses across these spheres of human activity. Free and fair elections stabilize markets, maturity is expressed through the ability to support oneself financially, religious affiliation and participation cannot be politically coerced, and self-help books constitute entire sections of bookstores. Promoting freedom in one sphere will influence other spheres and the meanings of freedom in one sphere creep into the meanings of freedom in others.

Because it is fetishized, it is difficult to explain why this creep is problematic: As much as such actions may violate our instincts, why shouldn't it be legal to sell one's vote or one's body? Why might it not be in society's political interest for wealthy parents to leave the whole of their estates to their children? How might a commitment to the providential work of God enable human flourishing? When might the pursuit of self-differentiation inhibit one's own happiness? Philosophers, theologians, and ethicists shape argument after argument and build case after case answering such questions, but it sometimes seems like the sheer preponderance of answers to such questions and the knotty problems that sit behind them only reveal the difficulty of thinking beyond freedom. Perhaps the modern metaphysicians doth protest too much.

One of the more insidious expressions of the fetishization of freedom comes in the way that it feeds an inequity between liberal political systems and late modern consumer capitalism. If liberal political systems are oriented toward a capacious future, consumer capitalism, with its relentless drive for acquisition and growth, insists on the urgency of the present: buy the latest, buy the biggest, and buy now. It is as if neoliberal economics rushed in to fill the vacuum intentionally left by classically liberal politics and its insistence that we will have time to deal with the problems that regularly constitute political life together.

Much has been written elsewhere about the dangers of conflating economic and political systems and the damage that comes with thinking as consumers in a market society rather than as citizens who live in market economy.[60] Much has been written on the tendency of capitalism to expand beyond the limits imposed on it by political systems: Political protections for economic institutions that are too big to fail only mask another set of dangers that come with failing by growing too big.[61] Much has been written about the dangers that neoliberal economic policies pose to liberal democratic systems.[62] And, over the past few years, much has been written about the relationship between economic growth and political inequality.[63] This is not the place to rehearse all those arguments. However, to the degree that those arguments stand up – and I believe many of them stand up very well – the avarice of contemporary neoliberalism is doubly dangerous.

On the one side, it disrupts and undermines the very political systems that may be necessary for the world's citizens to address the dangers and costs of environmental catastrophe. When people think of themselves as consumers rather than citizens (or, more insidiously, when they think that the primary practices of citizenship should be understood in consumptive terms), the scope of their attention to goods such as order and the pursuit of the common good shrinks. While liberty, as a political good, sits in constructive tension with other political goods, choice, as an economic good, is ill-fitted to maintain those same tensions when it is imported into political systems. Elections don't become clearer as the number of candidates on the ballot multiplies. Tax laws don't become fairer when taxpayers can choose where their taxes should go. Understanding political issues doesn't become easier when one can choose the news service that best fits one's prejudices. The result of emaciated political systems is weak environmental regulations at the very time when robust political systems are needed to address dangers to environmental (and, concomitantly, human) health.

60 See, for example, Michael Sandel, *What Money Can't Buy: The Moral Limits to Markets* (New York, NY: Farrar, Straus and Giroux, 2013).

61 See, for example, Joseph Schumpter, *Capitalism, Socialism and Democracy* (New York, NY: Harper and Row, 1950).

62 See, for example, Lisa Duggan, *The Twilight of Equality? Neoliberalism, Cultural Politics, and the Attack on Democracy* (Boston, MA: Beacon Press, 2003); Wendy Brown, *Undoing the Demos: Neoliberalism's Stealth Revolution* (Brooklyn, NY: Zone Books, 2015); and Bonnie Honig, *Public Things: Democracy in Disrepair* (New York, NY: Fordham University Press, 2017).

63 See, for example, Thomas Piketty, *Capital in the Twenty-First Century* (Boston, MA: Belknap Press, 2017).

On the other side, that avarice encourages persons to consume at an environmentally unsustainable rate. Studies suggest that human beings currently use up resources at two and a half times the rate that the earth can produce resources – and if everyone maintained a standard of living like those in the West, we would use up those resources at four times that rate. The rate of consumption in Western societies has tripled in the last forty years. Environmental economists stress the need for societies to decouple the pursuit of economic growth from the use of natural resources to drive that growth, but their calls for decoupling have not proven to be compelling to many in Western societies and have proven to be punitive (and overtly racist) to those in other nations whose primary sources of economic support are their natural resources.

There are ways to prioritize liberal political systems – and, for that matter, free-market economic systems – while still pursuing those practices that will reduce and/or mitigate the more damaging effects of the wicked problems associated with climate change. When governments support research into sustainable power systems rather than subsidizing fossil fuel exploration, extraction, and consumption, they help provide meaningful choices to their citizens about where they will get energy. When candidates for office include concern for the environment in their political agenda and news sources inform voters about the impact of environmental degradation and climate change, free and fair elections lead to greener candidates. When nations work together to establish and maintain systems for controlling and reducing greenhouse gasses, they benefit their citizens in myriad ways.

Yet prioritizing liberal political systems while pursuing environmentally sustainable practices will still mandate defetishizing freedom. Freedom is a great good, but it is not the only great good. And only as it is defetishized will its flow across spheres of human activity be restricted enough to allow those spheres to maintain their integrity and prioritize the distinct goods that sustain them.[64] Freedom is a great good; it is not, though, a good fetish. And the consequences of fetishizing it are especially costly for life in the Anthropocene.

Collectively, these three implications (confusing deferral with delay, blurring optimism and hope, and fetishizing freedom) complicate the process by which the Christian just war tradition will move into the

[64] I am, I suppose, arguing for something like Michael Walzer's political vision in his *Spheres of Justice*. See Michael Walzer, *Spheres of Justice: A Defense of Pluralism and Equality* (New York, NY: Basic Books, 1984).

future to address climate-shaped conflict. The costs of delay have grown exponentially. The credibility of optimism has diminished markedly. Confusions about the value of freedom proliferate. Moreover, these complications are not tangential to the way that Christian just war thinking will have to change because they are the products of Christian just war thinking. That is, as it moved into modernity, Christian just war thinking shaped the very values and concerns that have led to these implications. The US Constitution, as a Christian just war document, is evidence of this. Addressing these implications will necessarily catch the tradition up in projects not only of recognizing its complicity in shaping the problems we now face but also in providing new ways of thinking that can correct for the malformations of its current ways of thinking as we move into the Anthropocene. Those are heavy weights to bear.

Of course, bearing the weights of complicity and the malformation of thought is not a new problem for Christian just war thinking. It has always turned on questions about complicity and the malformations of thought under conditions of sin and finitude. It will continue to do so as it reshapes itself to answer questions about other, self, and state (i.e., the questions of theological anthropology) in a new age. And toward that end, it will need to attend to roads not taken during its development in early modern thought. Among such roads was the one having to do with the way climate and conflict interacted during the sixteenth to eighteenth centuries. After a brief chapter tying up the arguments of this and the previous three chapters about the way that the Christian just war tradition shaped modern thought, the remainder of the book maps that road.

5

Christian Just War Thinking and Modernity

> [T]he interrelationship [of productive capacity and military might] applied as directly and unequivocally as it did only after 1500, with the growth of modernity.
>
> Azar Gat, *War in Human Civilization*

> If Christian thinking about resort to war were as simple as leading spokespersons for the tradition would have us believe, summarizing its most salient features for comparative purposes would be easy. But traditions get complicated as they go along. And church officials and theologians are often engaged in disguised polemics when they make pronouncements about the essence of structure of their tradition's teachings. Thus there is reason to treat any simple account, above all an official one, with suspicion. It is a mistake to suppose that some such account must be accepted and that the problem is merely to decide which account to accept. Rather, doing justice to the diversity of a venerable moral tradition means coming to terms with competition among the various patterns of reasoning within it, each of which may present itself as *the* traditional pattern. So it is with Christian thought on resort to war.
>
> Jeffrey Stout, "Justice and Resort to War"

> These processes of ethical and cultural creation are especially difficult to think about or talk about because self and culture are reciprocal, mutually forming over time. We are in part the products of our language, but each time we speak we remake it. This reciprocity is so deeply embedded in our experience that it is hard to recognize it.
>
> James Boyd White, *When Words Lose Their Meaning*

In the introduction to this book, I claimed that the questions around which the Christian just war tradition coheres are those of theological anthropology: What are people like? What are their capacities and limits? What drives and constrains them? Who counts among them and how much does each count? How do they understand and relate to others and to the rest of the world? What types of systems do they put in place in order to clarify, constrain, and otherwise regulate their relations with each other? How do they communicate these things? These are questions of self, state,[1] and other and they distinguish just war thinking from, for example, pacifist, just peacemaking, and *realpolitik* thought.

The first four chapters of this book looked at answers to questions of self, state, and other given by important Christian just war voices; namely, Francisco de Vitoria, Hugo Grotius, and the US Constitution. Vitoria's vision of the "other" both restrained violence against indigenous persons in the Americas and West Indies and also shaped patterns of commodification and colonialization that have undergirded racism in the West and that drive a particular vision of the nonhuman natural world as a kind of other. Grotius' understanding of the self both allowed for the flourishing of international law and also shaped notions of the natural world as resource and of human agency that have complicated Western attempts to respond to the increasingly interlinked environmental crises and the needs of refugees. The US Constitution's just war vision of the state emphasizes both state sovereignty and citizen sovereignty and has also complicated the processes and possibilities of addressing the violence of nonstate players and nonstate-related violence. And the US Constitution's promotion of a politically secular state both created space for religious freedom and also manifested as a posture of temporal deferral that arose out of a Christian vision of time and that inhibits wise responses to the types of crises and violence we face in the Anthropocene.

Said differently, Vitoria shapes an instrumentalist vision of human engagements with the other; Grotius promotes a self that will come

[1] To again make explicit what I have noted already: I am using the term "state" in a capacious and primarily alliterative sense, rather than in a technical one. In this sense, any politically organized, publicly recognizable, and geographically bounded body of people could count as a "state," including bodies as large as empires and as small as tribes. And even this quasi-definition is riddled with problems around its edges: What about collections of politically organized groups (e.g., the Axis and Allies of World War II)? What about groups that are not internationally recognized (Myanmar after the ruling junta changed the name of Burma in 1989) or geographically bounded (Al Qaeda)? Or, for that matter, what of bodies that fit all three criteria but are clearly not states (e.g., the Presbyterian Church USA)? Dictionary and encyclopedic entries are of limited further help here, mostly noting that there is no consensus on the definition of "state" or agreement about a more basic term to substitute for "state." For now, I will simply trust in readers' charity and Wittgenstein's notion that within a language game, users can recognize a "family resemblance" of terms that so overlap in meaning as to be coherent and agreed upon by those users.

to be seen as autonomous; and the US Constitution emphasizes both universality in its approach to morality and immanence in its vision of the state. These four big ideas – instrumentalism, autonomy, universalism, and immanence – are hallmarks of modern thought. At minimum, the existence of these four big ideas within various components of the Christian just war tradition at the beginning of modernity reveal how quickly the tradition was shaped into its modern form. A more ambitious interpretation of the relation between the Christian just war tradition and modernity is that the former helped create the latter. It is this ambitious interpretation that I want to play out in this chapter. Let us begin with the meaning and significance of these four big ideas.

THE SHAPE OF MODERN THOUGHT: AUTONOMY, UNIVERSALIZABILITY, INSTRUMENTALISM, IMMANENCE

To get at the significance of the transition of just war thinking during the advent of modernity, it helps to make sense of the modern social imaginary. Toward that end, I begin with these four big ideas: autonomy, universalizability, instrumentalism, and immanence. Collectively, these four ideas not only define the modern social imaginary but also, through their interactions, shape the very conceptions of self, state, and other in need of reconsideration.

This is not to say that modernity is the product only of ideas. New technologies (the printing press, gunpowder-powered weapons, mechanical clocks), changing economic systems (the rise of trade and growing political power of the merchant classes), demographic shifts (the growth and expansion of cities, the increased mobility of persons, colonization), and new religious identities (Protestantisms), among other forces, all shaped modernity. Yet even as each of these other forces shaped modernity, they were also shaped by these four big ideas. As such, these ideas help make sense of the modern social imaginary, carrying representative weight for the breadth of the changes that led to and shaped modernity.

Perhaps no idea has been more closely identified with modernity in the West than that of autonomy. Whether derived from Rene Descartes' *cogito, ergo sum*, John Locke's emphasis on the individual freedom originally available in a state of nature, Jean-Jacques Rousseau's idea of the *tabula rasa* as a starting point for education, John Stuart Mill's maximal vision of liberty for the performances of selfhood, or Immanuel Kant's monumental solution to the problems of contingency under conditions of subjectivity (*Sapere Audi!*), autonomy – the freedom to act without the undue impositions of external authorities – has been the defining hallmark of modern thought.[2] It has shaped science, philosophy, theology, politics, and economics, undergirding many of the most significant advances in those fields over the past several hundred years. It has prevented, inhibited, or fought to overturn many of the worst abuses of heteronomous forces applied to people. And it has provided the basis for the advancement of human rights as basic universal goods.

Autonomy has not, though, been a force only for good in the world. It has reified the separations between persons and between persons and things. It has downplayed or disregarded the impact of the past on the present. It has elevated a particular form of abstract reasoning that is blind to the powers of emotion and habit to shape thought. It has hidden its own historically shaped and conceptually laden origins. It has not so much abolished conflicts as refashioned them and it has shaped resources – including just war thinking – that both limit and exacerbate those refashioned conflicts.

Likewise, universalizability – the ideas that all human beings share common qualities such as reasonableness, religiousness, and basic rights at their core; that people should, therefore, be able to give reasons for their actions that all persons would find acceptable; and that those reasons provide a basis for recognizing that the most basic human goods should be available to all persons – is a hallmark of modernity.[3] Driven by the fears arising out of the wars of religion (namely that particularity breeds injurious prejudice and that tribalism – whether religious or ethnic – gives rise to violence), universalizability functions as a lodestone for modern moral thought.[4] And, like autonomy, universalizability has given rise to important philosophical, theological, scientific, and

[2] Beyond reading Descartes, Locke, Rousseau, Mill, and Kant as drivers of autonomy in modern thought, one might further explore the significance and impact of the idea of autonomy through resources such as Isaiah Berlin's *Four Essays on Liberty* (Oxford: Oxford University Press, 1969); Gerald Dworkin's *The Theory and Practice of Autonomy* (New York, NY: Cambridge University Press, 1991); or, especially, Jerome Schneewind, *The Invention of Autonomy: A History of Modern Moral Philosophy* (Cambridge: Cambridge University Press, 1997).

[3] On universalizability, see R. M. Hare, *Freedom and Reason* (Oxford: Oxford University Press, 1977).

[4] For instance, even as he is moving beyond it, Kant shapes his first formulation of the categorical imperative in terms of universality: "Act only in accordance with that maxim through which you can at the same time will that it become a universal law." Immanuel Kant, *Groundwork for the Metaphysics of Morals*, trans. Allen W. Wood (New Haven, CT: Yale University Press, 2002), 37. Though the idea of universalizability is most closely connected to Kantian ethics, I would argue that it undergirds Leibnitz's concept of the

political goods. International law, modern democracy, and human rights (not to mention the scientific method) have all grown out of the conviction that universalizability is to be preferred to particularism.

And, likewise, universalizability has brought with it a range of problems and crises. The hegemonies of colonial powers and their suspicions of (or, worse, antipathy toward) indigenous populations; assumptions of similarity that have masked important differences in power and perception; the genocidal attempts to bring an end to entire groups of people who do not conform to an idiosyncratic vision of what should be included within the "universal"; and the seeming need to establish exceptions to moral norms during a state of emergency as a way of recognizing the universality of such norms even in the breach: all of these are products, in part, of the universalizing impulse.

In place of superstition and a world rife with noncorporal entities active in the world, modern thought has given preference to an explanatory view of cause and effect in which both causes and effects were physical, measurable, repeatable, and able to be discerned by anyone with proper training in the sciences. This is instrumentalism.[5] Beyond the relief that comes with being released from a world in which forces beyond our perceptions or control drive the movements of history, the benefit of such a vision has been the development and success of the sciences and a broader insistence on methodological clarity applied throughout scholarly work. The effect of this vision has been to reject not only the teleological assumptions operative in the premodern sciences, but also the larger teleologies that provided oppressive and unwarranted clarity about persons' places in the universe. Thanks to instrumentalism, the future, at least for moderns, became an essentially open-ended affair: free will dominated philosophy, just procedures replaced just outcomes at the center of law, kings lost their claims on "divine right," and God

monad, Grotius' vision of international law, Hobbes' understanding of the leviathan, Spinoza's understanding of nature, and Matthew Tindal's deism, among other ideas, and has expressed itself variously in R. M. Hare's utilitarian moral thought and Alan Gewirth's "principle of generic consistency." Further, I would suggest that the spirit of universalizability haunts modern forms of positivism and scientism, at least to the extent that positivism and scientism don't so much divide the world between "that which can be discussed through the use of universally acceptable methods and that which can't" as reject the consideration of any assertion that can't be generally verifiable.

[5] The scientific revolution inaugurated by Galileo and given perhaps its fullest expression in Newtonian physics both relies on and drives this kind of instrumentalism. Yet it hardly stops there; by the era of John Dewey, instrumentalism had pervaded even philosophies of education, political theory, and law.

retreated to a safely deistic distance. Given a choice ("choice" not only being given primacy of place but also simply being a given), most of us would prefer to live in such a universe.

Yet, like autonomy and universalizability, instrumentalism also brings with it deleterious consequences. Lacking an intrinsic moral shape, instrumentalism could be and was turned toward foul ends. Separating truth from meaning, process from purpose, and the thing studied from the one studying it, instrumentalism opens the door for nihilism philosophically, larger and increasingly indiscriminate weapons militarily, impotence politically, and degradation environmentally. Answers to the questions "Is this useful?" and "Is this good?" become increasingly difficult to distinguish, as is evident in growing and plaintive concerns about the loss of the ability to even think about the common good in a way that admits to the needs for national sacrifice. The power of wonder to shape not only inquiry but also the inquirer dimmed.

Immanence was shaped by and, in turn, shaped these emphases on autonomy, universalizability, and instrumentalism. Immanence – framed as it is by the rejection of a "beyond" that provides order and purpose, whether that beyond is manifest in Plato's idealism, Augustine's city of God, Kant's *Ding an sich*, or, weakly, Heidegger's *Sein* – insists on the priority of that which is present, available, and temporally fixed over a transcendence that exists beyond space and time. An important corrective to preoccupations with the hereafter and the insignificance of human beings, immanence disenchants the world and provides a vision of human beings and the world around them that is measurable on human-sized scales. Immanence displaces teleology. In the natural sciences, progress is assessed by attention to incremental growth from simplicity to complexity. In society, progress is assessed by humanity's ability to recognize more people as human and to desire their flourishing. In law, justice is measured by adherence to an implicitly agreed upon social contract between persons rather than correspondence to divine law. In economics, wealth is measured in fungible terms and fiscal health by market growth. In culture, the realm of the secular became ascendant. In ethics, decisions about right and wrong are shaped by rationally discerned rules and principles rather than the development of virtues through which one aspires to become something one is not yet. And in religion, faith was increasingly identified with the choices one makes – and eventually, therein, with the priority of morality over theology in adjudicating ecclesial and interreligious conflicts – rather than simply being understood as a foundation for and condition of existence.

IDEAS AND THEIR IMPACTS ON MODERNITY AND WAR

Autonomy, universalizability, instrumentalism, immanence: they are all concepts as much as goals; the building blocks of a social imaginary as much as ideals valued in it; means for criticizing a social imaginary as much as claims about it. Collectively, they constitute not only the philosophical bases for modernity but also exemplars of its benefits. I have named them here in perhaps greater detail than necessary (though in less detail than either a full-throated defense or a substantive criticism of them might warrant) for four reasons.

First, I wish to highlight the degree to which ideas have significant though bounded implications for social change. Over against a range of historians and philosophers who have argued that such changes are primarily the result of larger social forces, materialist causes, biological drives, or mere caprice, I am suggesting that human societies actually have been and can be shaped by the considered reflections of individuals whose thoughts then sway others and thereby shape societies. Larger social forces, materialist causes, biological drives, and random events certainly play roles in the movement of human beings through time, but they don't determine it in the absence of reflection on those forces/causes/drives/events. Instead, as ideas shape the ways we think about ourselves, others, and the rest of the world (and shape the ways we relate self, other, and the rest of the world), they shape that movement through time. More than that, though, ideas shape the habits and technologies of an age, which then can reshape not only those ideas, but those social forces, materialist causes, and even biological drives. In the context of the arguments in this book, giving attention to these ideas helps me clarify that though it is wrong to think of the Christian just war tradition as being framed only by ideas, it is also wrong to think that the tradition has moved through time as if ideas don't matter. They matter a great deal; they just aren't the only things that matter.

Of special note in the context of this argument is the degree to which autonomy, universalizability, instrumentalism, and immanence shape particular approaches to understanding and addressing war and its problems. As both Hans Joas and Charles Tilly have pointed out in otherwise quite different work, modern Western understandings of and approaches to war are distinct both from those that came before them and from war as it has been practiced in other parts of the world (at least until the twentieth century). Joas, on the one hand, has emphasized the degree to which Enlightenment/liberal thought, its growing appreciation of the social sciences, and, perhaps above all, its vision of progress have shaped modernity such that it struggles to make sense of – and therefore regularly ignores – reasons for and patterns of war. Modernity's struggles to make sense of war, though, have not so much eliminated war in the West as made it harder to constrain and, increasingly, also made it likely to be driven by persons whose primary motivations are not overtly political (e.g., those driven by religious fanaticism or economic gain).[6]

Tilly, on the other hand, has focused his attention on the degree to which modernity reshaped political forms and modernity was itself shaped by war. Whereas premodern and early modern Europe was littered with a remarkable range of political systems (city states, empires, diffused nations and national states, fiefdoms, colonies, territories, kingdoms, etc.), the forces of war gradually led to the consolidation of these systems into just a few political systems. These remaining systems are those we associate with modern nation-states, each of which is able to maintain its strength through its claim to have a legitimate monopoly on violence and a willingness to submit that monopoly to the rule of law.[7]

Neither Joas nor Tilly emphasizes the significance of ideas to the degree that I have here (though Joas comes far closer than does Tilly, whose argument looks more like "war shapes modernity" than "modernity shapes the idea of war"). Both of them, though, clearly lay out arguments that advance the claim that "war," "state," "self," and "politics" are not natural categories but social constructions. As constructions, they can also be reconstructed within and by the particular social imaginaries that are then shaped by those constructions.

It follows, second, that ideas have origins. In spite of covert appeals to permanence and universality driven by a preference for theory over history, law over contingency, and abstract reasoning over context, the

[6] Hans Joas, *War and Modernity*, trans. Rodney Livingstone (Cambridge, UK: Polity Press, 2003). See also John Mueller, *The Remnants of War* (Ithaca, NY: Cornell University Press, 2004), for the way war under conditions of modernity has changed – primarily, in his opinion, through the power of those ideas associated with pacifist movements such that conventional war has all but disappeared, to be replaced by irregular wars advanced by nonstate players who are seeking their own advantage by benefitting from the chaos that war brings.

[7] Charles Tilly, *Coercion, Capital and European States* (Cambridge, MA: Blackwell, 1990). See also the work of Max Weber as precedent-setting for Tilly's work by naming the modern state and its vision of having a legitimate monopoly on violence submitted to the rule of law. Which is not to say the two agree. Tilly is much more attentive to economic resources and drivers than is Weber, for instance. Among other sources, see Weber's essays in *From Max Weber: Essays in Sociology*, ed. H. H. Geerth and C. Wright Mills (New York, NY: Oxford University Press, 1946).

hallmarks of modernity have beginnings. The four concepts of autonomy, universalizability, instrumentalism, and immanence, in particular, rose to prominence at a distinct time and out of distinct circumstances in particular contexts, generally as a way of getting out from under the weight of heteronomous power used for particular and often nefarious ends under the mandate of divine purpose, especially as that power was expressed in the wars of religion in Europe. Where a premodern vision's emphasis on a hierarchically ordered universe led to the abuses of feudalism, autonomy gave people control of their own lives. Where a premodern vision's emphasis on static and ontological differences between persons led to social stratification, universalizability recognized the equality of persons regardless of social location. Where a premodern vision's emphasis on noncorporeal agents left people feeling vulnerable to forces not only beyond control but beyond resistance, instrumentalism invited people to express their own agency. Where a premodern vision's emphasis on the transcendent left people feeling insignificant, immanence shaped a world in which they had the power to bring about change.

Certainly, these four concepts all have precursors in premodern systems of thought, but none of those precursors functions so prominently and powerfully in the thought of premodern Europeans as these four have since the advent of modernity. Indeed (and paradoxically), one expression of the prominence and power of these four ideas is the degree to which they advance a vision of history in which ideas arise *de novo*. Moreover, since these ideas are shaped to be functional, it is worth observing that their power lies not only in the explicit ways they lay out a particular vision of the world but also in the way the world they have created is designed to emphasize their powers. That is, autonomy, universalizability, instrumentalism, and immanence not only rise to prominence during a particular time period; they also become the means through which to make a social world. They have made this world by providing recourse – sometimes implicitly and generally inchoately – to a guiding set of dominant ideas that shape practices and technologies as much as changing habits and new technologies refine ideas. Said differently, once introduced, these ideas participate in the production of the modern social imaginary even as they are products of that social imaginary. To the extent that these ideas arose out of the transition from the late medieval to the early modern forms of Christian just war thinking and then shaped modern practices and technologies, that tradition was instrumental in shaping the practices and technologies associated, for instance, with colonization and the doctrine of discovery, international law, modern constitutional forms of government, and contemporary environmental crises (not to mention our ability to separate all those things from each other).

Arising out of particular contexts, it also follows, third, that as both producers and products of the modern social imaginary, ideas have life spans. The dominant ideas of modernity came about, at least in part, as correctives to earlier ideas and the world that those earlier ideas shaped. Yet their corrective qualities are neither shaped for the coming context nor perfect for all contexts. Especially in a rapidly changing world, the very stability of ideas undermines their authority as they become less and less able to offer compelling arguments about how to address new problems, many of which they have contributed to. Where autonomy has given people control of their own lives, it has also treated people atomistically, thereby unduly separating them from each other. Where universalizability has promoted equality between persons, it has also papered over significant differences of power and perspective between persons and their respective needs. Where instrumentalism has invited people to express their own agency, it has also shaped a particularly mechanical approach to life that, when extended to questions about why we behave the way we do, has ironically undermined the very agency it intended to promote. And where immanence has shaped a world in which people have the power to bring about change, it has also restricted the scope of reflection and the power of wonder. Collectively, separation of persons from each other (and the rest of the natural world), blindness to power and its uses, mechanistic visions of natural behavior and cultural purpose, and the diminishment of vision and wonder have played powerful roles in creating and then exacerbating the environmental crises we now face.

Finally, these ideas have shaped a particular vision of time itself.[8] Let us begin with the emphasis on immanence. Though perhaps originally shaped as a response to a hypertrophied sense of transcendence, immanence gradually displaced transcendence such that appeals to the transcendent were increasingly cloistered in explicitly religious communities shaped by premodern views. The "whole" was quantifiable and, if not exactly touchable, at least open to examination through use of the five senses and whatever aids we could create to enhance those senses.

[8] See, again, Elizabeth F. Cohen, *The Political Value of Time: Citizenship, Duration, and Democratic Justice* (Cambridge, UK: Cambridge University Press, 2018).

The idea that there is a reality sitting beyond (or through or above or around) the observable one – a reality that demanded our attention – became increasingly quaint, suspect, or irrelevant. And as that vision of the transcendent disappeared, so did its implications for how to live in time. Life was not teleologically pulled into a future from a realm beyond but instead organically proceeded from the past. Progress meant evolutionary development over time rather than increasing proximity to otherworldly perfection. The future lay open as purpose became a project of achievement rather than discernment even as the past disappeared forever into a foreign country where things are done differently and with which we can only be vaguely familiar.[9]

Likewise, instrumentality and universalism favored visions of time that were measurable and consistent across space and through the ages. The use of mechanical clocks expanded dramatically as ways of ordering the day (rather than, e.g., sunrise and sunset), therein helping to regularize the days of merchants and city-dwellers. Chronometers were invented in the eighteenth century and helped those traveling across time zones to keep track of both time and location. The Gregorian calendar replaced the Julian calendar throughout much of Europe in the sixteenth century as a more accurate and universal way to keep track of days and years (not to mention a way that kept Christmas near the winter solstice and Easter in the spring). And as clocks and calendars changed and grew, they allowed persons to gain greater control not only over their lives but also over natural forces (e.g., those associated with planting seasons). Indeed, paradoxically, the very idea of a tradition as a linked set of ideas, practices, beliefs, and technologies that can find their origins in the past and help make meaning for particular communities at particular points in time is itself a modern construction that can exist only because of a modern vision of time.

And autonomy, universalizability, instrumentalism, and immanence were also shaped by and shape modern Christian just war thinking. Soldiers increasingly came from many classes rather than the knightly one and, as they did so, the processes for preparing them shifted from processes for training a specific few in particular skills to drilling masses in how to maintain discipline and morality in war.[10] Just war became a universalizable theory with standardized criteria for adjudicating decisions

[9] L. P. Hartley, *The Go-Between* (London: Hamish Hamilton, 1953), 1.

[10] This transition being but one expression of the transition away from a focus on virtues and toward a focus on rules that marks modern moral thought.

about whether to go to war and how to fight. Those criteria were themselves framed by wholly immanent moral considerations (ruling out, e.g., going to war because God told us to). The self-defense of sovereign states became paramount among just causes. As a result, the Christian just war tradition gained applicability even beyond the realms of Christendom and not only framed early international law but also resourced traditions such as pacifism in helping to explain why particular wars were unjust.

At the same time, as autonomy, universalizability, instrumentalism, and immanence shaped Christian just war thinking in its modern forms, their impact also led such thinking to downplay and/or ignore various components of the tradition (e.g., putting order in the service of just peace; the priority of legitimate authority among *jus ad bellum* criteria). They concealed various concerns and values that a differently formed tradition might have emphasized: the status and plight of refugees received little attention; moral obligations toward humanitarian interventions were put at cross-purposes to international laws affirming state sovereignty; and engagements with indigenous persons in Asia, Africa, and the Americas began with assumptions about the separability of those persons from their social and political structures. At least in comparison with *jus ad bellum* and *jus in bello* criteria, both *jus ante bellum* and *jus post bellum* considerations received very little attention in early modern just war thinking, even though the conditions that make new wars more likely can arise both before and after conflicts and there is nothing in the questions of the Christian just war tradition that precludes attention to what happens before and after wars. More importantly for this project, modern just war thinking not only treats wars as driven primarily by political concerns (as opposed, e.g., to being driven by humanitarian or resource concerns), but also ignores the role that natural events, including climatic changes, play in making wars more likely and worse.

THE CHRISTIAN JUST WAR TRADITION AND MODERNITY

That the Christian just war tradition took a modern form as it transitioned out of its late medieval form is unsurprising though still worth noting, if only to highlight the fact that traditions take on the characteristics of the historical contexts in which they find themselves. Traditions may transcend eras; they don't sit above them. There is a stronger hypothesis at least worth proposing here, though: that the Christian just war tradition played a significant role in shaping modernity. The evidence supporting this hypothesis is suggestive if not definitive.

The first piece of evidence supporting such a hypothesis turns on the very sources I've already laid out in this book and has to do with the fact that Vitoria, Grotius, and the US Constitution were all influential not only during the times they wrote/were written but after. Vitoria not only shaped some of the changes in Roman Catholic teaching that would lead to the Catholic Reformation but also influenced Grotius and initiated the Salamanca School as it shaped international law as well as reshaping Spanish approaches to dealing with indigenous persons in the Americas and West Indies during a time of exploration and conquest. Grotius shaped international law while also playing important roles in inter-Protestant theological disputes and interstate diplomatic concerns where his arguments and opinions were far-reaching, including in the shaping of the Peace of Westphalia. The US Constitution became one of the most if not the single most durable model of what such documents might look like in shaping a state and, therein, the forms that national polities would take. Collectively, they demonstrate influence beyond their years – and that influence pulses with Christian just war thinking as they were reshaping it.

A corollary piece of evidence in this regard is the fact that even as they exerted their own influence, they existed within networks that were aligned with them. Vitoria's thought connects not only to scholars such as Suárez and de las Casas but also to *Inter Caetera* and to larger movements within Roman Catholicism at a time when the papacy and the empire were redefining themselves in the face of the changing theological, geographical, and political landscapes of the time. Grotius was influenced by Alberico Gentili and influenced thinkers such as Samuel Pufendorf and John Locke. The US Constitution came about during an age of constitutions that were being written and implemented around the world. And, again, those in these networks were preoccupied not only with the problems of war but also with moral questions about when and how to fight; as they pursued these preoccupations, they did so in ways that promoted modern thought.

A second major piece of evidence is the degree to which modernity in the West was shaped by its preoccupations with war and violence. Beyond the more conventional but no less valid recognitions of the way Westphalia structured modern political systems as an attempt, in part, to get away from the so-called wars of religion; beyond the degree to which changing demographic, religious, and economic systems expanded the range of coercive practices available to function as surrogates for violence; beyond the advent and growing popularity of political ideas such as social contract theories that were described as ways of avoiding a "war of all against all"; beyond the range of developing technologies that made the horrors of war more visible and less acceptable; and, paradoxically, beyond the increased levels of violence that came with wars serving overtly political ends in the pursuit of new resources and greater wealth, wars of independence shaped by visions of self-determination, and technologies that made the horrors of war more horrible; beyond all these things, there is a compelling telling of the history of modernity that begins in questions about suffering and evil: the very questions that experiences of war raise. So, for example, Susan Niemann argues that "the problem of evil is the guiding force of modern thought ... Modern conceptions of evil were developed in the attempt to stop blaming God for the state of the world, and to take responsibility for it on our own."[11] In an age of immanence and instrumentalism, the types of supernatural forces that once not only shouldered the blame for horrors but also made those horrors an acceptable part of human life were no longer viable parts of the explanations of those horrors.

Niemann's argument is about the problem of evil, which is not the same as the problem of war. And she's much more focused on Enlightenment thinkers as representative of modernity than I am here. However, the gist of her argument applies across those differences. With the advent of modernity, the inevitability and naturalness of wars and the sufferings they produce no longer seem either inevitable or built into the structures of existence – and once they are both evitable and constructed (which is to say, once they can be thought of as the products of human decisions based on explicable motives), they must be thought about. Modernity in the West was shaped by its preoccupations with war and violence because the sufferings they produced became both more prominent as matters in need of address and more central to questions about how humans should make their way in the world.

A corollary piece of evidence for this is the way that moderns relocated the crises of suffering caused by violence away from divine will and toward other "inevitabilities": economic necessity makes chattel slavery necessary (and the division of people by race a natural category); political necessity makes *realpolitik* the way of the world (and a Machiavellian approach to political life justified by means of expediency); religious necessity makes the pursuit of purity necessary (and the

[11] Susan Niemann, *Evil in Modern Thought: An Alternative History* (Princeton, NJ: Princeton University Press, 2002), 3–4, 5.

separation of the righteous from the unrighteous, even if by force, the product of a utilitarian calculus about how to save and protect souls). This is the paradox of the modern preoccupation with war and violence: paying more attention to it, even for moral reasons, didn't make them less likely so much as mandate that they be reframed. Perhaps thinkers before Clausewitz might have thought of war as politics by other means; the success of Clausewitzian thought, though, is a modern thing.[12]

And since war and violence not only became preoccupying concerns in modern thought but also did so in a context that emphasized autonomy, universalism, instrumentalism, and immanence, the mandate to undergird the forms, causes, and pretexts for war and violence with moral language became preoccupying as well. The Christian just war tradition as it existed in the late Middle Ages provided an imperfect moral language through which to give voice to that preoccupation, and early modern thinkers such as Grotius (and, in a nascent form of modernist thought, Vitoria), not to mention Enlightenment philosophers such as Locke and Kant, picked up that imperfect moral language and transformed it to address the tensions that sat between a preoccupation with war and violence, on the one side, and a commitment to moral reasoning, on the other. That the US Constitution is simultaneously a paradigmatically Enlightenment document and suffused with just war thinking even as it has not been described as a just war document reveals the peculiarities that come with a preoccupying concern that will not be named as such. Little wonder that some of the ideas and movements we most associate with modernity (not only international law but, e.g., human rights, individual and national sovereignty, modern democratic polities, capitalism) are all defended on the basis of their varied abilities to avoid, minimize, or mitigate war: they have picked up the tasks that previously occupied Christian just war thinking.

This second piece of evidence is analogous to the arguments of political theology in Chapter 4 only written in a smaller font. As much as they promote new intellectual beginnings and non-tradition-based reasoning, modern thinkers still needed to use the tools at their disposal in

[12] This, in part, has been the heart of Hans Joas' argument: that modern theorists have promoted ideas that have been simultaneously pacifistic and progressive even as the ways the wars are carried out have become more violent and less likely to admit to visions of moral progress. The inability to integrate coherent and evidence-based ideas about war into modern thought shows just how much the intersections of war and morality have been preoccupying: those intersections are the sites where reality keeps poking at conceptual coherence.

shaping their thought. In matters having to do with war and morality, the tools at their disposal were, especially, those associated with late medieval just war thinking. They needed to do significant work in translating the ideas of one social imaginary into the contexts of another, but such is always the work in such transitions. The overtly theological (or at least God- and teleologically directed) language of the earlier social imaginary may have disappeared but the values, virtues, and visions of the world that undergirded that overtly theological language remained, shaping new expressions about obligations to neighbors, the relation of justice and peace to law and order, and the capacities of human beings to behave in moral and immoral (as well as wise and foolish) ways. Part of what makes Vitoria, Grotius, and the US Constitution so interesting is that each of them shows these processes of translation in their texts.

Finally, a third suggestive piece of evidence about the significance of the impact of Christian just war thinking on modern thought manifests as an absence. In early modern thought, attention to the interpenetration of the human and nonhuman natural worlds is not present because (as will be discussed in the next chapter) that interpenetration as it existed in its late medieval form was necessarily bound up in matters of transcendence, and matters of transcendence are the one thing that early modern thought cannot discuss. Recognizing that arguments from silence are of limited value, it is nevertheless worth noting that such arguments gain value when the silence stands in contrast to surrounding noise. In this instance, the early modern Christian just war tradition's silence is notable for its multiple contrasts. So, for instance, it is silent where both Jewish and Islamic just war thinking give explicit attention to the natural world in their thought. Likewise, late medieval/scholastic Christianity and culture more broadly are attentive to the natural world, as was late medieval just war thinking. As John Grim and Mary Evelyn Tucker have noted, "[s]cholastic thought from the twelfth through fifteenth centuries reflects the influence of classical ideas of a divinely ordained origin, plan, and order to creation. God as manifest in God's works became identified with the views of the Divine Artisan ... In this spirit, many religious and philosophical thinkers depicted nature as similar to Scripture that reveals the mysteries and mind of the Creator."[13] St. Francis and the order that took his name, for instance, celebrated a sacramental connection to the natural world (in contrast, e.g., to the Cathars). As Richard

[13] John Grim and Mary Evelyn Tucker, *Ecology and Religion* (Washington, DC: Island Press, 2014), 48.

Hoffmann has pointed out in his magisterial text *An Environmental History of Medieval Europe*,[14] human and natural history during the period were wholly interactive with each other: the former shaped the latter and vice versa. For early modern Christian just war thinking to pay so little attention to the natural world in the face of its interactions with other parts of the faith and other just war traditions is peculiar.

Modern thought left this human and nature in interaction paradigm behind for a variety of reasons, not the least of which were the rising attention to human capacities to bring about change and deep suspicions of any theology that treated events such as earthquakes and plagues as part of a divine order.[15] The emphases on autonomy, universalizability, instrumentalism, and immanence that drove modernity built on each other and built away from a vision of interactivity with the nonhuman natural world and toward one of separability from that world. Again, though, it did not need to be this way: moderns were encountering developed cultures in other parts of the world that did not think this way and even some of the benefits that modern thought provided could have been enhanced with closer attention to the interactions of human beings and the nonhuman natural world. Perhaps this is the case because the preoccupying questions of the Christian just war tradition on self, state, and other have also been the preoccupying questions of modernity (though only to the extent that "the other" was framed exclusively within the human realm). It is worth noting in spite of all this, though, that the natural world refused to drop away entirely from modern lines of vision: romanticism and transcendentalism, for example, would soon offer exalted visions of the natural world but neither played much of a role in shaping just war thinking. Again: arguments from silence are weak arguments – but they are, nonetheless, arguments worth noting, especially when placed in concert with other supporting arguments. In this instance, the disjunctions between late medieval Christian just war thinking and its early modern forms with regard to their perspectives on the relationship between the human and the nonhuman natural worlds are worth noting.

These three arguments (the influence of Christian just war thinkers on modernity; the preoccupations of modernity with war and violence

[14] Richard Hoffmann, *An Environmental History of Medieval Europe* (Cambridge, UK: Cambridge University Press, 2014).

[15] See John Aberth, *An Environmental History of the Middle Ages: The Crucible of Nature* (London: Routledge, 2013).

and the need to offer moral arguments about them; and the disjunctions between late medieval Christian just war thinking and modern thought) are only meant to be suggestive of a possible conclusion that medieval Christian just war thinking played a significant and underexplored role in shaping modern thought. It is not the purpose of this chapter nor of this book to offer a more rigorous defense of that hypothesis. It is, however, to note one possible conclusion that might follow from the arguments that I have laid out in this chapter and in this book.

This book's purpose is, instead, to attend to the way the transition from one particular social imaginary to another might help us make sense of the challenges we face as we transition into a new social imaginary that is shaped by environmental concerns. And to do that, it helps to recognize how the modern social imaginary has helped to create the environmental crises that are driving the current transition.

MODERNITY, VIOLENCE, AND THE ENVIRONMENT

Going forward, I want to do more than describe the way the just war tradition's answers to questions about self, state, and other changed as the tradition entered modernity. I want to suggest that those changes exemplify the way that a modernist sensibility has had a significant and deleterious effect on the nonhuman natural world and, eventually, on the human one as well. For all the advances it offered to humanity – and there are many – modernity has not been kind to the wider natural world. As we enter into the Anthropocene, we are confronted with an irony: many of the advances in thought and technology that have so helped us over the past few centuries are now contributing to the harms we face as they are made more likely and more destructive through those very advances. Said differently, autonomy, universalizability, instrumentalism, and immanence, as they have been promoted in modern thought, processes, and technologies, have unduly distinguished the human and nonhuman natural worlds, ignored the complex interrelations and variability within the larger natural world, treated the nonhuman natural world in ways that cause harm, and cut us off from ways of thinking about the natural world that might have prevented or mitigated such harm.

Christian just war thinking has played a role in this harm when it has funded disregard for the natural world, the reduction of the nonhuman natural world to "resource," and a posture of opposition toward respect for natural systems. In spite of some attention to interactions with the nonhuman natural world in late medieval just war thinking, the larger tradition's theological foundations in the chastened eschatology of the third- and fourth-century church, its political foundations in Augustine's city of God/city of man distinction, and its moral foundations in neighbor-love all shaped a vision in which restraints on war circled around concerns for the protection of human life and the preservation of culture without embedding human life or culture in larger natural systems. Asking questions about self, state, and other has, for the most part, effectively excluded questions about creation.

Modern Christian just war thinking, via its contributions to the shaping of modern thought, has made things worse by adding modernist modes of inattention and opposition to the natural world to its own engrained habits of focusing on the protection of human life and preservation of culture. Indeed, to the degree that Christian just war thinking played a significant role in shaping modernity, the sins of its historic inattention to nature have been visited upon the modern West and reinforced the modern West's own sinful tendencies vis-à-vis the natural world. The particular configuration of answers to questions about self, state, and other given by modern Christian just war thinking have had the effect (if not the intent) of causing violence to the environment and shaping environmental changes that help to induce violence.

Contemporary Christian just warriors, therein, face a Catch-22: on the one hand, if they claim that the Christian just war tradition has played a significant role in shaping modernity – as they do when they note the impact of the tradition in shaping modern international law – then they must also shoulder significant blame for the causes of environmental destruction. On the other hand, if they attempt to insulate themselves from the causes of modernity's environmentally destructive tendencies, they must also relinquish their claims about the significance and impact of Christian just war thinking on modernity.

Here, I would argue, it is better to accept blame and move forward in light of that responsibility than to avoid blame (and significance). Not only is this a more accurate reading of the history of the relation between the Christian just war tradition and modernity, but it also opens the tradition up to face its own limits, failings, foibles, prejudices, and evils. Or, rather, a more accurate reading of the history of the Christian just war tradition and its impact on modern thought, systems, and technologies as they have shaped environmental destruction, are of a piece. On the one side, reading history and impact together highlights the ways that the tradition has moved through time: it brings the qualities of irony, paradox, tragedy, and hope expressed during discontinuous change into relief. On the other side, reading impact and history together provides diagnostic force to proposals about the way that the tradition may change as it enters the Anthropocene. Its attention to the tradition's complicity in environmental destruction locates the tradition in a place where it must take up climate-shaped conflict and where it is also motivated to do so.

THINKING THEOLOGICALLY ABOUT CONTEMPORARY AMBIGUITIES OF THE CHRISTIAN JUST WAR TRADITION

Said more theologically, accepting blame and moving forward in light of that responsibility not only locates the Christian just war tradition in a larger narrative about the way God's works of judgment, redemption, and salvation manifest themselves in history but also opens the tradition up to the resources that come packaged with Christian understandings of confession in the face of those works. As confession is that peculiar type of project that simultaneously recognizes sin and trusts in grace, shaping a story about the movement of the Christian just war tradition through time that is fundamentally confessional in orientation resonates with the larger story of the movement of Christians through time.

Treat this book, then, as a kind of confessional project (in the sense that I embed it and myself within a particular creedal expression of Christianity that is shaped to draw from a history of formal theological claims while at the same time being open to new theological visions). I do not presume to speak for the whole tradition nor for all those who align themselves with it. I do, though, count myself as among those who are persuaded by the arguments the tradition makes about the ambiguous goods of peace and justice and the nondemonic qualities of force and violence. As such, this text performs a kind of liturgy – a testimony that recognizes that mine is not the only perspective (from either within or outside Christian just war thinking); that all I am doing is making my arguments as visible and transparent as I can so that they can be weighed; and that these chapters are a kind of expression of imperfect faith and reveal both convictions about the intended object of that faith and the imperfections of those convictions.

This is no easy thing to do. Not only can the language of sin and grace sound strange within the context of war (while perhaps there are no atheists in foxholes, the horrors of war have driven any number of thoughtful persons away from many of the basic convictions of Christian faith), but that language is not especially welcome in modern thought. Where modern thought emphasizes autonomy, the language of sin and grace begins in a recognition of helplessness and dependence. Where modern thought emphasizes universalizability, the language of sin and grace recognizes that sinners and the sinned against may not share the same moral space. Where modern thought emphasizes instrumentalism, the language of sin and grace recognizes that there is nothing mechanical about divine–human (and, therefore, human–human) relationships. Where modern thought emphasizes immanence, the language of sin and grace begins in the conviction that this world is not all there is.

The lack of hospitality to such language on the part of modernity has created any number of tensions (and even more responses to such tensions) within modern and Christian thought. Mostly, modernity attempts to address such tensions by resolving them, notably by advancing an idea of secularism in which all great needs can be addressed through resources available within a secular frame. Does religion cause violence? Then keep religious language out of the public sphere. Does religion get in the way of healing? Then rely on the resources of modern medicine and psychotherapy. Does religion distract people from the pursuit of the common good? Then raise a great wall of separation between church and state. Does religion promote dominionist understandings of the relationship between human beings and the nonhuman natural world? Then reject the unenlightened thinking of the faithful. And, mostly, Christianity has responded to such tension through various projects of retrenchment and accommodation, either digging in to fight modernity or modifying itself to conform to it – and in either case, Christianity's responses are its own way of attempting to resolve those tensions, which may reveal just how beholden to modern thought modern Christianity actually is.

Yet the language of sin and grace would suggest that such tensions are not to be resolved so much as borne. Resolution turns on a set of assumptions not only about human agency and control but also about the progressive power of human thought to achieve permanent and wholly virtuous change. As such, a preoccupation with resolution ignores the limits of human agency, the failings of human control, and the realities of mixed motives, ambiguous goods, and unanticipated consequences. Treating such tensions as things to be borne, however, begins in a position in which humans are only some of the agents in the world (and never fully in agreement with each other, anyway); in which there will always be forces beyond their control; and, this side of eternity at least, in which human beings will pursue goods imperfectly and imperfect goods. They are still agents, they still have power to bring about change, and they still pursue goods – but they do none of these things in ways that are beyond improvement, reproach, or the need for transformation.

Among the chief pieces of evidence of human beings' limited agency, partial control, and ambiguous wisdom in the twenty-first century is that their pursuit of the very best of goods in modernity, including an end to violence, the elimination of the caprice of natural forces, the increase and spread of wealth, and the equalizing distribution of political power, has also led to environmental destruction at unimaginable levels (the term "Anthropocene," after all, is used to describe both a time of environmental crisis and an epoch in which human actions are the dominant forces reshaping the planet). The goods of modernity are now under threat as a consequence of the benefits of modernity. A posture of resolution has not only failed to resolve; it has also exacerbated the very problems it was intended to address. Life in the environmental age will require both modesty and a willingness to live with tension.

One aspect of this focus on modesty and living with tension is that it resists overclaiming either the tradition's impact (for weal and woe) on the past or its resources for the future. Christian just war thinking was hardly the only source of modern thought. Indeed, if anything, getting at the complexity of sources and their interactions is a precondition for making sense of the movements of human projects through time at the beginning of the Anthropocene. The simplification of ideas about causation and human agency is but one of the problems of modern just war thinking. And even if modernity had looked very different, it is quite likely that some of the environmental crises we face – crises that threaten to exacerbate violence and threats of violence – may well have arisen. Forces that have existed since far before modernity and whose increase is incidental to modernity (e.g., human population growth and migration patterns) could still have played roles in shaping contemporary environmental crises.

As the Christian just war tradition moves into the Anthropocene, then, its primary purpose ought not to be to provide answers to questions about self, state, and other that will be good for all time, but to provide answers that will be helpful for a particular time. Although I will not pursue that purpose in this book, keeping that purpose in our peripheral vision will remain important: *principiis obsta; respice finem.*[16] My goal in this book, then, isn't so much to get the story of the movement of the

16 "Resist the beginnings; consider the endings."

Christian just war tradition through time right as it is to shape a story that is functional for the wars of the Anthropocene. This project is not only about recognizing that contemporary just warriors have tended to tell a modern story about the Christian just war tradition; it is also about renarrating the tradition in a way that will prove useful as the tradition exits modernity. In all this, I take paradoxical comfort in the resources of the Christian just war tradition, which may not only find new footing (provided it returns from its far-flung travels in secular theory) but also offer guidance, or at least a fresh witness. The tradition has, after all, long traded in modesty and the tensions that come with living in morally ambiguous (and, from a theological perspective, eschatologically incomplete) time.

6

Historical Roots and Roads Not Taken

An Environmental History of (Part of) the Christian Just War Tradition

Christianity, in absolute contrast to ancient paganism and Asia's religions (except, perhaps, Zoroastrianism), not only established a dualism of man and nature but also insisted that it is God's will that man exploit nature for his proper ends.

Lynn White, Jr., "The Historical Roots of Our Ecologic Crisis"

Already in the twelfth but especially from the thirteenth century onwards, acknowledged public authorities intervened variously vis-à-vis both private lords (owners) and users (collective or individual) of especially non-arable resources. In halting steps and against continual resistance, what a later age would call resource management policy or even environmental legislation emerged and sought to constrain behaviors of both sorts of private interests.

Richard Hoffman, *An Environmental History of Medieval Europe*

Here lies another irony: a worldview given over to the ontological difference between culture and nature produces historical experiences in which both are destroyed in terms of what are supposed to be their mutual interests.

Edward E. Antonio, "Ecology as Experience in African Indigenous Religions"

CHRISTIANITY, THE NONHUMAN NATURAL WORLD, AND LYNN WHITE'S THESIS

In 1967, Lynn White, Jr., published his enormously influential article, "The Historical Roots of Our Ecologic Crisis," shaping decades of debates about religion, cosmology, and the natural world and effectively inititing the field of environmental ethics. In that article, White, a historian of the medieval period, argued that the long, slow processes that led to the ecological crises of the twentieth century began with and were driven by a Christian narrative about the relationship between human beings and the natural world. More than perhaps any other historian of his day, White recognized that "[w]hat people do about their ecology depends on what they think about themselves in relation to things around them. Human ecology is deeply conditioned by beliefs about our nature and destiny – that is, by religion."[1] Examining the history of the technological changes in, especially, Western Europe over the past millennium, White traced the way a particular version of Christianity shaped "what people do about their ecology" and, therein, "our ecologic crisis."

The starting point for White's argument was his conviction that, "[e]specially in its Western form, Christianity is the most anthropocentric religion the world has seen"[2] and that this anthropocentrism led human beings not only to identify with a supernatural divine other but also to resist attempts to identify themselves with a natural other, pursuing instead the conquest of "paganism" and its animisms. The dualism between human beings and nature that White saw from the Book of Genesis forward eventually shaped attitudes in Western Europe toward science and technology and led to a point when "no creature other than man has ever managed to foul its nest in such short order" and for which "Christianity bears a huge burden of guilt."[3] The combination of anthropocentrism and an instrumentalist view of nature shaped an unsustainable way of living; Christianity gave us both. While Christianity could have gone a different direction (White raises both Eastern Orthodoxy and St. Francis of Assisi as offering different narratives about how to relate human beings to the nonhuman natural world), it didn't. As a result, "the present increasing disruption of the global environment is the product of a dynamic technology and science which were originating in the Western medieval world ... [and their] growth cannot be understood historically apart from distinctive attitudes toward nature which are deeply grounded in Christian dogma."[4]

The arguments of this chapter (and, indeed, this book) will reveal that White was partly right, but only partly so – and that the reasons

[1] Lynn White, Jr., "The Historical Roots of Our Ecologic Crisis," *Science* 155.3767 (March 10, 1967), 1205.
[2] Ibid.
[3] Ibid., 1204, 1206.
[4] Ibid., 1207.

he was right and the reasons he was wrong are both important if we want to address wars in a warming world. In brief, White was right to link contemporary environmental crises to the way Western Christianity developed, but his timing was off because of his essentialist assumptions about the nature of Christianity. Where White's argument relied on his conviction that there is, within Christianity, an essential human/nature dualism, mine is built around the concatenation of a number of forces – including natural ones – that shaped the way Christian just war thinking transitioned from its late medieval to its early modern forms (and, in turn, Christian just war thinking shaped understandings of the natural world). Where White expressed a unidirectional sense of causality (Christian idea X leads to environmental crisis Y), I argue that ideas and effects interpenetrate: Ideas are shaped by experiences of events in the natural world and, in turn, shape understandings of that world. Moreover, because not everyone in Christian Europe experienced the same events (let alone interpreted events in the same way), there are many understandings of the natural world; this is one of the reasons that animisms, senses of the supernatural in the natural, and medieval versions of natural theology (expressed, e.g., in allegories from nature and arguments of God's grace evidenced in human collaboration with nature) lived on.

The purpose of this chapter, though, is far larger than a refutation or an appreciation of White's work; that work has been done elsewhere.[5]

[5] Because White's essay has drawn criticism from so many sources – including, especially, arguments from Christian theologians and biblical scholars about misreading the Book of Genesis as well as Christian theology and history – it is worth noting the range of the strengths of White's essay. He recognizes that "[a]ll forms of life modify their contexts ... [and ever] since man became a numerous species he has affected his environment notably." White, "Historical Roots," 1203. He addresses the way that human beings modify their contexts through advents in technology, especially those having to do with agriculture but also namely having to do with war. He links scientific and technological changes to particular religious perspectives. He attends to the way that the human modification of their contexts have had unintended and deleterious consequences for both the human and nonhuman natural world. He raises suspicions about the viability of historically naïve attempts to address those consequences. He provides a new resource that offers a hopeful new direction to pursue in addressing environmental crises: Francis of Assisi as "patron saint for ecologists." White, "Historical Roots," 1207. In providing this resource, he even tacitly recognizes that the resources to help us must be linked to the causes of our problems. To do all this in five short and well-written pages is a remarkable feat.

There are, though, weaknesses to White's essay, many of which have been named elsewhere (see, e.g., Willis Jenkins, "After Lynn White: Religious Ethics and Environmental Problems," *Journal of Religious Ethics* 37.2 (2009): 283–309; and Ben A. Minteer and Robert E. Manning, "An Appraisal of the Critique of Anthropocentrism and Three Lesser Known Themes in Lynn White's 'The Historical Roots of Our Ecologic Crisis,'"

Instead, the larger point of this chapter is to highlight the way Christian just war thinking, as it transitioned from its late medieval into its early modern forms, was shaped by human interactions with the natural world. In earlier chapters of this book, I have argued that Christian just war thinking shaped modernity; in this one, I argue that human interactions with the nonhuman natural world (e.g., experiences of climatic events and changing climates, technological developments related to the use of natural resources and legal systems put in place to adjudicate that use, changing human perceptions of and language about the natural world and human beings' place in it, encounters with the nonhuman "new world" and the impact of those encounters) shaped Christian just war thinking during its transitions between the sixteenth and eighteenth centuries. These were, not coincidentally, the very centuries in which attention to the natural world receded from the focus of the Christian just war tradition; ironically, the consequences of those transitions blinded us to many of their causes.

THE SHAPING OF ENVIRONMENTAL HISTORY

In the not too distant past, when historians spoke of "history," they meant human history such that the natural world served, at best, as a backdrop for the far more important matters of interhuman interactions

Organization and Environment 18.2 (June 2005): 163–176, as examples that summarize some of these weaknesses and develop new criticisms of that essay). Here, momentarily, I want to note merely that even as White recognizes that all forms of life modify their contexts, he virtually ignores the way "contexts" modify all forms of life. The result is its own problematic binary: dynamic (human) life vs. passive context. This binary ignores the degree to which all forms of life may be responding to changes in their context when they act to modify it: There are feedback loops in play that lead to movements in all sorts of unanticipated directions. As he addresses the way human beings modify their contexts through technology, he presents an accretionary narrative about how human beings move through time. His is a weak myth of progress. By contrast, I want to highlight the interactive qualities of the nonhuman natural world, human perceptions of that nonhuman natural world, and the social world human beings construct around them. Attention to such interactions means resisting narratives of progress and, for that matter, any essentialized cosmologies that fund such narratives by reinforcing straightforward causal links (i.e., "this worldview leads to this set of practices/type of technology"). In contrast to many critics of White, I want to argue that White did get Christianity (or, more carefully named, a particular historical expression of Christianity) right; he got historiography wrong. That he refers to an "ecologic *crisis*" (singular) that arises from a single (essentialized) root cause rather than many different overlapping crises that emerge out of many causes is one bit of evidence in this regard. He shares this problem, by the way, with many of his most vociferous critics.

and the consequences of those interactions for human beings.[6] Beginning with Roderick Frazier Nash's 1967 book *Wilderness and the American Mind*, however, the burgeoning field of environmental history began to find its place within history departments.[7] One of the chief contributions of environmental historians has been to trouble the nature/culture distinction that has shaped the modern social imaginary. That distinction has been replaced by a recognition that "all environments are hybrid ... [such that] ... the field has largely given up on a pristine, ahistorical nature as a baseline against which to measure human-induced change ... [while also implying] ... that human history and culture cannot be easily isolated from environmental forces and circumstances."[8] Attending to hybridity not only helps us recognize the variety and interactivity of forces at work in human–nonhuman interactions (while also raising intriguing questions about causation, power, and agency), but it also relieves us of the malformations that come with the implicit myths of decline or progress that tend to shape conventional narratives of human–nonhuman interactions. Human beings neither corrupted a previously idyllic natural world nor successfully emerged from and then gained mastery over a chaotic one.[9]

The starting point for joining environmental history to the history of the Christian just war tradition as it transitioned from its late medieval to its early modern form involves getting a better sense of the way late medieval Europeans interacted with the natural world. Such work will, necessarily, involve some reasonable speculation. Environmental history is a fairly new field of study and little has been done thus far to link it to war, let alone to the ideas that shape whether, when, and how we fight. And while there are a number of interesting texts connecting environmental history to war, almost all of them deal with wars fought over the past 200 years where information about both the wars and the

[6] "Natural history" – the study of the nonhuman natural world and its interactions in its environments – wasn't really treated as history and was taught in science departments, not history departments. For a brief but articulate overview of the entrance of the nonhuman natural world into the study of history, see Sarah Maza, *Thinking about History* (Chicago, IL: University of Chicago Press, 2017), 108–117.

[7] Roderick Frazier Nash, *Wilderness and the American Mind*, 4th ed. (New Haven, CT: Yale University Press, 2001). For an overview of environmental history, see, e.g., J. Donald Hughes, *What Is Environmental History?* (New York, NY: Polity, 2006).

[8] Paul S. Sutter, "The World with Us: The State of American Environmental History," *Journal of American History* 100.1 (June 2013): 96.

[9] Along the way, environmental history's efforts to hybridize all environments has helped to undermine the racism and Eurocentrism implicit in geographical determinism.

environment is plentiful.[10] As such, joining environmental history to the history of the Christian just war tradition as it transitioned from its late medieval to its early modern form will involve adapting wisdom from these texts and joining it to books on medieval and early modern environmental history in Europe more broadly.

Among the best resources on European medieval environmental history is Richard C. Hoffman's *An Environmental History of Medieval Europe*. A deeply interdisciplinary work that treats "medieval history as if nature mattered,"[11] Hoffman's book covers a thousand years of history across a continent (and, toward its end, European encounters with life on other continents) with special attention to the way that people colonize natural ecosystems, modifying the natural world for their varied uses.[12] How, Hoffman asks, do "material nature, living and non-living, and human communications (symbolic culture) join in an inter*active* and reciprocal relationship by human material life"?[13] His answer: "Over time culture and nature co-adapt; they engage in co-evolution."[14] Toward explicating this answer, Hoffman identifies and explores three overarching themes: environmental influences on human existence and

[10] Among such books, I would include Lisa Brady, *War upon the Land: Military Strategy and the Transformation of Southern Landscapes during the American Civil War* (Athens, GA: University of Georgia Press, 2012), Brian Allen Drake, *The Blue, the Gray, and the Green: Toward and Environmental History of the Civil War* (Athens, GA: University of Georgia Press, 2014), and chapter 6 of Paul D. Escott, *Rethinking the Civil War Era: Directions for Research* (Lexington, KY: University of Kentucky Press, 2018), on the US Civil War; Chris Pearson, *Mobilizing Nature: The Environmental History of War and Militarization in Modern France* (Manchester: Manchester University Press, 2016); and Judith Bennett, *Natives and Exotics: World War II and the Environment in the South Pacific* (Honolulu: University of Hawaii Press, 2009).

[11] Richard C. Hoffman, *An Environmental History of Medieval Europe* (Cambridge, UK: Cambridge University Press, 2014), 3. John Aberth's similarly named *An Environmental History of the Middle Ages: The Crucible of Nature* (London: Routledge, 2013) is also impressively researched and encompassing but suffers from being too generalized in its treatment of European history (so, e.g., Aberth proposes four periods of European environmental history: an eschatological period from the third to fifth centuries; an adversarial period from the sixth to tenth centuries; a collaborative period from the eleventh to fourteenth centuries; and a (new) adversarial period from the beginning of the fifteenth century on) and geography (Europe may not be a large continent, but it and its people are varied in ways that Aberth's four periods don't account for). That said, Aberth's book is rich in anecdotes and pieces of literature that help get at the many ways that medieval Europeans encountered, made sense of, engaged, and were engaged by the nonhuman natural world.

[12] Hoffman, *An Environmental History*, 8ff.

[13] Ibid., 9. Emphasis his.

[14] Ibid., 10.

actions, human attitudes toward nature, and human impacts on the natural world.[15]

While the range of foci, disciplines, concerns, and questions within Hoffman's work far exceeds anything I can manage to adequately summarize (let alone apply) in these pages, those three themes offer a helpful starting point for exploring the intersections of environment and late medieval Christian just war thinking such that its transition into its modern form was, if not inevitable, at least likely. What were the environmental influences that shaped the Christian just war tradition prior to and during its transition to its early modern form? How did those in the Christian just war tradition think about the natural world? And what were some of the impacts of the tradition on the coevolving natural and human world?

THE COEVOLUTION OF THE NATURAL WORLD AND LATE MEDIEVAL CHRISTIAN JUST WAR THOUGHT

Late medieval (European) Christian just war thinkers entered a non-human natural world that had been colonized for over a millennium. Over the centuries, land cultivation in Western Europe manifested like a slow-moving tide. When the climate and other natural forces allowed for it (as was the case during the Roman Climate Optimum from the third century BCE to the end of the fourth century CE and then again during the Medieval Warm Period of the tenth to thirteenth centuries, when cereal production helped the population of Europe to nearly triple), more and more land was transformed for agricultural use. And when the climate turned against it (e.g., during the Early Medieval Cold Period of the fifth to eighth centuries) or natural forces such as the Justinian Plague of the sixth and seventh centuries and the Black Death of the fourteenth century led to population collapses that prevented land from being farmed,[16] the land went feral.[17] The environmental history of the period reflects these tidal forces and reveals itself in changes

[15] Ibid., 11.

[16] In the course of a half-decade in the mid-fourteenth century, the Black Death afflicted up to 80 per cent of the population of Europe and of those, 75–90 per cent died from the disease. Almost overnight, Europe was all but emptied of people.

[17] Interestingly, Roman imperial practices associated with its conquests had shaped the first great colonization of much of non-Mediterranean Europe. Toward uniting the empire and maintaining its military might, Rome developed a sophisticated all-weather road system through Europe (many of which paved roads continue to be used today) and

to arable land, woods, and waters; in mining and smelting activities; in the spread of diseases, famines, and refugees fleeing them; and in the creation of "second natures" associated with the development of cities.[18] All these interactions with the environment shaped not only European culture (and vice versa) but the practices of war and how Christian just warriors thought about war.

LAND, WOOD, AND WATER

Let us begin with arable land. Throughout medieval Europe, natural forces shaped food sources and practices (so, e.g., when populations fell, cultures were more likely to turn toward herding as it required less labor than working the land), but the predominant foods of the age were cereals. The cerealization of medieval Europe manifested in several ways pertinent to just war thinking. A growth in monocultures led to more calories per person being available (making wars possible, as wars carry significant metabolic costs). Increased areas of arable land shaped medieval social structures that included minor nobles gathering small militaries to defend those lands and the peasants who worked them. Land maintenance structured both agricultural practices such as crop rotation and increasingly complex laws concerning land ownership, use, and protection, all of which would shape ideas about colonization in the new world. Famines led to conflicts, especially where they drove the creation of refugees and the search for new resources.[19] And, over time, crop success led to the growth of those great metabolic sinks, cities (with all the

placed military fortifications along those roads, particularly along the empire's northern boundaries. Peasants in the area then cleared patches of forest along these roads and near these fortifications for agricultural use. Over time, villages grew up around the cleared areas, where peasants worked, among other reasons, to provide food and other resources for the garrisons stationed in the area. See Richard P. Tucker and Edmond Russell, eds., *Natural Enemy, Natural Ally: Toward an Environmental History of Warfare* (Corvallis, OR: Oregon State University Press, 2004), 20–21. As Rome declined during the Early Medieval Cold Period of the sixth to eighth centuries, many cultivated lands were abandoned as growers faced increasingly cold and wet seasons that were inimical to crop success.

18 "… some scholars see the entire fourteenth century and through to the mid-sixteenth as characterized by irregular cycles of weather extremes, with concomitant regional bouts of catastrophic weather." Hoffman, *An Environmental History*, 324.

19 "Indeed, failures of agricultural output in the late fourteenth and early fifteenth century are offered as one reason why Portuguese turned their attention overseas. What they sought as they sailed south along the African coast were cereal supplies or commodities they could there acquire and use in exchange for cereals elsewhere." Ibid., 172.

implications associated with urban living: increased vocational specialization, social stratification, and political organization; the rapid spread of diseases; increased trade and the movement of peoples; reshaped metabolic patterns because cities consume more calories than they produce; and, eventually, the use of fossil fuels), as well as siege warfare around those cities. Indeed, the explosion of cities in the sixteenth century fundamentally altered natural Europe: "If human numbers grew two- or three-fold [during the sixteenth century], the movement of plant and animal biomass grew twenty-fold, a huge increase in the flow from the geographical center of Europe to its western fringes ... more people now ate what they had never seen alive."[20] The increasing abstraction of the natural world in the late medieval/early modern mind was well underway and shaped the abstractions of early modern just war thinkers when they addressed conflicts in the new world.

Unlike most of the wars arising during and after the early modern period, medieval wars were fought over land and food and governed – to the extent that they were governed – according to rules shaped by those conditions. As brutal as wars during the period could be, *jus ad bellum* criteria such as legitimate authority gained prominence not only because social systems were stratifying (making legitimate authorities easier to discern) but also because adjudicating the costs and benefits of wars over land and food became increasingly pressing even as the calculus in measuring them became clearer (or at least clearer than wars that did not have, as their centers, basic material goods). And a *jus in bello* criterion such as discrimination became more significant both by way of its affirmations (chivalric codes to avoid targeting peasants were established, in part, so that someone could work the land after conflict) and it its absences (siege warfare targeting whole cities). The Peace of God (beginning in 898 CE) was intended to protect not only the clergy but also the agricultural lands owned by the church. The Truce of God (beginning in 1027 CE) was intended to restrict conflict to only certain days each week not only as a way of recognizing the fact that wars are horrible and that those who share the body of Christ should not be at war with each other, but also because of the unrelenting obligations associated with filling bellies and larders. And while both the Peace of God and the Truce of God may seem, in retrospect, naïve and more likely to be recognized in their breach than practiced, they nevertheless shaped patterns of chivalry and systems of land maintenance for hundreds of years.

[20] Ibid., 357.

Beyond matters of arable land, two other expressions of land maintenance pertained to wood and water. As forests were razed to create cropland, the value of forests and wood grew. Nobles claimed forests and the creatures that inhabited them for themselves not only as commodities to control but also as signs of privilege (including having time to go on hunts and eating more meat). Conflicts between various nobles, especially as the political power of the church waned (via, for instance, the Concordat of Worms in 1122 CE), led to a gradually growing political infrastructure in Western Europe that would eventually become the modern nation-state. This infrastructure was shaped to negotiate conflicts over property and property rights as laws were developed to adjudicate such conflicts.[21] In continuity with its predecessors, this infrastructure generally ignored the place and claims of peasants, which, in turn, shaped tensions and, eventually, conflicts between the classes: "By the later Middle Ages in most parts of Europe commoners were barred from hunting and the activity reserved for nobles. Even if the social monopoly was never complete or well enforced, it still provoked conflicts on the land as, for instance, when peasants found their lord's protected deer coming out of the park to eat in their grain fields or meadows. This grievance, among others, helped trigger the German Peasants' War of 1525."[22] Moreover, access to timber for building became increasingly important, including in the building of navies as coastal nations took to the seas in search of power and new lands with fresh resources.[23]

Property laws regarding water were especially complex. Because it was nonarable, access to water – especially inland water – was controlled via a mix of private and collective rights, especially where that water was needed for agricultural purposes. The complex mix of such rights, shaped as they were by regional geological, cultural, and political differences, meant laws about water in Holland, Germany, England,

[21] So, for instance, laws were put in place to prevent the poaching of trees by banning saws in woodlands because they are so much quieter than the sounds made by an axe as a tree is chopped down. See Ibid., 187.

[22] Ibid., 192.

[23] "Early ecological damage outside Europe reflected the navies' needs for construction timber and naval stores. By the 1700s European navies began cutting the hardwood and white pine stands of northeastern North America, the coastal hardwoods of Brazil, and later the teak forests of monsoon Asia, to find substitutes for depleted English oak and Scandinavian conifers." Richard P. Tucker, "The Impact of Warfare on the Natural World: A Historical Introduction" in Tucker and Russell, *Natural Enemy, Natural Ally*, 25–26.

and Spain were both byzantine and quite different.[24] As each of these countries pursued colonization outside Europe, they carried their laws about water with them; so, for instance, when Francisco de Vitoria adjudicated matters having to do with rights to water in the new world in his *Relectiones*, it is worth noting that he was not so much creating new laws as carrying old laws into new lands (and, therein, shaping justifications for conflicts with indigenous persons in those lands).

ORES AND METALS

Perhaps the most obvious role that the nonhuman natural world played in shaping late medieval European Christian just war thinking had to do with the mining of ores and the increasingly sophisticated processes involved in smelting and casting metals to be used for a wide variety of purposes. Especially after the development of the blast furnace (sometime around the thirteenth century), the sheer volume of metal available to be used for household, agricultural, civic, and military purposes multiplied dramatically.[25] Blast furnaces required greater amounts of wood for fuel even as mines required timbers for pit props, thereby contributing to the shaping of property laws about forests and increasing tensions between nobles intent on securing both wood and metals.

Metal, in turn, shaped other engagements with the nonhuman natural world. Durable metal plows became more common and allowed more land to be cultivated; greater volumes of cereals followed. At the same time, mines cleared land of surface vegetation and led not only to erosion but also to land and water contamination, all of which reduced the amount of arable land around the mine while leaving plants, humans, and other animals vulnerable to toxic contamination that would endure for generations, thereby reshaping the natural world in ways that not only restructured its value for human use but also oriented medieval persons to increasingly think of the natural world in instrumental ways

[24] The "Tribunal of Waters" has been meeting on the steps of Valencia's cathedral every Thursday since the Middle Ages to address conflicts related to water in the canal systems in that part of Spain.

[25] "One modern scholar estimated that an average late twelfth-century peasant household possessed only a few kilograms of iron and one or two kilograms of other metals, notably brass and bronze. But with a great expansion of the use of these materials, by the late fourteenth century the descendants of that same family would possess somewhere between twenty and a hundred kilograms of metal." Hoffman, *An Environmental History*, 215.

as a collection of fungible resources.[26] So, for instance, Georg Agricola's 1556 text *De re metallica* includes his argument that while mining destroys land and affects plants and animals in the area, "arable land and metals obtained by clearance of woodlands and destruction of the land surface were worth far more than the trees or mountains they had replaced, while fish, birds, and animals could be purchased to replace those lost. Humans could see nature, make use of nature, and, if they disliked what had become of it, counterbalance loss with economic values gained."[27]

Clearly, though, the greatest impact of mining on late medieval just war thinking had to do with the way metals and gunpowder could be brought together to make weapons. Between the mid-fifteenth century and the early sixteenth century, the amount of metal being used by European armies rose at a geometric rate.[28] Though small arms were so costly and unreliable even during the Late Middle Ages as to be relatively rare in combat, the arrival of cannon onto the fields of battle beginning in the thirteenth century would dramatically reshape war. Where sieges might previously have been literal wars of attrition, lasting for months as the resources of the besieged were gradually depleted, cannon made it possible to end sieges in only days or weeks. Neither protracted sieges nor short intense ones favor the *jus in bello* criterion of discrimination; the latter, though, do contribute to shaping *jus ad bellum* concerns that had been highlighted by Thomas Aquinas about the obligation of attacking parties to publicly announce their intentions. Where the leaders of a besieged city may judge that it has sufficient resources to wait out attackers (who may either tire of siege or face resource shortages of their own), those same leaders' calculus will change if they anticipate a fairly quick and violent end to the siege.

Moreover, the economic and environmental costs associated with wars fought via metal and gunpowder were so high that they contributed toward shaping the political economies of early modern nation-states and reshaping the natural world and human relations to it:

> Soaring capital and resource costs of war drove states to contrive ever more elaborate institutions and grounds to interfere in their subjects' relations with nature.

[26] That money was made from metal certainly contributed to notions of the fungibility of natural resources.

[27] Hoffman, *An Environmental History*, 356–357.

[28] Ibid., 362.

New cannon-resistant defence schemes laid waste the immediate environs of cities. Early modern officials entered private cellars to scrape from the walls blooms of saltpeter needed for more gunpowder. Pursuit of this natural resource meant seizing control over the geochemical process itself. Firearms not only increased the ability to kill humans and other organisms, they embodied a collective shift of cultural patterns. And most had to do with power.[29]

Paradoxically, as metal came to drive the machines of war, war became less about natural resources (at least in Europe) and more an expression of "politics by other means."[30] Christian just war thinkers would have to change the way they thought about war in order to address the new types and understandings of war that were arising, in part, due to the coevolution of human beings and the nonhuman natural world.

Sitting behind food, cities, disease, forests, water, and metals is a history of climate change in medieval Europe. Climatic change in Europe over the centuries since Roman conquest had shaped patterns of planting and of famines; cycles of population growth and decline; and periods of sedentary living mixed with times of significant movement of peoples. It would play a significant if not easily calculable role in the transition from late medieval to early modern times in Europe. So, for instance, "1450–1500 in central Europe and 1500–40 in the Netherlands became the longest runs of mild winters either would see again until the mid-eighteenth century."[31] These were years of plenty – which was good, as, with the onset of the Little Ice Age beginning around 1560, Europeans would draw from those stores and rely on them as they pursued resources on other continents. Moreover, such large-scale projects reinforced the importance of a particular kind of political system capable of organizing them: "By the waning of the Middle Ages in the fifteenth and sixteenth centuries, the state had become an important determinant of what natural elements did belong to its subjects and what they could do with that possession."[32]

29 Ibid.

30 Contrast, for instance, R. C. Smail's claim that "[l]and was the essential basis of Latin dominium. The policy of the Latin rulers centered on its possession and war, as an instrument of policy, was employed in the interests of territorial expansion or defence," in R. C. Smail, *Crusading Warfare, 1097–1193*, 2nd ed. (Cambridge, UK: Cambridge University Press, 1995), 24, with Clausewitz's argument that "[w]ar is not merely a political act, but also a real political instrument, a continuation of political commerce, a carrying out of the same by other means," in Carl von Clausewitz, *On War*, trans. J. J. Graham (NP: Value Classic Reprints, 2017), 14. Politics can be about land but land never exhausts politics.

31 Hoffman, *An Environmental History*, 327.

32 Ibid., 278.

Thereafter, the nation-state would be the dominant player in the politics of Christian just war thinking.[33]

These, then, were many of the environmental influences and factors shaping the late medieval Christian just war tradition as it started its transition into its modern form. Yet how were these environmental influences and factors understood by late medieval Christian just warriors? Or, asked differently: How did late medieval European Christians understand nature and how did they express those understandings?

LATE MEDIEVAL UNDERSTANDINGS OF NATURE IN CHRISTIAN JUST WAR THOUGHT

To be fair, the question of how late medieval European Christians understood nature is unanswerable if we are looking for a single unambiguous answer. Temporal, geographical, and social locations (not to mention personal experiences, educational training, and individual psyches) would all shape answers to that question and not all those answers would align. So, for instance, Thomas Aquinas reinforced the common idea that an obligation to shape property rights having to do with the control of fields, forests, and rivers was a consequence of the fall, but a generation later, John of Paris argued that property rights are of a piece with human labor and human obligations to work precede the fall. Both men lived in the same city, were theologians of the same faith, were shaped by the same teleological perspectives, and were – at least on this matter – interested in the same questions. That they might come to different answers only hints at the range of late medieval attitudes toward nature and the expressions of those understandings.

This is not to say, I hasten to add, that the range of attitudes toward nature is unrelentingly vast. To the degree that late medieval Western Europe was suffused with Christendom, certain basic presuppositions so grounded reflections on the natural world as to be all but beyond consciousness, let alone argument. God, who is not nature, has created all that is out of nothing and nothing that exists sits outside of God's providential work. There is an order to all that God has created and although that order might be distorted by human sinful behavior, it is

[33] One bit of evidence in this regard: No less a just war light than Paul Ramsey would all but concede the obligation to address *jus ad bellum* questions to the state (not the church, not private armies, not insurrectionist groups, and certainly not the individual) in his major text *The Just War: Force and Political Responsibility* (Lanham, MD: Rowman & Littlefield, 2002).

never replaced by chaos. The things created by God and the order in which they exist are good and although that goodness is not perfect, neither can it be wholly removed from creation. That order and the things of it are not static but instead participate in a telos in which all things will be made new. Within that order, human beings sit in a privileged location: higher than the rest of material creation but not so high as the angels (let alone the triune God); the spiritual is higher than the material. With that privileged location, human beings have both dominion over the rest of material creation and responsibilities for it. The occurrences of horrendous events (whether earthquakes, famines, wars, or epidemics) complicate how we understand God's providential activity but don't negate it. They do, though, remind human beings of just how tenuous their dominion is. By God's grace, all that is necessary for continued existence has been provided. The term to contrast with "natural" is "supernatural," not "cultural." These presuppositions constitute what Julian N. Hartt has called "the theological consensus from Origen to Calvin" – that is, from the second century to the seventeenth.[34]

Such a consensus set the boundaries within which late medieval Christians made sense of the natural world and shaped the work of just war thinkers of the time, including orienting convictions about the very problems and possibilities of war that make just war thinking necessary. Within this broad consensus, late medieval Christian just war thinking on matters having to do with the nonhuman natural world coalesced around four themes that found expression at theoretical as well as concrete/situational levels: nature as resource in fighting, nature as brake on fighting, nature as opponent to fight, and nature as sign through which to make sense of fighting.[35]

[34] Julian N. Hartt, "Creation and Providence" in Peter C. Hodgson and Robert H. King, eds., *Christian Theology: An Introduction to Its Traditions and Tasks*, 2nd ed. (Philadelphia, PA: Fortress Press, 1985).

[35] These four themes relate, loosely, to what Hoffman has called "[t]hree less abstract, more firmly demonstrable, if not mutually compatible, threads in medieval thinking about nature – its hostility to humans, its value as a sign, the possibilities of human collaboration with a friendly personified *Natura*" (Hoffman, *An Environmental History*, 87). Since just war thinking is focused on matters having to do with when and how to fight, its greater specificity – and the greater utilitarian sensibility within it – helps to explain differences. Parenthetically, these three foci – and the tensions between them – help explain both the power of Lynn White's thesis (each of them being, in some way, preconditions for the framework that gave rise to "our ecologic crisis") and its inadequacy. The idealism and determinism in White's thesis (i.e., that certain Christian ideas about God, human beings, and the nonhuman natural world led to environmental destruction) ignores the complexity, multiplicity, and entanglements of those ideas and the variety of

It is important to note that making sense of how medieval Christian just war thinkers linked just war thinking and the nonhuman natural world involves a degree of speculation based on anecdotal evidence (often in the form of analogies), on direct references to the nonhuman natural world that were separated from more systematic reflections on war, and on arguments from absence. For instance, when Honoré Bonet asks, "[w]hether in time of war the ass should have the privilege of the ox?"[36] he is raising a question about whether an ass should be protected during time of war as canon law prescribed oxen should be. His answer: "[W]e must consider for what reason oxen are privileged. Certainly the only reason is that they plough the soil. Now if a poor man has only one ox, and harnesses his ass along with the ox to plough, then, considering that the ass does the work of an ox, I hold that he must have the same privilege, for he takes an ox's place."[37] Asses gain immunity where they work more like farm animals (oxen) than war animals (horses and mules); the reasoning is by analogy. Or when Thomas Aquinas argues that once victory against foreigners is assured, the victors "should use moderation in pursuing the advantage of victory, by sparing women and children and by not cutting down the fruit-trees of that country";[38] the closing note on fruit trees is almost an aside (although it does connect him to the scriptural injunction of Deuteronomy 20:19 against cutting down fruit trees when laying siege to a city). Note, though, that Thomas' argument about saving fruit trees occurs in I.II of his *Summa Theologica* although his systematic discussion of war occurs hundreds of pages later in II.II.

Or when Christine de Pizan argues that a wise and just military leader, upon recognizing that his enemy's forces lack food and money, should avoid engaging in battle but, instead, "he will remain quiet as if he were not paying attention, and as quietly as he can he will set out to bar the ways of escape. Thus he will surprise the enemy if it is at all

practices made manifest in those ideas. There was no inevitability that these ideas would combine as they did in early modern thought to shape the environmental crises funded by modernity (after all, Christianity in other parts of the world – not to mention most Christians in Europe for the bulk of medieval history – did not join those three foci in the way that the early modern Christians did).

36 Honoré Bonet, *The Tree of Battles*, cited in James Turner Johnson, *Ideology, Reason, and the Limitation of War: Religious and Secular Concepts, 1200–1740* (Princeton, NJ: Princeton University Press, 1975), 70.

37 Ibid. That Bonet ignores the prohibition against yoking an ox and an ass together in Deuteronomy 22:10 may be a sign worth interpreting at some other time.

38 Thomas Aquinas, *Summa Theologica I.II*, Q. 105, A. 3.

possible,"[39] she signals both the importance of basic human resources and the value of natural cover as a way of reducing casualties in war. Where resources are scarce and an enemy's force is demoralized, she seems to argue, it is better (and comes at less cost to all involved) to ambush them rather than announce your intentions and enjoin battle even though ambushes involve the use of deception and such use can raise moral concerns. Although she is not explicit in her reasoning about preferring ambush, that reasoning – especially in the context of her wider writings on war – suggests a link between fighting well and attending to the material conditions of human existence: The absence of natural resources necessary for an enemy's human well-being (as opposed to his or her military well-being) seemingly shapes how just war thinking is applied. As the link between natural resources, military victory, and just fighting isn't explicit, though, this is an argument from silence. What Barbara A. Hanawalt and Lisa J. Kiser note regarding written records about the natural world in medieval and early modern periods pertains fully to late medieval just war writings:

> Despite the ubiquity of nature's presence in the artistic and literary cultures of these periods, it is also true that overt discussion of it is seldom found. Until the sixteenth century, when scientific writing began to take as its primary subject the close observation of nature, responses to nature were often recorded only in the course of an investigation of some other subject ... In other words, medieval and early modern writers, unlike their modern counterparts, seldom sat down to write essays or treatises on nature in and of itself. Indeed, for many writing in these early periods, "nature" was arguably not even a discursive category; it simply went without saying.[40]

THE NONHUMAN NATURAL WORLD AS RESOURCE

That the nonhuman natural world provided resources in premodern warfare could also go without saying. However, drawing out some of

[39] Christine de Pizan, *The Book of Deeds of Arms and of Chivalry*, trans. Sumner Willard (University Park, PA: Penn State University Press, 1999), 54. At times it seems that as much of Book I of Pizan's *Book of Deeds* is occupied with questions about how military leaders should negotiate the natural world as it is with how they should negotiate enemy forces and that as much of Book II of *The Book of Deeds* is occupied with the securing and use of resources in the natural world in anticipation of battle as with battles themselves. Pedantic as Pizan's writings might be, they also display recognition of the focus that (at least premodern) military leaders must give to the nonhuman natural world. This is, in no small part, due to the influence of Vegetius' *De Re Militari* on her work.

[40] Barbara A. Hanawalt and Lisa J. Kiser, "Introduction" in Hanawalt and Kiser, eds., *Engaging with Nature: Essays on the Natural World in Medieval and Early Modern Europe* (Notre Dame: University of Notre Dame Press, 2008), 2.

the ways that the natural world provided resources for warfare begins to signal the complexity of the relationship between war and nature. Bows made of English yew proved stronger and more durable than alternatives such as ash and hazel and shaped the power of English archers during the Hundred Years War – yet Peter the Chanter did not question the morality of using yew for bows even though his opposition to the use of crossbows was so fierce that it led him to argue that crossbowmen did not deserve to be saved (and this in spite of the fact that in some battles – such as that of Crécy in 1346 – English longbows were better weapons than French crossbows).[41] Horses were not only necessary for a cavalry but also aided possibilities for military communication across distances. Of course, they also served nonmilitary purposes, which made their status complicated (as seen in the quote from Bonet above). As he prepared to lay siege to Auberoche in the middle of the fourteenth century, the Earl of Derby forbade his troops from foraging in France lest their excursions upon the land give away his army's presence and disrupt the possibilities for surprise: Preservation of the resources of the land was justified not because they were good in themselves but because they served military purposes.[42] Archbishop William Courtenay used the Dover earthquake of 1382 during a meeting of the Blackfriar's council in support of his argument that the church should attack John Wycliffe and the Lollards, for, "just as the earth lost its evil vapours [via the earthquake], so would the church gain from expelling the wicked."[43] In 1415, Henry V published an ordinance governing his army's conduct with a special focus on restricting pillaging, including the regulation that "no manner of man beat down housing to burn, nor apple trees, pear trees … nor no other trees bearing fruit."[44] Henry's preoccupation with the (im)morality of pillaging had, in part, to do with the prominence of wars over land in the Middle Ages, the growing focus on defense of land as legitimate *casus belli* (such that, by the seventeenth century, Grotius would include it

[41] See Fredrick H. Russell, *The Just War in the Middle Ages* (Cambridge, UK: Cambridge University Press, 1975), 243.

[42] For all the claims about chivalry and the shape of proper fighting, a willingness to hide behind natural features in the land and to use the protections offered by natural resources such as forests in order to pursue guerrilla-type warfare were not only common but also seem to have been an acceptable way of fighting. See David Whetham, *Just Wars and Moral Victories: Surprise, Deception and the Normative Framework of European War in the Later Middle Ages* (Leiden: Brill, 2009).

[43] Hoffman, *An Environmental History*, 309.

[44] Henry V, quoted in John A. Haymond, "Laws of War: The Origins of Restraint," *MHQ: The Quarterly Journal of Military History* 31.4 (Summer 2019): 18.

among just causes for war), and the way pillaging both harmed the land and destabilized social and political systems.[45] Whether through its use in weapons, its complex mix of military and civilian uses, its providing of cover/protection for military activities, its use in shaping the rhetoric of war, or contestations over it as the actual cause for war, the nonhuman natural world featured regularly in the way late medieval Christian just warriors thought about nature. Natural resources resourced military arguments that were freighted with both moral and strategic purposes.

THE NONHUMAN NATURAL WORLD AS BRAKE

Implicit in the previous paragraphs is the idea that the stuff of the nonhuman natural world also served as a brake on war – as a cause for restraint in war and, sometimes, the prevention of war – for late medieval Christian just war thinkers. While obligations to protect the nonhuman natural world lack the prominence in Christian just war thinking that they have in both Jewish and Islamic just war thinking, Deuteronomy 20:19, with its prohibition against destroying fruit trees while laying siege to cities, influenced just war thinking in the Middle Ages (as the quotes from Thomas Aquinas and Henry V above suggest).

More strategically, in the same way that the pursuit of natural resources justified war (politically, if not morally), it also shaped an implicit brake on war: If the costs associated with going to or staying in war were substantially greater than the value of those natural resources, the *jus ad bellum* criterion of proportionality could be applied in strategically straightforward ways. So, for instance, although the area that would become the Kingdom of the Netherlands in 1581 was the site of battles going back to the Bronze Age, during the Middle Ages it was relatively free from conflict, at least in comparison with immediate neighbors such as Germany and Flanders (as well as France and neighbors across the sea such as England). While there are a variety of reasons for this (and, to be clear, the Netherlands was politically linked to other countries – including Flanders – where conflicts were more frequent),

[45] At the end of the fourteenth century, Honoré Bonet argued that though one could not morally wage war in order to recover lost property in general, "I may recover it by force of arms if I do it in heat, that is to say on the instant, before I undertake anything else." Honoré Bonet, *The Tree of Battles*, cited in G. W. Coopland, *The Tree of Battles of Honoré Bonet: An English Version with Introduction* (Cambridge, MA: Harvard University Press, 1949), 140. By the early seventeenth century, Hugo Grotius treated defense of property as a generally legitimate *casus belli* in *The Rights of War and Peace*.

one reason was certainly that its land was seen as less valuable – whether for agriculture or as the basis for a port – than its neighbors' lands. The Dutch spent their premodern energies fighting back the seas in order to grow cereals on small plots – a task complicated by the fact that as peat bogs dried and were walked on by farm animals, they sank, which mandated new drainage systems. Of course, the combination of technological development used by the Dutch to address their needs and a comparative lack of natural resources also helped drive both economic growth and the search for new resources during its heyday as a colonizing power.

Though the Netherlands mostly avoided English incursions during the late medieval period – and especially during the Hundred Years War (1337–1453 CE) – its neighbors were less fortunate. Both Belgium and France were repeatedly raided and invaded during that war, and the battles of that war have come to symbolize the brutality of wars during the period. Among the most notable manifestations of violence during that war were the English *chevauchées*: plundering raids in which English forces burned and pillaged the countryside in order to reduce its productivity and weaken their enemies. Clearly, these *chevauchées* did great harm to the land; indeed, it would not be too much to say that the English waged war on the land itself during their *chevauchées*. Yet the purposes of the *chevauchées* were not necessarily oriented toward destruction and harm per se; instead, they were often intended to highlight the inability of a government and its forces to maintain control of that land. During a period in which one's ability to control land was the chief expression of one's right to that land, lack of control was tantamount to losing property rights. As such, the *chevauchées* not only revealed the limits of authority over land but also tacitly expressed the rights of the invading force to that land and, by extension, to the other lands that the defending forces had claimed.[46] Their intended effect was, among other things, to make long and damaging sieges unnecessary. Indeed, the *chevauchées* fell out of tactical favor as siege warfare changed. Destructive as they were – and as much as they drove the flight of refugees to cities – they were also a way of preventing or inhibiting the potentially even worse violence associated with siege warfare.

46 "In a period that had a very positivist conception of rights, if one was unable to defend something, one did not have a right to it. Therefore, a *chevauchée* could be used to, very publicly, call into question the legitimacy of the position of the impotent lord and demonstrate the justice of one's own cause at the same time." Whetham, *Just Wars and Moral Victories*, 12.

The codes of chivalry also recognized the value of the natural world (and the resources it provided) as a reason to limit violence. So, for instance, there were codes that opposed pillaging, thereby – at least in theory – limiting destruction of the land. In *The Tree of Battles*, for example, Bonet asks: "Can a man-at-arms who goes to war for the sake of pillage legally ask for wages?" Having laid out even optimal conditions for pillaging (the men-at-arms are serving a legitimate authority and have otherwise performed "good deeds of arms" in winning towns and doing their duty well), Bonet argues that if they also pillage, they lose their good grounds to ask for wages because "according to written law no obligation can be incurred by what is dishonest, condemned or wicked."[47] At another point in *The Tree of Battles*, Bonet asks whether serfs can be compelled to go to war at the command of their lords. His answer is that they can – with the proviso that "[t]here is another kind of servitude called libertine serfdom, which involves obligation to perform certain tasks in the fields: for instance in the corn-fields or vineyards. Such serfs would not be required to take part in war, for, considering the nature of their state, a new servitude cannot be imposed on them."[48] While the meaning of "libertine serfdom" is uncertain, the passage is clear that at least some fieldworkers are protected from participation in war in order to maintain those fields because their serfdom is to the land as much as it is to the lord. Even though Bonet rarely discusses the nonhuman natural world in the context of his book, these two examples, combined with the example Bonet gives of protecting an ass used for plowing fields mentioned above, are consistent in suggesting that attention to the nonhuman natural world shaped rules of chivalry that limited violence.

THE NONHUMAN NATURAL WORLD AS ENEMY

This is not to say that late medieval Christian just warriors saw the nonhuman natural world as serving only in beneficial ways; at times, that world was, itself, an opponent to be battled. No less than other thinkers, late medieval Christian just warriors aligned themselves with the miner in Paulus Niavis' 1492 poem "The Valley of Beauty," in which a miner is put on trial by the gods for desecrating mother earth as he digs deep into her bowels and damages the woods and rivers around his mine. Though guilty of such crimes, the miner defends himself by arguing that he is

[47] Bonet, *The Tree of Battles*, in Coopland, *The Tree of Battles of Honoré Bonet*, 144.
[48] Ibid., 168.

justified in doing so in order to produce goods – including luxury goods such as silver – that can benefit human beings. While the poem recognizes that eventually the earth will strike back against such behavior, it also names the reality of a conflict between human beings and the natural world.[49] In a presecular world, natural calamities (i.e., "acts of God") and human-caused ones were closely connected. During the rise of insurance plans in the Middle Ages, for instance, insurers didn't distinguish between severe storms and earthquakes, on the one side, and war and piracy, on the other: the four horsemen of the apocalypse (famine, disease, war, and death) rode together. The language of war was common in late medieval "battles" against fickle and capricious natural forces.

Natural forces also impacted actual battles. Early in the Hundred Years War, storms apparently rendered the French crossbows ineffective against English forces in the Battle of Crécy; later, the French knights, wearing heavy armor, needed to make their way through thick mud after heavy storms to confront the English at the Battle of Agincourt (1415 CE). Historian Juliet Barker reports that many French soldiers, once knocked to the ground, were unable to stand to rejoin the melee and some actually drowned in the mud, having been held under by the weight of their armor.[50] Food shortages and famines complicated the ability of rulers to feed their armies during war. Diseases – particularly airborne diseases, which spread rapidly through armies – were a continued drag on military strength and preparedness. And terrain was understood as playing a significant role in shaping battles since Vegetius' *De Re Militari*, a fourth-century text on Roman war that was treated as a textbook for warfare throughout the Middle Ages. In the midst of battling opposing forces, militaries of the time were also always battling the elements. That late medieval just war thinkers do not write on such matters probably has less to do with their absence as a concern than the reality of such problems being so obvious as to not warrant mention.[51]

49 For a brief overview of the poem, see Ulrich Grober, *Sustainability: A Cultural History*, trans. Ray Cunningham (Cambridge: UIT Cambridge Ltd., 2012), 45–46.

50 Juliet J. Barker, *Agincourt: Henry V and the Battle that Made England* (London: Little, Brown Book Group, 2005), 300.

51 It is worth noting, tangentially, that during the early modern period of colonization, battling the elements was an especially significant and fraught endeavor. During the Little Ice Age, European explorers/colonizers found North America, particularly, inhospitable. Indeed, one of the reasons that South America was colonized more aggressively than North America during the sixteenth century was that Europeans simply found it easier to colonize. For the difficulties of colonizing North America, see Sam White, *A Cold*

A final connection between just war thinking and nature as opponent manifested in the way that persons from non-European lands (including during the Crusades and in early explorations of Africa and the Americas) were viewed as animals – that is, as part of the nonhuman natural world – as a way of justifying attacking them. And while some Christian just warriors grew increasingly leery about the moral legitimacy of crusading and colonizing efforts, the equating of enemies with the nonhuman natural world reveals not only the biases of Western European thinking about non-Europeans but also the way they thought about the natural world.[52]

THE NONHUMAN NATURAL WORLD AS COLLECTION OF SIGNS

A fourth way that late medieval Christian just warriors understood nature was as a collection of signs to interpret in making moral sense of the world. It is no small task for twenty-first-century interpreters to make sense of the way the nonhuman natural world functioned as a system of signs in the late medieval period, let alone how it functioned in the context of just war thought; our immanent frame inhibits such projects. In a world bursting with the supernatural, though, discerning something of God behind materiality was a significant project within human existence. Assumptions about a world behind the world, regularly manifest in allegorical interpretations that start with the natural world and move beyond it – whether in illuminated manuscripts, bestiaries, stained glass windows, statues, or elsewhere – shape difficult questions for contemporary environmental historians to address: "At what point, if any, do these meaningful notions of the natural world actually intersect with materiality? When and how, if at all, do they arise from experience that is becoming representation? How, why, if at all, do these ideas serve to inform programmes for work?"[53] What is true about the way nature functioned as sign in medieval thought generally is no less true – though rarer, perhaps – in medieval just war thought.

Welcome: The Little Ice Age and Europe's Encounter with North America (Cambridge, MA: Harvard University Press, 2017).

52 For the way medieval and early modern thinkers linked "outsiders" (e.g., Africans, Jews) to animals, see Jeffery Jerome Cohen, "Inventing Animals in the Middle Ages" in Hanawalt and Kiser, *Engaging with Nature*, 39–62.

53 Hoffman, *An Environmental History*, 101.

There are symbol-rich references to nature as it relates to war in the writings of late medieval just warriors, though these are scarce and tend toward the metaphorical rather than the allegorical. So, for instance, Christine de Pizan notes that when soldiers enter into the territory of an unprepared prince, "[t]hey should not be like wolves, who are not satisfied with one sheep when they enter the fold, but must kill the whole flock ... as if they were true ravishing wolves without consciences, as if there were no God or they should never die."[54] Likewise, there are occasional visual cues in drawings and illuminations of war as they appear in various settings and reinforce allegorical interpretations of the world: the power and subservience of the horses ridden by royalty that are allegories for the power and subservience of their armies (such as the pictures that appear in the front pieces of translations of Vegetius' *De Re Militari*); the majesty of eagles or the strength of lions on military crests to signal nobility in how an army fights; mythic animals such as griffins or unicorns on royal coats of arms that signal, at some level, rulers' exoticism and freedom to live beyond the everyday experience of most persons and so reinforce the legitimacy of the rulers' authority; trees nourished by the blood of those who die in battle appearing in stained glass. Indeed, in a world interpreted by picture as much as by text, not only does nature show up symbolically in war and in reference to war but also war with nature shows up symbolically in reference to well-lived military lives: pictures of chivalrous knights engaging in combat not only with other knights but also with beasts – some of them demonic – that represent moral and spiritual enemies to be conquered and battles to be won.[55]

Certainly, though, the most prominent way that nature offered signs for late medieval just warriors to interpret was via the dominant form of moral inquiry of the time: natural law thinking. Beginning with the assumption that there is a moral order to the universe and that such order can be discerned through observation of and reflection upon the world – including the nonhuman natural world – natural law theories guided all just war thinkers of the day.[56] It provided coherence behind the chaos of battle and morality in the face of the immoralities of war. Such thinking

[54] Pizan, *The Book of Deeds of Arms and of Chivalry*, 166.

[55] See, for example, The *miles spiritualis* displayed in Aden Kumler, *Translating Truth: Ambitious Images and Religious Knowledge in Late Medieval France and England* (New Haven, CT: Yale University Press, 2011), 81.

[56] Natural law was so prominent among just war thinking in late medieval thought that even those who did not write specifically on just war (and perhaps were not just warriors themselves) nevertheless found it necessary to pursue natural law reasoning for their own

was prominent in the work of these scholars, and particularly in those who followed after Thomas Aquinas (e.g., Cajetan, Suárez). If anything, such scholars emphasized Thomas' version of natural law thinking even more forcefully than Thomas had, repeatedly inserting it into questions about war.

To make sense of these insertions, though, two caveats warrant attention. First, "natural law" is not a single approach to making moral sense of the world but, instead, a family of approaches that share resemblances with each other.[57] So, for instance, Richard Gula identifies two major strains of interpretation within natural law thinking during the High Middle Ages: "One, the 'order of nature,' identified with the Stoics and Ulpian, focused on the physical and biological structures given in nature as the source of morality. The other, the 'order of reason,' identified with Aristotle, Cicero, and Gaius, focused on the human capacity to discover in experience what benefits human well-being."[58] Thomas, it is worth noting, used both forms at various points in his *Summa Theologica*. Moreover, multiple distinct streams of natural law reasoning flowed

purposes. So, for instance, in his book on usury, *De usuris,* the Franciscan theologian Peter Olivi (1248–1298 CE) argued that although it seems that usury is "contrary to nature," the Book of Deuteronomy allows for usury in the case of interest charged to strangers. Since divine law as found in Scripture is perfect and cannot contradict natural law, natural law must, therefore, allow for usury. Olivi then applies this same mode of thinking to just war thinking: because the divine law as found in Scripture is perfect and cannot contradict natural law, and since Scripture affirms just war thinking, therefore the laws of nature must affirm just war thinking. This form of reasoning shaped a problem for Olivi as he was opposed to usury. His solution: that God had given permission for the Hebrew people to charge usurious rates when they were entering the promised land in much the same way that God had given them permission to pillage and massacre – but those permissions no longer applied. On Olivi's reasoning, war in Europe was problematic because the theological justification for war in the Old Testament did not pertain in that context. See Steven A. Epstein, *The Medieval Discovery of Nature* (Cambridge, UK: Cambridge University Press, 2012), 139–141.

57 For that matter, Charles E. Curran has written, "it has been pointed out that the word nature had over twenty different meanings in Catholic thinking before Thomas Aquinas." Unfortunately, he does not cite the source who has pointed this out. See Charles E. Curran, "Natural Law in Moral Theology" in Charles E. Curran and Richard A. McCormick, S. J., *Readings in Moral Theology No. 7: Natural Law and Theology* (New York, NY: Paulist Press, 1991), 253.

58 Richard M. Gula, S. S., *Reason Informed by Faith: Foundations of Catholic Morality* (New York, NY: Paulist Press, 1989), 223. Gula goes on to argue that Thomas Aquinas inconsistently applied both forms of natural law reasoning, thereby complicating not only natural law thinking but also the way his successors claimed to follow him. My own opinion is that Thomas chose the interpretation of natural law that he felt best supported his argument at the time. When referring to sexual ethics, he referenced the order of nature; when referring to war, he referenced the order of reason.

through the Middle Ages, including canon law and Roman law, each shaping premodern just war thinking. Canon law, shaped by Augustinian thought and given specificity through the Decretists as they followed Gratian, viewed war as justified if it was in the immediate defense of self or in order to punish wrongdoing after the fact (hence its emphasis on and approaches to just cause). Roman law, on the other hand, followed from the recovery of Justinian's 534 CE *Corpus juris civilis* in the eleventh century and was framed by attention to the rights of authorities to wage war as a manifestation of the laws of nations in which war was a natural phenomenon and needed only limited justification (so they emphasized legitimate authority).[59]

The second caveat is that the temporal distance between, for example, Thomas (d. 1274) and Francisco Suárez (d. 1617) is enormous. The three and a half centuries between the two men cover a long and productive history in the development of natural law thinking, including developments in the idea of natural rights. There is a tendency in just war thinking to jump from the thirteenth to the sixteenth century without noting the way just war thinking changed in the intervening centuries (let alone the way interactions with the nonhuman natural world shaped those changes). Both Thomas and Suárez rely most obviously on what Gula referred to as the "order of reason" version of natural law when discussing war, but their assumptions about how the order of reason expresses itself differ not only because the political systems they were engaging differed but also because their perceptions about the relationship between the human and the nonhuman natural world differed.

Recognizing these caveats, it is nevertheless the case that there are a few generalizable claims to make about the role of natural law thinking in late medieval just war thought. First, just warriors of the period relied almost entirely on "order of reason" thinking to conclude that wars are natural things. Observers of the nonhuman natural world did not look out and see actual wars being waged in that world even if they did see predation, the protection of the vulnerable, and a hierarchical order to things. Instead, they inferred from first principles (there is an objective good in the universe and people desire it) to right actions in that universe (the good is to be done and evil avoided) to law (pursuing that good means seeking justice through rules) to behavior in war

[59] See Ryan Greenwood, "War and Sovereignty in Medieval Roman Law," *Law and History Review* 32.1 (February 2014): 31–63. There was also Germanic law, which had been shaped by encounters with both canon and Roman law but was different from both.

(those who break those laws – including at the level of the state – should be punished). In this, they had not only Scripture and theologians such as Augustine and Thomas but also centuries of earlier just war/natural law thinking to guide them. Indeed, to read through Honoré Bonet or Christine de Pizan (or, for that matter, Suárez, Vitoria, or Grotius) is to read through a history not only of military strategy but also of martial rectitude on the part of generals and soldiers from the time of the Roman Empire forward. As Nigel Biggar notes, premodern Christian "just war theory ... took its bearings from morality, natural and revealed, not from positive law. Its paradigm was the defence of the innocent against injustice, not the defence of the (national) self simply."[60]

This is not, though, to say that the "order of reason" is an order set apart from the natural world. Reasoning within a natural law framework both reveals and shapes the way human beings understand their relationship to the nonhuman natural world. Human beings may be in a privileged place in the world, but they are not set apart from it. Instead, they inhabit a unique and distinct position in a hierarchy of being that places them above the nonhuman natural world and below more "spiritual" entities. And not only are they embedded in a hierarchy, but also their ability to reason reveals how they are embedded in it. Because they are both subject to and participants in its processes, they are capable of using it for their own purposes but only within limits. Because the natural order of things reveals a hierarchy, their own political configurations should be hierarchical. Because they are subject to laws both divine and natural, they are ordered toward making and maintaining laws themselves. Because human beings participate in the natural order of things, they can turn their faculties of reason upon themselves as creatures in order to make sense of how to behave in the world. Or, said differently, as medieval just war thinkers were making arguments about when and how to fight, they did so by interpreting the (natural) signs that they were to themselves as those signs revealed the natural law to which they were bound.

This is one of the primary reasons that late medieval just war thinking is so oriented around laws and codes, including codes of chivalry. So, for instance, Bonet "did not claim to be creating a new body of rules to govern warfare. Rather he was attempting to interpret or 'find' the law that he believed was already present, even if it was not always

[60] Nigel Biggar, *In Defence of War* (Oxford: Oxford University Press, 2013), 159.

adhered to."[61] Over time, this natural law emphasis diversified and then receded as Christian just war thinking both shaped and was shaped by the stirrings of early modern thought even as such shaping was, in part, a response to climatic changes in Europe and North America. With the arrival of modernity, nature as sign played an increasingly diminished role in Christian just war teaching.

More broadly, these late medieval understandings of the natural world (as resource, brake, enemy, and sign) would also change as a variety of practices – including those associated with war – emerged out of the interactions between human beings and the nonhuman natural world. The natural world would not only continue to provide resources for war but also increasingly be understood as a collection of resources. Whatever brakes it placed on war were weakened, first as the whole of the natural world became "resources" and as new resources were discovered and then as wars increasingly focused on the acquisition and maintenance of political power. Nature was still to be battled, but those battles became increasingly one-sided as traditionally organic interpretations of the natural world shifted to mechanical metaphors.[62] Nature offered signs, but those signs pointed only to other parts of the nonhuman natural world. Practices shaped ideas, which reshaped practices. New ways of interacting with the nonhuman natural world shaped practices, which reshaped understandings of the nonhuman natural world. And, over time, human beings – including Christian just war thinkers – ceased to see themselves as participants in the natural world as they shaped new practices and new understandings of old practices, all of which helped them live – and fight and die – in the new world they were creating as it was also creating them.

SILENCES AROUND THE NONHUMAN NATURAL WORLD AFTER 1500

Earlier chapters in this book have focused on the way that Christian just war thinking, as it transitioned from its late medieval to its early modern form, shaped the conditions that led, eventually, to contemporary

[61] Whetham, *Just Wars and Moral Victories*, 53.

[62] See, e.g., Francis Bacon's language about conquering nature by forcing it to reveal its secrets or Rene Descartes' treatment of animals as biological machines (recognizing that both Baconian and Cartesian understandings of the natural world are arising out of complex experiences of that world and are more complex than the stereotypes of their positions on such matters suggest).

climate-shaped conflicts and to our difficulty in making sense of and responding to those conflicts. This chapter has thus far focused on interactions between the nonhuman natural world and late medieval Christian just war thinking. Left underexamined up to this point, though, is the question of the way Christian just war thinking shaped the nonhuman natural world and human experiences of it during that complex transition into early modernity.

Each of the preceding chapters could serve as a starting point toward answering that question. So, for example, as a response to wars in Europe, international law of the sort given shape by Grotius is understandably preoccupied with intrahuman matters – though even that preoccupation is peculiarly shaped, as the conceptual lacuna where refugees should be makes clear. Almost entirely left out of Grotius' work on international law is attention to the nonhuman natural world and human interactions with it, and where Grotius does give attention to it, it is almost exclusively in the context of treating the nonhuman natural world as a site of resources. Even Grotius' version of natural law fails to show much interest in the nonhuman natural world.

Such an omission is unsurprising. The focus on sovereignty and intrahuman violence from an immanent perspective doesn't offer much of an entry point into thinking about interactions between the human and the nonhuman natural world. Nature doesn't respect boundaries and, within an immanent frame, lacks agency.[63] The very fact that there continues to be so little law – international or otherwise – to lend structure to human interactions with the nonhuman natural world (e.g., the language of "rights" only awkwardly applies to other species; there aren't really any effective international judicatory bodies to address environmental degradation; indigenous cultures with different ways of relating

[63] This is not to say that actions of the nonhuman natural world ceased to be treated as driven by will and purpose. It is, though, to say that when will and purpose were imputed upon actions in the nonhuman natural world, they entered the realm of the supernatural (and so evaded assessment). Indeed, there is scholarship suggesting that times when the nonhuman natural world is most hostile to human beings are also times when attention to/preoccupation with the supernatural is most prominent. See, for example, the work of Wolfgang Behringer and Emily Oster, who argue that the worst periods of the Little Ice Age were the very times that witchcraft persecutions in early modern Europe were most pronounced. Wolfgang Behringer, "Climatic Change and Witch-Hunting: The Impact of the Little Ice Age on Mentalities," *Climate Change* 43 (1999): 335–351; and Emily Oster, "Witchcraft, Weather and Economic Growth in Renaissance Europe," *Journal of Economic Perspectives* 18 (2004): 215–228. Both are mentioned in Hoffman, *An Environmental History*, 339.

to the nonhuman natural world struggle to get the status necessary to appeal to law on behalf of that world) only highlights this.

That omission carries over to the development of modern constitutions. The Constitution of the United States of America, for instance, includes no mention – whether in its original articles and the Bill of Rights or any amendment – of the nonhuman natural world and it has only been in the last few decades that the nexus of environmental law and constitutional law has begun to be filled out at state levels.[64] And that constitution is the norm, not the exception. That many of us would think of inclusion of environmental law into constitutional law as odd is, itself, a sign of just how much the nonhuman natural law receded from visibility during the advent of early modern thought. The nonhuman natural world may have been the backdrop for all the interhuman matters that shaped modern thought, but it stayed in the background. It is, then, no surprise that many of the most significant implications of the Christian just war tradition's engagements with the nonhuman natural world during its transition from its late medieval to its early modern form have proven to be simultaneously invisible and far-reaching.

THE NONHUMAN NATURAL WORLD'S ROLES IN COLONIZATION AND CONQUEST

Undoubtedly, the most significant exception to this invisibility is found in the European exploration, conquest, and colonialization of the rest of the world. As Richard Hoffman notes,

> [a]fter the Columbian voyages, no human and no European experienced the natural sphere as they had before. While much nevertheless remained unaltered (limited energy resources, the immanence of spirit, the centrality of agriculture), coincident natural and cultural mutations reset key features of European environmental relations at a scale comparable to the late antique discontinuity which had delineated a medieval European from a classical Mediterranean civilization. Narrative arcs contained within European culture and its own natural world would thereafter be transcended.[65]

What, though, did these transformations look like and how are they related to Christian just war thinking? Given the centrality of Christian just war thinking within the projects of colonization, that question might

[64] See, in this regard, James R. May, ed., *Principles of Constitutional Environmental Law* (NP: American Bar Association, 2013).

[65] Hoffman, *An Environmental History*, 19–20.

be asked more simply: What is the environmental history of European colonizing projects?

The history of colonization has often been told as if the nonhuman natural world functioned only as the setting for the stories of human endeavors: courageous explorations, brutal oppressions, brave refugees, wondrous discoveries, the globalizing nation-state, the dawn of Enlightenment philosophy, the pluralization of religious sensibilities, and the conflicts – some bloody, some less so – associated with each of those endeavors. Though those endeavors warrant the extensive attention they are being given, treating the nonhuman natural world primarily as a backdrop for them does significant injustice to the causative power of the interactions between the nonhuman natural world and those endeavors.

Let us begin with the obvious: Colonization was pursued by those with access to natural resources and driven by the desire to get more resources. Interactions with the nonhuman natural world in the fourteenth and fifteenth centuries shaped the conditions for colonization. Favorable climates had increased cereal yields and helped multiply European populations, sent them to cities, and shaped and expanded markets and trade. Epidemics of the Black Death had largely subsided. Issues of control over waters and woods reshaped European political and legal systems. Developing technologies increased access to and the use (and exhaustion) of natural resources in Europe. Even calamitous events favored Europe: "Some historians," Hoffman writes, have argued that "Europe's relative immunity to large-scale natural disasters gave it an advantage over other advanced pre-industrial civilizations. While between 1000 and 1550 some twenty-four earthquakes are recorded to have killed more than ten thousand people world-wide, only three of these occurred in western Christendom."[66]

This is not to say that life in Europe was easy during these centuries. In places such as Holland, limited access to natural resources drove technological innovation in order to take better advantage of those resources. In others places such as Portugal, diminished access to natural resources drove the pursuit of natural resources in new places. Throughout Europe, contested access to natural resources drove conflict as well as the displacement of groups of people from one place to another. It is, though, to say that the conditions for expansion, shaped by both plenty

[66] Ibid., 312.

and need, were connected to interactions with the nonhuman natural world and were, at the dawn of modernity, ideal for projects of exploration and colonization.

War – and Christian just war thinking – played its own role in these projects, shaping and being shaped by them. By the time Vitoria was advising the pope on how the Spanish conquistadors should treat indigenous persons in the Americas and West Indies (and, less overtly, how to think about access to natural resources in those places), just war thinkers were shaping and adjusting to the new shape of just war thinking. By the time Grotius was writing *On the Rights of War and Peace*, wars had moved onto the oceans – an activity that both required natural resources (metal and gunpowder for cannon, lumber for ships) and was shaped by attempts to control access to and shipping lanes for natural resources. By the time the Westphalian nation-state was emerging out of the chaos of bloody European wars, European countries were already involved in proxy wars over control of resources in the new world. Indeed, the transformation of the Christian just war tradition as it shifted from its late medieval to its early modern form was, perhaps, the most momentous shift in its history – and it was dramatically shaped by colonization and attempts to control natural resources in colonized spaces (which is ironic, given the way attention to the natural world during that shift would then largely drop out of consideration in answering questions about whether, when, why, and how to fight).

That the conditions for colonization were right, though, doesn't explain why so many of the projects of colonization were successful. After all, earlier and concurrent failed attempts at colonization (e.g., Norse attempts to colonize Greenland until the Little Ice Age, Spanish attempts to colonize the areas that would eventually become the southern United States in the first half of the sixteenth century) might have diminished the colonizing impulses of European states had those failures not been dwarfed by successes in the sixteenth and seventeenth centuries. The need for resources had led to repeated conflicts within Europe prior to the early modern period and the romance of some foreign lands (and the resources available there) had long stimulated European imaginations. It's not as if Europeans had spent their premodern history successfully colonizing places they did know about, such as the Middle East or North Africa. The scale and success of colonization in the sixteenth century reveals the power of some relatively new ideas as they were joined with new technologies and practices in the early modern period, making this period of colonization stand out within the long arc of European history.

Toward answering the question of why European colonization was so successful once it was launched, Alfred W. Crosby has offered a provocative thesis:

> [T]he areas that the Europeans most successfully colonized are far from Europe in distance but have climates similar to hers, and European flora and fauna, including human beings, can thrive in these regions if the competition is not too fierce ... Perhaps European humans have triumphed because of their superiority in arms, organization, and fanaticism, but what in heaven's name is the reason that the sun never sets on the empire of the dandelion? Perhaps the success of European imperialism has a biological, an ecological component.[67]

Crosby's argument, then, is that European nations succeeded in their colonizing adventures not simply because they were armed with guns rather than spears or because they were more fanatical in their pursuits than the indigenous persons they found when they arrived in new lands, but because of the parts of the nonhuman natural world they brought with them (sometimes intentional and often not), which then outcompeted indigenous species. They brought not only dandelions and other weeds but also rye and oats to harvest, clover to feed their animals, and fruits that proliferated in the temperate areas in the Americas. They brought not only domesticated animals such as horses, oxen, dogs, pigs, and rabbits (some of which went feral, much to the detriment of native species), but also pests such as rats and insects such as honeybees (beneficially) and mosquitoes (not so beneficially).[68] And, above all, they brought diseases including typhus, malaria, yellow fever, and smallpox that wiped out entire civilizations. In a little over 100 years, diseases reduced indigenous populations in the Americas tenfold, from around sixty million to closer to six million. Almost all at once, the continents "were emptied, creating what Europeans perceived as primal wilderness where once had flourished anthropogenic landscapes shaped by hunter-gatherers, agriculturalists, and indigenous urban cultures."[69]

There were, obviously, also species that made their way back from the new world to the old: plants such as potatoes, tomatoes, and tobacco; animals such as turkeys; diseases such as syphilis. Many of these imports

[67] Alfred W. Crosby, *Ecological Imperialism: The Biological Expansion of Europe, 900–1900* (Cambridge, UK: Cambridge University Press, 1986), 7.

[68] In a fascinating exploration of the impact of the nonhuman natural world on colonization, J. R. McNeill has explored the impact of mosquitoes on the Americas (and, particularly, the Caribbean) in reshaping not only natural systems but political ones as well. See J. R. McNeill, *Mosquito Empires* (Cambridge, UK: Cambridge University Press, 2010).

[69] Hoffman, *An Environmental History*, 365.

would reshape Europe. But the sheer levels of invasion and impact were hardly commensurable. By the beginning of the twentieth century and the end of Europe's colonizing projects, the "success of the portmanteau biota and of its dominant member, the European human, was a team effort by organisms that had evolved in conflict and cooperation over a long time."[70] As much as Europeans were interested in acquiring resources from the nonhuman natural world in their colonized lands, the parts of the nonhuman natural world that they brought with them were instrumental in shaping the success of their colonizing projects.

Between what the colonizing powers imported into the Americas (including not only nonhuman nature but also human beings in the form of enslaved Africans, who brought their own agricultural practices and parts of the nonhuman natural world of Africa with them) and what they took from colonized lands (including timber, agricultural products such as cotton and sugar, spices, precious metals, fish, furs, and whale and seal oil), the metabolic and commerce systems of the entire world were transformed.

THE ROLE OF THE MILITARIES IN ENVIRONMENTAL IMPERIALISM

That transformation required a military presence to protect those in the colonies from indigenous persons and serve as labor in physically clearing spaces for colonists, but even more to protect colonial powers from encroachments by each other. Militaries that served such purposes could no longer be composed of a single class of persons trained in the skills and mores of war; they required increasingly large numbers of persons from a much wider range of classes with a wider variety of skills using weapons that required less training. The era of the mass army, initiated during the European wars of the sixteenth and seventeenth centuries, bloomed in full, and with it the transformation of just war thinking of the sort described in earlier chapters. Much has been written about the roles that the modern nation-state, the advent of political liberalism, the rapid expansion of capitalist economies, and the arrival of the secular age have played in shaping modern just war thinking. Relatively less has been written about the role that colonization played in shaping modern just war thinking. And almost nothing has been written about the role that the nonhuman natural world played in shaping it. Yet to make sense

[70] Crosby, *Ecological Imperialism*, 293.

of how and why just war thinking changed, we need to attend to the way all these forces have played out as they interacted with each other.

Especially as we enter the environmental age, we ought to give special note to one of the great ironies that came with the arrival of early modern warfare: At no point in the history of human conflict had the nonhuman natural world played a larger role in shaping the ideas, practices, and technologies of warfare, including those associated with the Christian just war tradition. Yet in spite of its magnified role in shaping such things, the attention that Christian just war thinkers gave to the nonhuman natural world was much reduced. Not only did they pay less explicit attention to it, but they also paid attention to it in fewer ways: late medieval conceptions of nature as brake on violence and nature as sign to interpret receded from prominence in the way Christian just warriors encountered and made sense of the natural world.

The nonhuman natural world need not be a brake on violence. Finding a new world rich in resources, early modern Christian just warriors gave little to no thought to the idea that the nonhuman natural world might provide a brake on fighting. Why protect the fruit trees when one can simply go find that fruit somewhere else? Pursuing technologies that increased crop yields and reduced the threat of famine, they increasingly turned their attention to technological innovation rather than crops. Why pillage neighboring countries when there are enough cereals at home? Living in cities, their encounters with the natural world (including their own food) became increasingly grounded in abstractions. Why focus on how to grow grain when one could buy bread? Christian just war thought turned toward theories in which costs and benefits were measured in human-sized quantities.

Nor need the nonhuman natural world offer signs in need of interpretation. Pursuing written ways of meaning-making, they gradually left allegorical interpretations grounded in attention to the natural world behind. Why look to the natural world in order to make sense of the divine when the Bible was available? Reducing their teleologies to the idea that God had provided abundantly for them, they turned away from looking to the nonhuman natural world for signs about whether, why, and how they should fight. Why pursue the insights of natural law when positive law seemed more pertinent for dealing with conflicts across cultures? Encountering more religious perspectives, whether from other Christian communions or from other parts of the world, they shaped natural law theory in increasingly immanent ways (as described in Chapter 1). Why maintain the complex metaphysical infrastructure of medieval natural law thinking when simpler forms of natural law thinking would suffice? Separating themselves from the conceptual hierarchies that situated human beings vis-à-vis the nonhuman natural world, they increasingly shaped philosophical systems that started with questions about what it meant to determine their own fates. Why shape political systems around a cosmically established order when politics ought to be about how human beings structure their own systems of organization? Christian just war thought turned toward theories grounded in an immanent framework oriented more toward pursuing human flourishing than maintaining divine order.[71]

Many of the changes that came about through the transformations of thought, practices, and technologies that occurred during the transition from the late medieval to the early modern period have been enormously beneficial – including even (if perhaps only eventually) in the reasons and ways we fight. The people of the world are simply better off when fewer people have faced the threat of starvation due to failed crops and when more people can enjoy the benefits of "life, liberty, and property," as John Locke advocated. These benefits, though, have come with payments that we are only recently coming to recognize even as they are increasingly coming due.

CLIMATE CHANGE AND EARLY MODERN EUROPEAN IMPERIALISMS

Operating behind the scenes in the changes brought about by these transformations were changes in climate happening on a global scale. Even as the Little Ice Age was beginning to exert its power in the fourteenth

[71] Hoffman suggests that "between the mid-fourteenth and the mid-sixteenth century Europeans did initiate and undergo seminal cultural developments which inflected basic ways they perceived and represented their experience of the natural world and then set plans and priorities for material action in it. Much had autonomous origins in a symbolic cultural sphere itself." Among these changes, he includes "empiricism, quantification, the historic synergy between vernacular literacy and emerging print culture, Renaissance humanism, and ever-pressing claims from markets and from states over use of natural objects and materials." Hoffman, *An Environmental History*, 351. His argument here seems largely correct and in alignment with the ones I'm making about transitions in Christian just war thinking throughout this book, with the single exception of his use of the word "autonomous." I would suggest that the natural and cultural spheres so interpenetrate each other (including in the way both are engaged symbolically) that the very idea that one of the spheres could actually operate autonomously is problematic. Given his emphasis on an interaction model for making sense of the way humans relate to the nonhuman natural world, I'm more than a little surprised he would use the word.

century, it was nevertheless the case that, as noted above, "1450–1500 in central Europe and 1500–40 in the Netherlands became the longest runs of mild winters either would see again until the mid-eighteenth century."[72] These mild winters, combined with the decline in epidemics of the Black Death and relative seismic stability and volcanic inactivity, shaped periods of good crop yields as well as significant population growth and technological advancement in Western Europe. By the time the impacts of the Little Ice Age accelerated after these periods, Western Europe was not in the same condition it had been in the two preceding centuries. Rather than fall victim to those impacts, the continent had both the human resources and technological savvy to move forward: By the end of the sixteenth century, the Dutch and the English – both countries which had been vulnerable to earlier climate changes and their effects – had become political, military, economic, and colonizing powers. Their resilience manifested itself in the colonies they established and the resources they imported over the next century. So, for instance,

> England, largely by virtue of its maritime strength, became a world power with colonies and possessions and trading interests on a global scale ... British consumers benefitted from a growing influx of butter, meat, and staves from Ireland; silver and gold from the New World; saltpeter, spices, dyes, and textiles from the East Indies; copper and iron from Sweden; timber, potash, tar, and pitch from North America and Siberia; coffee from Arabia; tea and porcelain from China; sugar from the Caribbean; and fish from Icelandic and North American waters. British producers benefited from the market created by transplanted compatriots – a quarter million strong by 1700 – who offered a protected market for all manner of manufactured goods. Power and trading efficiency conferred material prosperity and security.[73]

Moreover, after working (albeit uncomfortably) with the English to fend off a series of attacks by the Spanish at the tail end of the sixteenth century, the Dutch increasingly exerted their economic and political muscle on the world stage, including initiating a series of military innovations to build a modern state. They "built a large professional standing army, reliably paid and strictly disciplined, that made near-optimal use of the most advanced firearms and fortification technology of the time."[74] They maintained that discipline, in part, by publishing "a code of military conduct in 1590, which was read out to recruits and to all

[72] Hoffman, *An Environmental History*, 327.

[73] John F. Richards, *The Unending Frontier: An Environmental History of the Early Modern World* (Berkeley, CA: University of California Press, 2003), 194.

[74] Ibid., 42.

troops at the beginning of each campaigning season. Special military judicial officers enforced these rules in each garrison. There were severe penalties for theft, affrays, and assaults, but death by hanging was the penalty for abduction and rape. So effective were these measures that towns actually applied to have garrisons assigned them."[75] That code preceded the Lieber Code by 250 years and reveals the shape of just war thinking as it turned to address the new shape of the military and military actions in the modern world.

This, then, is a second great irony that came with the arrival of early modern warfare: Increasingly harsh climatic conditions during the Little Ice Age may have helped to stimulate colonization during the early modern period and, with colonization, to stimulate the Christian just war tradition's transition into its early modern form. The universalization, theorization, and secularization of Christian just war thinking – whether expressed in the context of international law or revised visions of domestic law; whether manifest in a preoccupation with *jus ad bellum* and *jus in bello* criteria or a growing suspicion about the role that religious thought can or ought to play; whether emphasizing the modern nation-state as the basic political body or the mass army as the basic military one – are not simply expressions of the shift toward thinking of war in modern political terms. They are consequences of natural forces and a vast array of human reactions and responses to those forces, many of which seem otherwise unrelated to matters of war and reflections on war.

I am not suggesting that climatic events and changes determined how colonization happened or how colonization reshaped Christian just war thinking. Such a suggestion would fly in the face of the complex interplay of nonhuman natural forces and human engagements with the world, not to mention undermine arguments about causation I made earlier in this book. And, to the degree that the term "Christian" carries any weight in the phrase "Christian just war thinking," such a suggestion would also close off theological questions about the possibility that there are larger, providential, forces at work.

I am, though, suggesting that the pattern of a period of beneficial and stable environmental conditions in Europe followed by entrance into the Little Ice Age, when occurring simultaneously with a concatenation of particular intellectual, technological, political, and demographic changes in Europe, shaped not only European desires to explore and colonize non-European lands but also how that exploration and

[75] Ibid., 43.

colonization occurred and the successes (and failures) that resulted from colonization. And since colonization played such a significant role in reshaping Christian just war thinking, ignoring that pattern to focus exclusively on the concatenation of those other changes – or, worse, to focus only on political changes of the time – dramatically misnarrates how the Christian just war tradition moved through the time between the fourteenth and eighteenth centuries.

REDUX: REREADING (AND REVISING) LYNN WHITE

All of which returns us to Lynn White's essay, "The Historical Roots of Our Ecologic Crisis." If my argument above is at all compelling, it highlights a number of flaws in White's argument. The first, as I noted early in this chapter, is a matter of timing: White locates the start of the growth of those roots in the early church and suggests that it was just a matter of time before those roots, grafted to the technologies of early modernity, would nourish ecological crises. Nature-friendlier theological visions such as those of St. Francis and of the Eastern Orthodox church may offer alternative ways the church could have, but did not, go. Such ways, though, are, for White, clearly "alternatives" not only in the sense of being another way but in the sense of being an expression of alterity; they are, to White's mind, not nourished by the taproot of Christian thought.

Viewing even late medieval Christian just war thinking through the lenses of environmental history begins to reveal the way that even late medieval Christian thought was replete with a wider range of theological visions of nature than early modern Christian thought (let alone modern Christian just war thought) has provided. Francis is not an expression of alterity from the way most of Western Christendom thought about the natural world so much as he is simply one among a wide range of perspectives about how to understand the nonhuman natural world. And if this is the case, the pertinent question should not be the comparatively ahistorical, "Is Christian thought at the root of our ecologic crisis?" but the more nuanced and grounded, "What happened to Christian thought as it transitioned from its late medieval to its early modern forms such that a particular way of understanding the nonhuman natural world within the wider Christian tradition gained prominence?" One of the things that attention to the Christian just war tradition reveals is that the projects of exploration, colonization, and conquest played a much greater role in advancing such an understanding than "Christian thought," generically conceived, ever could have.

This begins to get at a second flaw in White's argument. On White's reading, technologies (such as plows and cannon) and practices (such as those associated with the scientific revolution and measuring time) interact with human thought, but in subsidiary ways. It is as if human thought provides the boundaries for such interactions: If that thought favors exploitation, it is only a matter of time before human beings develop technologies and practices through which to exploit. Yet the relations between ideas, technologies, and practices are far more complex than this and those relationships are, themselves, shaped by unforeseen (and often unforeseeable) events. Traditions of thought are far less stable and far more open to change and the influence of other forces than White recognizes. Viewing the Christian just war tradition through the lenses of environmental history reveals how half a century of beneficial weather, coming along at an opportune time in European history and followed by the advent of the Little Ice Age, both drove and shaped colonization and, therein, our modern ecological crisis.[76]

And this, in turn, gets at a third flaw in White's argument: that even as he was, helpfully, attending to the nonhuman natural world, he was still treating it in (modern) inert ways. The nonhuman natural world is acted upon rather than functioning in quasi-agentic ways, interacting with the human world such that they coadapt and coevolve. Indeed, read through the lenses of environmental history, the very term White so derogates, "anthropocentrism," begins to lose its meaning. Treating human beings as the center of things makes assumptions about center and periphery that are no more supportable in White's thesis than they are in modern thought more generally. The human world – its ideas, practices, and technologies – changes as it both shapes and is shaped by its interactions with the nonhuman natural world. War may be among the most "human"

[76] One of the perennial risks associated with attending to links between the nonhuman natural world and the human/cultural one is that such links have been used, via the arguments of environmental determinism, to reinforce claims about the superiority of some European racial/ethnic groups in comparison with all others. This chapter argues quite the opposite: Not only are such deterministic claims fallacious because they are deterministic, but also the very resources used to defend such claims actually suggest that deficiencies in European cultures arising during their transition from the late medieval to the early modern period drove the very projects of conquest, colonialization, and exploitation that environmental determinists implicitly and so troublingly valorize.

of things that *Homo sapiens* does,[77] but – as this chapter shows – even the practices of and reflections upon war are profoundly and perpetually shaped by interactions with the nonhuman natural world. White may have initiated the projects of environmental history, but those projects have since gone in ways that offer piercing correctives to his work.

The point of these criticisms is not simply – or even primarily – to challenge White's thesis and its impact. Instead, I want to suggest that the flaws in White's thesis are shared by modern interpretations of the history of the Christian just war tradition. Like White's thesis, those interpretations are insufficiently grounded in the actual and complex history of how that tradition moved through time. Like White's thesis, those interpretations are too determined by a history of ideas and insufficiently attentive to the complex ways that ideas, practices, technologies, and events interact with each other and shape the movement of that tradition through time. And, like White's thesis, those interpretations treat the nonhuman natural world as a backdrop for history rather than an actor in it, shaping and being shaped by its movements. Having focused for most of this book on one particular period in the history of the Christian just war tradition, I will turn in the final chapter toward shaping a narrative of the movement of the Christian just war tradition through time that corrects for these deficiencies.

Before doing so, however, I want to offer two further thoughts about the implications of this chapter. First, it is important to recognize how the lexicon of modern just war theories – replete with economic, political, cultural, and military terms – has served as the basis for making sense of modern war. That lexicon has enriched our capacities not only to make sense of modern wars but also to make judgments about them. It also distinguishes the modern lexicon from that of late medieval just war thought, which relied on a wider and richer – if anecdotal and occasional – language for linking war and the nonhuman natural world. Moreover, absent a deeper attention to the interactions between the nonhuman natural world and human activities, this modern lexicon lacks the terms necessary for making sense of the origins of modern just war theory, the wars of the Anthropocene, and the nonhuman natural world-related forces that have shaped those wars.

77 Recognizing that our closest primate relatives, chimpanzees, also seemingly wage war against each other. See Michael L. Wilson, "Chimpanzees, Warfare, and the Invention of Peace" in Douglas P. Fry, ed., *War, Peace, and Human Nature: The Convergence of Evolutionary and Cultural Views* (New York, NY: Oxford University Press, 2013), 361–388.

This absence has gone unnoticed during the last several centuries, largely because modern just war thinking functioned perfectly adequately in the wars of this period. Yet its functionality has diminished rapidly over the past several decades. Attempts to address this diminished functionality through, for example, the development of *jus post bellum* thinking, just peacemaking practices, giving closer attention to non-Western just war traditions, and arguments for humanitarian intervention, may all serve valuable roles in reshaping the tradition. Any attempts to increase its functionality or reshape the tradition that ignore the degree to which the political world is tethered to the natural world, though, will repeat the errors of modern just war thought. At the entrance into the environmental age, attention to the interactions between the human and the nonhuman natural worlds – especially as those interactions provoke and shape conflicts – must be a necessary feature of any responsible and responsive version of just war thinking, including Christian just war thought.

Second, there are a range of resources within the Christian just war tradition for thinking about the interpenetration of matters having to do with war and those having to do with the nonhuman natural world. This is not to suggest that the tradition should – or even can – revivify premodern ideas in order to solve the problems of the environmental age. Not only would this kind of conceptual revanchism fail to achieve the goals toward which it aimed (as revanchism always does), but also such an attempt repristinates the modern project that the just war tradition needs to get away from: separating ideas from practices and technologies and then focusing only on the former. Practices and technologies (and, therefore, ideas) change over time in unidirectional ways. Trying to move backwards in time – as romantic as it may seem – is not a viable direction to travel. Early modern just war thinkers were preoccupied with problems other than the interactions of the human and nonhuman natural world and were shaped by changing interactions between ideas, practices, technologies, and the nonhuman natural world. Their solutions don't match our problems. Likewise – and perhaps even more obviously – those solutions don't match the preoccupations and influencing forces of late medieval just war thinkers. Whatever the Christian just war tradition becomes in the environmental age, it will look like a new thing rather than an earlier one.

Instead, a primary benefit of attending to the interactions between the nonhuman natural world and the Christian just war tradition as it transitioned from its late medieval to its early modern form is that such attention highlights the way the tradition changed over time. This and earlier chapters have focused on one particularly interesting period during which the tradition flexed and shifted in response to a range of forces and events, giving special attention to the way that interactions with the nonhuman natural world shaped the tradition that shaped modernity. The next, and last, chapter will offer a briefer account of the way that interactions with the nonhuman natural world have always shaped the tradition even if the tradition has not been aware of such interactions. The tradition neither can nor needs to go backward. The more self-aware it can become, though, the better able it will be to go forward.

7

Renarrating the Christian Just War Tradition

> The praxis of mortal combat is not destructive to human sociality as such; it is simply a moment at which human sociality regroups and renews itself.
>
> Oliver O'Donovan, *The Just War Revisited*

> … nearly all academic studies, models and prognoses regarding the phenomena and consequences of climate change have been in the *natural sciences*. In the social and cultural sciences, it is exactly as if such things as social breakdown, resource conflict, mass migration, safety threats, widespread fears, radicalization and militarized or violence-governed economies did not belong to their sphere of competence.
>
> Harald Welzer, *Climate Wars*

> The context of widespread environmental collapse, along with the suffering of the most vulnerable communities and the specific problems associated with climate change, has provided the occasion for public declaration by religious leaders calling for greater moral responsibility … However, other than calling for technologically advanced societies to adopt simpler lifestyles and improve energy efficiency, there was no specific attention paid to the relatively drastic nature of the social and political changes that are necessary in order to address climate change.
>
> Celia Deane-Drummond, "Public Theology as Contested Ground"

Over the past chapters, I have explored one particular period in the history of Christian just war thinking. Along the way, I have tried to make two related arguments. First, the Christian just war tradition as it transitioned from its late medieval to its early modern social imaginaries played a significant role in influencing modernity in the West. The structures of modern thought, shaped as they have been by emphases on autonomy, universalizability, instrumentalization, and immanence, have inhibited our recognition of the significance of the tradition's role by downplaying the significance of religious perspectives, the pliancy of ideas in interaction with each other, the specificities of cultures, and the value of traditions in general. Under-regarding the tradition's significance has not only distorted the tradition but also hidden from view the influence of the Christian just war tradition on modernity. That is, early modern Christian just war thinking not only mirrors the patterns and emphases in modern thought more generally but also shapes those emphases. If we want to understand modernity – and, especially, modern understandings of war and its problems – it will help to give closer attention to what happened to the Christian just war tradition during this period.

Second, I have suggested that understanding the way the Christian just war tradition transitioned from its late medieval social imaginary into its early modern one may help us not only understand the environmental crises we currently face and their attendant connections to violence, but also begin shaping resources for addressing war in a warming world as we transition into the environmental age. Among those resources that will come with a closer attention to environmental history will be an increased recognition of the way that the nonhuman natural world interacts with the world of war. We are already bumping up against the limits of modern understandings of the just war tradition as we face conflicts that it is poorly shaped to understand, let alone respond to. At a time when we need a wide range of diagnostic tools at our disposal to address the new types, causes, exacerbators, and understandings of war we face in the twenty-first century, shaping such tools will be vital to meaningfully responding to such matters.

Implicit in these two related arguments is a third: that attending to one specific transition by the just war tradition as it moved from one social imaginary into another may help us make sense of the movements of the tradition, especially when it is in flux. This chapter moves toward advancing that argument. So where the previous chapters have taken up the first two arguments through attention to one particular time period, this chapter steps back to explore the more general implications of those arguments for narrating the history of the Christian just war tradition writ large. If those first two arguments are compelling (or at least sufficiently provocative), they motivate this step back. The former argument funds an attempt to narrate the process through which traditions move through time in order to better understand the way traditions – and the Christian just war tradition, in particular – change. And the latter argument provides at least one practical reason to renarrate the tradition; namely, to find construction materials that are useful during a time of conceptual bricolage at the entrance into an environmental age.

Along the way, this collection of arguments suggests the need to revise the way human beings understand their relationship to the nonhuman natural world (and, concomitantly, the way the nonhuman natural world relates back to us). At the end of the day, in fact, I want to suggest that any dividing line – no matter how bright or how dull – between the human and the nonhuman natural world is far harder to draw than we might think.[1] Human beings are, to use a word promoted by James M. Gustafson, best understood as "participants" in systems that are far larger than themselves: Human beings "are participants in larger spheres of interaction, and even in the development of the natural world. When we attend to participation in interaction with wholes, we acknowledge the prior reality of arenas of activity to which persons and institutions respond and exaggerated notions of human creativity (as if we somehow created the worlds of which we are a part *ex nihilo*) are avoided."[2]

Gustafson's huge project – to describe what it means to be a participant in a world that is best viewed from a theocentric perspective – reaches far beyond what I am trying to take up in my own large project of thinking about the implications of climate change for wars in the twenty-first century. Nevertheless, his vision of human being as participant is congruent with the argument I am making here: that a proper (that is,

[1] Perhaps the effects of pollution, the impacts of climate change, and the catastrophic loss of biodiversity, all treated as the inescapable residues of the "civilizing process," help us see how much human beings are currently reshaping the world. Just as significantly, scientists are increasingly naming the degree to which the nonhuman natural world is shaping us. So, for instance, conservative scientific estimates suggest that perhaps half the cells that constitute each human body do not share that body's DNA and that many of those "foreign" cells that compose the bacteria in our gut, the protozoa in our blood, and the mites on our skin are either harmless or beneficial to us. Said differently, giving attention to our microbiomes reveals that we always carry the nonhuman natural world with (and within) us. We are literally teeming with life. Moreover, especially as the global human population now exceeds seven and a half billion, travels at unprecedent rates, and relies on the wonders of chemistry in shaping its ease, how we collectively treat our respective microbiomes can make a huge difference in our health – and right now, many of our eating, sanitation, and medical habits are harming the wilderness within as well as beyond us.

[2] James M. Gustafson, *Ethics from a Theocentric Perspective*, Vol. 2: *Ethics and Theology* (Chicago, IL: University of Chicago Press, 1984), 13. Not only does Gustafson's two-volume *Ethics from a Theocentric Perspective* make this argument, but also much of the work he did after it (e.g., *Intersections: Science, Theology, and Ethics* [Cleveland, OH: Pilgrim Press, 1996]; *An Examined Faith: The Grace of Self-Doubt* [Minneapolis, MN: Fortress Press, 2004]) built constructively from it.

historically more accurate, theologically more coherent, and morally more functional) telling of the history of the Christian just war tradition recognizes the interpenetration of human and nonhuman forces in shaping that tradition. Christian just warriors are not simply participants in a tradition; they are also participants in "larger spheres of interaction, and even in the development of the natural world," and they ought to see themselves as such.

Likewise, they should see their tradition as also part of those "larger spheres of interaction, and even in the development of the natural world." If in earlier chapters I have argued for the importance of a particular stretch of time (that during which the Christian just war tradition transitioned from its late medieval to its early modern form) for understanding modernity and the modern just war tradition's lack of attention to the nonhuman natural world, this chapter partially inverts that work by asking what it would mean for the Christian just war tradition to understand itself as embedded in those "larger spheres of interaction."

Such a question may seem destabilizing. If traditions are understood as accumulations of culture (that is, if they are composed wholly of the products of human interactions and the artifacts that human beings have brought into existence), suggesting that traditions interact with the nonhuman natural world rather than simply draw resources from it is strange, and countering such an understanding of tradition may threaten to introduce a kind of "nature vs. culture" debate into historiography that has long proved unhelpful in other disciplines. If traditions are to be understood as internally coherent collections of ideas, customs, artifacts, and institutions within particular communities, then situating those ideas, customs, artifacts, and institutions within a wider range of social and natural forces that are interacting with them highlights how porous traditions actually are and countering such an understanding of tradition may threaten to introduce equally unhelpful essentialism/constructivism debates into historiography. And if traditions are to be understood as manifesting stability and continuity across time, then recognizing their transience and discontinuities may threaten to introduce yet another unhelpful debate – that between nominalism and realism – into historiography. In short, treating traditions as fundamentally conglomerate, porous, and dynamic would seem to undermine all the qualities that make traditions worth our attention in the first place.

Yet should it? Might treating traditions as (at least relatively) homogeneous, impermeable, and stable reveal more about continued Western beholdenness to Platonic sensibilities than actual attention to the way traditions exist and function? In what other natural or social science is there not at least as much attention to questions about how things are shaped and changed as there are to what constitutes and identifies a thing? Giving attention to the conglomerate, porous, and dynamic realities of traditions isn't so much a project of undermining their meaning as a way of including them as the very types of things worth researching, and recognizing such qualities doesn't so much dissolve the idea of "tradition" as help to locate it within wider spheres of scholarly engagement.[3] If anything, such attention-giving offers traditions credibility and weight. And, at least from a certain theological perspective, it suggests that they too exist contingently within the providential work of God.

The above few paragraphs obviously don't begin to address the complicated questions that recognizing the complexities of conglomerate, porous, and dynamic traditions entail: How do traditions begin? What drives their change? What features of traditions and/or their component parts stabilize and destabilize them? How does one go about doing comparative work on traditions? To what might we look in naming more and less identifiable and perduring features of traditions? How do we measure their rates of change? Can traditions change in both continuous and discontinuous ways? How are spatial (center/margin), temporal (origin/consequence), relational (enmeshed/autonomous), and linguistic (argument, conversation) metaphors clarifying and distorting in making sense of traditions? Which social sciences (e.g., anthropology, history, sociology) and schools of thought within those social sciences are especially attentive to and/or helpful in making sense of traditions? Do all traditions move through time in recognizably similar ways? These are big questions.

Perhaps no scholar has given as much concentrated attention to the concept of a tradition as Edward Shils. In a series of books and articles – most notably *Tradition* – the distinguished service professor in the Committee on Social Thought and in sociology at the University of Chicago not only offered a capacious if functional definition of the term ("Tradition – that which is handed down – includes material objects, beliefs about all sorts of things, images of persons and events, practices,

[3] Brandom locates not only the source of attention to the dynamic qualities of the Western philosophical tradition but also the beginning of the way that our understanding of a tradition affects the dynamics of that tradition in Hegel, after whom both dynamism and reflexivity are foundational for understanding any history. I am, in a minor way, attempting to do the same in this chapter. See Robert Brandom, *Tales of the Mighty Dead: Historical Essays in the Metaphysics of Intentionality* (Cambridge, MA: Harvard University Press, 2002), esp. 14–15.

and institutions")[4] but also explored how traditions endure and why they change. Particularly pertinent to my project and this chapter is his recognition that traditions change due to endogenous and exogenous factors. "A tradition does not change itself," he writes. "It contains the potentiality of being changed; it instigates human beings to change it."[5] Endogenous forces include those driven by the creative power of those within a tradition to imagine it differently or to see something in it that had hitherto been overlooked; the significance of particularly charismatic figures to bend a tradition to their purposes; and resistance to either particular aspects of a tradition, the whole of a tradition, or the very idea of "tradition" to reshape arguments about a tradition. Exogenous forces include societal encounters with other traditions, changes in the centers of power within a society that cause traditions to expand or contract, the expansion of traditions into other societies, and changes in circumstance.

Shils' pluralization of the forces that reshape traditions is helpful but limited in three ways. First, his method primarily involves drawing from particular historical events and movements within particular traditions to make claims about traditions more generally. That is, he makes his argument by accumulating examples. But, at least in this instance, the plural of anecdote may not be data: Shils' approach has the potential to ignore counterexamples or conceal patterns. Second, even as he attends to the forces that can reshape traditions, Shils' approach offers little insight into the directions that such forces typically push. Much as he attends to human agency, for example, he leaves human psychology almost wholly unexamined (which, I suppose, isn't wholly surprising: many schools of psychology reveal just how little agency we have over even our own thoughts and actions). Why do societies so often retain particular practices associated with a tradition over time even though the meaning of those practices according to those who engage in them may change? Why are some persons within a tradition more likely to cling to it than others and how do their location(s) and power(s) within a society impact their commitment to the tradition? How do those within particular traditions understand which charisms are worth valorizing such that charismatic figures can change traditions? Third, Shils, like most modern authors, views traditions almost wholly from within an anthropocentric framing: Natural events – whether sudden ones such

[4] Edward Shils, *Tradition* (Chicago, IL: University of Chicago Press, 1981), 12.
[5] Ibid., 213.

as earthquakes and tsunamis or gradual ones such as climate change and species extinctions – simply do not register as significant exogenous forces even though such events can bring about dramatic change.

These limits do not necessarily undermine Shils' work. If anything, they highlight directions for future research on the stability and mobility of traditions. Yet almost three and a half decades after the publication of Shils' book, Elisa Grimi can still legitimately write: "It must be observed that there are few studies and little research still underway on the issue of tradition and innovation."[6] I hope neither to address all the questions that Shils takes up – let alone those he leaves unanswered – nor to fill the lacuna Grimi points to. Leaving much of that work to other scholars,[7] I want at this point only to highlight a heuristic claim I made in *Christian Pacifism for an Environmental Age*: that questions exert a kind of centripetal force on traditions whereas the answers given to those questions exert a kind of centrifugal force. That is, the questions central to a particular tradition (about what has meaning and value, about how to understand and approach those outside the tradition, about how to

[6] Elisa Grimi, ed., *Tradition as the Future of Innovation* (Newcastle upon Tyne: Cambridge Scholars Publishing, 2015), 2.

[7] Among other texts from which I have learned (if not with which I have always agreed) in exploring the idea of a tradition beyond those named in the Introduction (i.e., work by Alasdair MacIntyre, Jeffrey Stout, and Robert Brandom), I would include Charles Mathewes, *Evil and the Augustinian Tradition* (New York, NY: Cambridge University Press, 2001); Stephen H. Watson, *Tradition(s): Refiguring Community and Virtue in Classical German Thought* (Bloomington, IN: Indiana University Press, 1997) and *Tradition(s) II: Hermeneutics, Ethics, and Dispensation of the Good* (Bloomington, IN: Indiana University Press, 2001); and even Josef Pieper, *Tradition: Concept and Claim*, trans. E. Christian Kopff (Wilmington, DE: ISI Books, 2008), not to mention a wide variety of texts and book series in which questions about what makes a "tradition" have been raised in fields as varied as history, anthropology, philosophy, political theory, and religion.

Among the more invigorating theological reflections on the idea of tradition are Peter Casarella's "'A Healthy Shock': Tradition and the Epiphany of Beauty" and John Milbank's "After and Beyond MacIntyre," both in Grimi, *Tradition as the Future of Innovation*, 220–242 and 242–254, respectively. Milbank's arguments, growing out of his use of Rémi Brague, continue his ongoing argument with modernity and his radical orthodoxy apologetic for Christianity. As such, the argument invigorates but suffers from the same frustrations and failings of his other work. Casarella's argument, though, suggests that discerning the coherence and significance of traditions has more to do with a kind of aesthetic/theopoetic understanding of them than a logical or moral one and, as such, potentially aligns better with the way that Stout's and Brandom's neopragmatist/ Wittgensteinian approach finds stability in a kind of family-resemblance understanding of what links potentially disparate things and their ways of reconstruing traditions rather than disassembling them (not to mention my use of Charles Taylor for similar purposes).

encounter and engage change) perdure over time even as the tradition itself changes. Questions stabilize a tradition.

Answers to those questions given by members of a tradition, meanwhile, are more closely bound to the contexts in which those questions are answered and, as such, less likely to maintain their viability and significance as contexts change. Answers drive change. So whereas tradition-specific questions reveal continuity within that tradition and help to stabilize it, the variability within the tradition's answers to its questions reveal its discontinuities and direct its change. So, for example, in that book I argued that questions of fidelity are central to the Christian pacifist tradition even if understandings of the one to whom they were faithful, what it means to be faithful to that one, and how to practice being faithful to that one changed over time.

The Christian just war tradition, I would argue, is oriented less toward questions of fidelity than questions of justice; hence, its name.[8] How shall we advance political ends while also maintaining moral commitments? Who counts and how much do they count in the weighing out of the costs and benefits that accrue in pursuing actions that can involve the use of coercive force? What do the laws of the land (or of many lands) demand in this context? If the Christian pacifist tradition tends to orient its questions of fidelity around its understandings of what God is like and what he asks of believers, the Christian just war tradition orients its questions of justice around its understandings of what people are like and what they can ask of each other. Answers to such questions – as I hope this book has made clear – can change, but the questions remain central.

This heuristic about centripetal questions and centrifugal answers is inadequate if not couched in larger arguments of the sort that Shils makes about the range of forces that cause traditions to endure and change; it is shallow if not linked to deeper philosophical arguments of the sort that Brandom and Stout make about how we recognize traditions as they change; and it is simplistic if it refuses to recognize comparatives and

[8] There has been a kind of glissading in the language of just war thinking over the past century in which the word "just" in the phrase "just war thinking/tradition/theory" increasingly came to bear the connotations of "justified" as much as "justice." While questions of justification certainly need to be addressed, and etymologically the words "justify" and "justice" are linked, it is worth keeping the two terms separable and recognizing that the word "just" truly does orient itself around questions of justice. For an exploration of why this is (that is nonetheless attentive to the kinds of practical reasoning through which to make judgments about war), see Oliver O'Donovan, *The Just War Revisited* (Cambridge: Cambridge University Press, 2003), esp. ch. 1.

matters of degree (so, e.g., we need to note that Christian pacifists care about justice and Christian just warriors care about fidelity). It does, though, achieve three things. First, it helps make sense of the kinds of questions that participants in the Christian just war tradition (such as Vitoria, Grotius, and the US Constitution) were at least implicitly asking during a time of dramatic change: How do we treat others, how do we understand ourselves, and what kind of enduring society can people like us put together? Second, it helps to position the way continuities and discontinuities within the Christian just war tradition have located themselves vis-à-vis each other as the tradition has moved through time: Answers to such questions may have changed dramatically and even in ways that feel discontinuous, but the questions endure. And, third, it helps to open the tradition to an attentiveness to its interactions with the nonhuman natural world as it has moved through time: Interactions with the nonhuman natural world during times when that world was changing shaped new answers to abiding questions.

Keeping these arguments and caveats in mind, the latter half of this chapter takes up the project of redescribing the history of the Christian just war tradition in just such a frame. First, though, I want to offer a theological gloss on these arguments and caveats. This is, after all, an attempt to think about the *Christian* just war tradition and, as such, warrants at least some attention to the question of how such a redescription might be identifiably Christian.

JUST WAR THINKING AS MORAL PRECIPITATE: SHAPING A THEOLOGY OF "TRADITION"

In the Introduction to this book, I offered a conventional narrative about the history of the Christian just war tradition and then disassembled it, suggesting that such a narrative is problematic on a number of historical and theological grounds, not the least of which is its reliance on accretionary and "great man" ways of telling history.[9] This chapter reassembles that narrative for reasons having to do with historical accuracy, theological acuity, and moral utility. I want to more truly portray the

[9] To be clear: Much as in previous chapters, this chapter makes theological arguments not to replace arguments that are accessible to a wider range of readers but to supplement such arguments for those who, like me, find theologically tinted lenses to be a useful and faithful way of seeing the world in its complexity. It is hoped that even those readers who are not especially interested in the wars of the Anthropocene will find this renarrated history of the tradition helpful and instructive.

way in which the tradition has actually moved through time, to offer a narrative that coheres with the way that at least some Christians can make sense of this part of their own history, and to do so in a way that reveals the continuing relevance of the tradition for the wars of the Anthropocene. In the twenty-first century, the Christian just war tradition may find space to be both a *tradition* and *Christian* in ways that it has not been fully able to do during modernity.

Repeating the argument I made in the Introduction, my starting point for telling the history of the Christian just war tradition is this: Some form of just war thinking – however limited – is endemic to all stable societies (i.e., societies that persevere by advocating an idiomatic and dynamic mix of order, justice, security, and freedom, which are the values through which they can maintain both stability and flexibility over time). Correlatively, just war traditions consist of the arguments, practices, technologies, institutions, and their interactions with each other and the nonhuman natural world that such societies make regarding the prudential and moral considerations about when and how to fight.[10] As such, just war traditions must always be understood in the plural because (a) there has always been more than one stable and relatively secure society in the world (at least since the agricultural revolution); (b) we have no evidence that there has ever been – or ever will be – a permanently stable society; and, (c) societies' arguments change organically over time as the way they understand and order social values change and as those changes are shaped by forces beyond the powers of their collective intellects, all of which lead to the Reformation and/or creation of new societies. Most of the time in stable societies, these values and their ordering are assumed. They float below public consciousness, shaping not only the laws of the land but also the way citizens engage each other and understand the world around them. They are formed by and anchored in their social imaginaries, to use Charles Taylor's phrase.

[10] I make no claims about the ontological origins of just war thinking such as natural lawyers might make. It isn't beyond my imagining that some societies never engage in just war thinking. My claim is more constrained: Societies that endure over time have some tradition of just war thought because the mix of prudential and moral considerations that stabilize and dynamize those societies can be applied to questions about how to engage other societies. Rule of law, a broadly shared sense of place in the world, an ordered way of creating and distributing material goods, the means to communicate matters of social importance across distances and generations: where such things exist, questions about whether, when, and how to fight will follow.

Stable societies change over time, though, and are more likely to change in dramatic ways that are followed by long stretches of stability than to change in constant and gradual ways. During periods of dramatic transition, the social imaginaries that shape societies become less coherent and are, therefore, more likely to rise toward public consciousness and drive endogenous forces for change. They are analyzed, reshaped, and given concrete expression, therein producing new (or significantly modified) social imaginaries.[11] Such periods of transition from one social imaginary to another, which I am calling "axial periods,"[12] are the occasions during which the arguments that partially constitute a just war tradition become more explicit *as* arguments and, through those arguments, a particular just war tradition – including its practices, structures, and technologies – is reframed. These axial periods – which are, in modern thought, unduly associated with the scholars who live through and participate in shaping them – do not drive the advancement of a just war tradition so much as they do its modification through argument. That is, just war thinking evolves in light of the evolution of the societies in which it is embedded and, in particular, the evolution of ideas about self, state, and other in those societies as those ideas shape and are shaped by practices, structures, and technologies and interact with the nonhuman natural world.

The concrete expressions of just war thinking – things such as modern just war criteria – exist as precipitates of a tradition arising during a particular time and within a context of particular social imaginaries. Moreover, as social imaginaries change, the precipitates increasingly reveal their historical context and become less able to adequately address the concerns of new societies about whether, when, and how to fight. Some precipitates (such as just cause) continue to exert a significant pull over a series of transitions in the social imaginary, even if the way they are understood changes. Some precipitates (such as legitimate authority)

[11] The analog within evolutionary biology is, perhaps, the theory of punctuated equilibrium rather than gradual change over time; the analog within philosophy of science might be found in Thomas Kuhn's structure of scientific revolutions (or, better, within the work of Imre Lakatos, since everything doesn't change all at once), and the analog within philosophy of language might be found in the work of James Boyd White, particularly *When Words Lose Their Meaning: Constitutions and Reconstitutions of Language, Character, and Community* (Chicago, IL: University of Chicago Press, 1984).

[12] My use of the term "axial periods" needs to be distinguished from the "axial age" that Karl Jaspers coined to describe the flowering of civilizations between 800 and 200 BCE. The shared component between my term and his – axial – simply refers to a historic pivot point when societies move in new and unanticipated directions.

recede from priority. And still others (such as discrimination) grow dominant. All of them, though, will eventually be reshaped or displaced by new thoughts from new contexts as those thoughts are shaped by contexts in which interactions with and perceptions of the nonhuman natural world are likely to play an important role.

Most of the points I make here – that societies change over time in nonuniform ways; that ideas inhabit societies rather than standing above them; that traditions are never univocal; that arguments about what matters in societies are more vocal when the core ideas of a society are felt by its members to be in flux; that social flux has to do not only with tensions between ideas but also with the developments of new technologies and new social practices; that when societies change, the ideas that inhabit them or the structures and practices that make those ideas coherent can change as well; and that interactions with the nonhuman natural world play a role in those changes – have been developed elsewhere, including in scholarship not related to matters of war and peace and in my last book. This chapter functions as an invitation to imagine another way of thinking about Christian just war thought in light of these points and it has, as its eventual end, the goal of situating this book's arguments about war and the nonhuman natural world within a larger narrative about how a specific set of moral and prudential questions might be understood and applied over the next several decades and their wars.

I have, to this point in the chapter, been emphasizing the discontinuous/revolutionary character of how traditions change over time. I have done so for three reasons. The first reason, as noted in the quote from Grimi above, is that discontinuity has been underemphasized by many scholars writing on the idea and history of "tradition" – which may reveal the degree to which their arguments, though critical of modernity's approach to understanding history, participate in that approach by emphasizing the accretionary, discrete, and pluralized characteristics of traditions.[13] The second reason for emphasizing the disjunctive character of how traditions change over time is historical. This approach does a better job of accounting for the way history – and, especially, the history of discrete societies – moves forward in spurts and starts as a result of the advent of external pressures (new conflicts, new neighbors, changing environmental conditions, etc.) and internal changes (new technologies

[13] To say that differently, much of the criticism directed at MacIntyre's concept of "tradition" (an argument extended through time) had to do with his claim that reason was tradition-dependent. My criticism, meanwhile, is that he never adequately explained how he thought arguments worked over time.

and practices, the creation and dispersal of cultural memes and particular schools of thought, the arrival of new charismatic leaders on the scene, etc.). Labeling particular ages – the Hellenistic, the medieval, the industrial, the information, and so forth – and the trends/ideas that characterize them is meant to do more than provide useful mnemonics to help schoolchildren learn history; it is meant to suggest that there are identifiable characteristics that mark certain periods of time as congruent with each other in a way that they are not congruent with the periods that come before and after them. Things change, but they do not change at a constant rate and the changes do not allow succeeding periods to be wholly congruent with their preceding ones.

Third (and finally), I have emphasized the discontinuous/revolutionary character of how traditions change for theological reasons: because I believe such an emphasis better recognizes the way God works in time, as attested not only in Scripture and the history of the church but in a theological vision of time marked by disruption and revelation, tragedy and hope, the transience of all things and the permanence of God. Whether one personally believes in God and the work of God in history or not, a tradition such as the Christian just war tradition that binds itself to a religion ought also to bind itself to the way that religion thinks and talks. As such, I offer this theological reason for emphasizing the discontinuous/revolutionary character of the movement of traditions through time because it coheres with the way Christianity understands and expresses itself. And Christianity thinks in terms of transformation. The character of the transformation brought on by God includes loss and gain, death and rebirth, the sufferings of sin and the joys of divine presence, all as crucifixion and resurrection reverberate through time. For Christians, God's saving work can feel so indirect and ironic because the redemption of time looks less like progress or decline than like death and resurrection, and time's end is the promise of eternity – albeit an eternity that continually impinges on the present. To live in history is to live in light of a longing for something that is never fully realized while being regularly and surprisingly touched by blessing and caught by loss. That is, history, theologically construed, is on a kind of pilgrimage, and pilgrimages include journeys onto peaks and into valleys, surrendering things that are precious and receiving gifts along the way. Traditions are marked by discontinuity and irony because, short of the consummation of all things, everything is marked by discontinuity and irony.

Likewise, this is also a better way for the Christian just war tradition to understand how God works in space. Troublesome and frequently ambiguous biblical attestations of theophanies aside, most of Scripture and the vast majority of human experience suggest that God's actions are neither obvious nor clear and they are almost always mediated through other agents and forces. The words of Isaiah 55:8 are, in almost all instances, wholly appropriate: "For my thoughts are not your thoughts, nor are your ways my ways, says the Lord." Moreover, divine action is always interpreted action; as such it is to be pondered, doubted, argued about, and otherwise examined. Those who would affirm that God does, indeed, act in particular spaces can only faithfully do so in ways that are provisional and alert to the possibilities for new interpretations/revelations. Theological interpretation, because it is interpretation, is always at least partly apophatic; even in its most prophetic forms, it still carries within itself a whispered "and yet these words describe neither the divine nor the actions of the divine." And while it may be easy to succumb to the temptations of easy negations – a kind of apathetic apophatic, as it were – such capitulations are but the shadow side of falling into the even easier and more dangerous temptation of rejecting the possibilities of negation built into transformation.

Partly because divine actions are always interpreted actions, the quality of interpretations improves as interpreters are able to take more pertinent data into account. As such, attention to God's activity in particular spaces means attention to as many of the forces through which divine action may be mediated as possible (including forces of the nonhuman natural world). And partly because divine actions are always interpreted actions, the quality of interpretations improves as interpreters attend to the ways God seems to have acted in the past and to the possibilities that come with God acting in new ways in the present and future. As attested in Scripture and in Christian theological interpretations of human experience, the same God creates and redeems, judges and saves, governs and transforms – and God does these things in mediated ways, including through the ministrations of human beings. It is for this reason, I think, that the Christian just war tradition has been and should continue to be unwilling to absolutely reject acts of violence. For this tradition, the issue isn't whether such acts are antithetical as a response to God's call that we live holy lives; it is whether we can preclude the possibility of a kind of divine violence in the kingdom of God, given the complexities of God's actions and the reality of their being mediated through human actions.[14]

[14] For a more extended set of reflections on divine violence, see Ted A. Smith, *Weird John Brown: Divine Violence and the Limits of Ethics* (Palo Alto, CA: Stanford University Press, 2014).

Moreover, interpretations of divine action within such a range of possibilities must include attention to the possibility that God's actions are being mediated through events in the nonhuman natural world, therein giving serious – if fraught – attention to events that have euphemistically been termed "acts of God."[15] If nothing else, attention to the interpreted qualities of divine action can at least highlight the way the forces that interact in particular spaces during particular points in time are always multiple, complex, and dynamic. Contra the claims of most modern thinkers, it is secular approaches to life in the world that reduce, simplify, and rigidify. As God works in time and space (linked as they are, perhaps, in a kind of super-physical theory of special relativity), the modes, expressions, durations, and purposes of such work manifest as both constancy and change, mediated through both human and nonhuman actions.

These last several paragraphs are intended to say neither that history consists of a series of restarts in which traditions disappear at axial periods in history, nor that attention to divine action must be a part of all assessments of the forces that shape particular times and spaces. At the same time that attention to the forces that shape traditions and their boundaries highlights the distance between when ideas, practices, and technologies come into existence and when they're being applied, traditions also serve as a conservative and stabilizing influence during times of transition. They do so by serving as a primary resource for finding, shaping, and laying the building blocks out of which new ideas and new practices emerge. Indeed, this is the great strength of MacIntyre's understanding of traditions as arguments extended through time: The very multivocality of traditions makes them sufficiently malleable to be reshaped for use in new times and places (and, as is often the case, to beget new subtraditions where those participating in the arguments prefer leaving to winning).[16] They may not look or sound the same,

[15] Such attention leads, inevitably, to questions of theodicy: how to interpret divine actions in the face of suffering. Whole libraries' worth of books and articles have been written on such questions. This is not the place to lay out my own thoughts on the matter (I'm fairly Augustinian in approach); I have done so, reservedly, in *Confessing Christ in the 21st Century* (Lanham, MD: Rowman and Littlefield, 2005), particularly in the chapter on sin in that book. I would note here, though, that modern attempts to address theodicy questions are ill-suited for the kind of slow-moving tragedies that constitute our transition into the environmental age. For a few brief reflections on this, see my "Theodicies for Slow-Moving Tragedies," *Sci-Tech†: The Journal of the Presbyterian Association on Science, Technology, and the Christian Faith* 28.3 (August 2019): 1–4.

[16] Alasdair MacIntyre, *After Virtue*, 2nd ed. (Notre Dame, IN: Notre Dame University Press, 1984), 222.

but Tradition 2.0 (or following) bears sufficient resemblance to its predecessor to be demonstrably part of the family. All of which suggests that traditions perdure precisely because they are capacious enough to contain and shape differences, though it is this very capaciousness that makes possible their disjunctive character when viewed over long swaths of time.

There are also both historical and theological reasons for attending to the stability of traditions over time. The historical reasons need no rehearsal here; most of those who think about and write on traditions attend to them because they manifest themselves in broadly coherent ways across time and we value them precisely because they function as a stabilizing force throughout. The theological reason to recognize and value traditions is that they are signs that the God who acts in history is One and God's desires for the world's welfare are consistent. Indeed, part of what this stability suggests is that the disjunctions, ironies, and indirections that constitute history's pilgrimage have more to do with our blindness about and resistance to that welfare than to any sense of capriciousness on God's part. God does not leave us bereft of resources or of Godself. Stability, as much as discontinuity, marks the indirect way in which the faltering and impure movements that constitute human projects are contained within and given momentum by the saving work of God.

Indeed, to get a sense of how traditions work, we need to attend to both discontinuity in and stability over time – and to do so in the kind of bifocaled way that lets us see both with only the slightest shift in our angle of vision. Only in this way can we keep clear in our minds that traditions are not so much *things* we study as *activities* in which we participate and, since it is we who are participating in them, we are carrying our own sensibilities, visions, possibilities, and failings into them – all the while remembering that they are activities that are simultaneously shaping us. A bit of honest self-reflection on our parts will help us be alert to the discontinuities and constancies, the ironies and possibilities, in our own lives. It will also, though, highlight our need to treat our lives with both tenderness and rigor, so we might learn to treat traditions as in need of both judgment and grace. As Charles Mathewes notes,

> [w]e inherit our ability to reflect from our elders, and we come to see that their tools only imperfectly fit the problems we face. Hence, unless we wish totally to jettison our minds, we find ourselves compelled to work through that inheritance, accepting its imperfections (and, furthermore, accepting that our own solutions will cause our descendents similar problems), and this acceptance

involves, indeed just *is*, in part, forgiveness. This is not false piety; it is simply the condition of being responsible in ones thought to one's predecessors.[17]

Learning to forgive traditions, we learn how to read them, to appreciate them, and to not hold them too tightly. And, at least from a Christian perspective, we learn a bit more about how God works in history so that we can learn something about how God works in us as well. Toward that end, though, we treat traditions as shaped by the twin trajectories of discontinuity and stability.

INTERPRETING THE SIGN THAT IS THE "GREAT MAN" MYTH OF HISTORY

How do the twinned trajectories of discontinuity and stability within traditions undermine the conventional narrative of the Christian just war tradition? And how does the inclusion of attention to the nonhuman natural world within the multiplication of forces shaping traditions at any given time manifest itself in a more accurate and helpful telling?

Let us return to the "great man" myth that animates the conventional narrative. While that narrative is multiply problematic, its very success – at least heuristically – is a sign worth interpreting. Why has it been so successful? Assuming there is more going on than individual brilliant minds at work, what is it about those great thinkers that makes us return to them so persistently? Aristotle, Cicero, Augustine, Gratian, Thomas, Vitoria, Grotius, Ramsey (and, in several instances, a few of their contemporaries): even recognizing their brilliance and the continued power of their thought, perhaps paying closer attention to what is going on around them may not only help make sense of why we have repeated a problematic myth but also help us think about how to use their insights in the future.

First we should note the way each of them is being shaped by their changing contexts as much as they are shaping them, living during disjunctive periods within the history of the Christian just war tradition over which they have little to no control, and trying to use the increasingly stale language of that tradition to say something new. Note, also, the way such "great thinkers" regularly lived in contexts in which climatic forces and shifts in human interactions with the nonhuman natural world were on display. All of these thinkers, their ideas, and their society's practices,

[17] Charles Mathewes, *Evil and the Augustinian Tradition* (Cambridge, UK: Cambridge University Press, 2001), 69.

structures, and technologies, as well as interactions with nonhuman natural forces, could warrant individual chapters in a different book (much as I gave closer attention to only Vitoria and Grotius in this book), so what follows will do justice neither to the subtlety of their ideas nor to the complexity of their contexts. Yet perhaps the following paragraphs can at least make the case that the "great thinkers" of the just war tradition lived during periods when their worlds were in the throes of especially significant changes, including relationships between the human and the nonhuman natural worlds, and that their "great thoughts" are products of their time as well as their wisdom. In what follows, then, I am not so much trying to undermine the significance of just war thinkers within the history of the Christian just war tradition as to challenge the way the conventional narrative of that tradition understands them and, concomitantly, itself. That is, I want to reinsert them into a history that is rife with social and natural forces impinging on each other and on them and, therein, argue for a way of narrating the tradition that better recognizes its complexity, highlights its dynamism, locates its significant claims, and, primarily, funds its utility. Following (and slightly paraphrasing) Robert Brandom, I want to attend to the way that "cultural formations such as the Christian just war tradition, like self-conscious individual selves, exhibit the peculiar freedom that consists in having what they are *for* themselves be an essential element of what they are *in* themselves" because "[t]he way we understand and conceive what we are doing affects what we *are*, in fact, doing."[18] That is, in what follows, I am intentionally reconstruing the tradition because making sense of its past is tantamount to shaping it for our future.

ARISTOTLE AND GREEK PERCEPTIONS OF THE NONHUMAN NATURAL WORLD

Aristotle wrote late in the period of the flowering of Greek thought, following not only directly after Socrates and Plato but also about 200 years after Thales and the Pre-Socratics founded Greek philosophy, contributing toward reshaping Greek society. Although it is true that Aristotle's political thought is recognizably provincial, it is also the case that his thought brought together metaphysics, aesthetics, ethics, physics, and logic into a newly (and nearly) coherent whole, therein paving the way for something that not only could recognizably be called "Greek

[18] Brandom, *Tales of the Mighty Dead*, 15.

culture" but also was exportable to the rest of the Mediterranean – as Aristotle's student, Alexander the Great, proceeded to do.

During the time of Aristotle, Greek political thought and Greek culture became integrated: The *demos* not only arose but also provided the mix of stability and dynamism that allows for cultural flourishing – which, though generally celebrated, was still something of a mixed bag as it carried within it not only assumptions about the superiority of Greek culture in comparison with the surrounding cultures but also a kind of evangelical (and, with Alexander, martial) zeal to reshape the world in the image of Greece. As Sheldon Wolin notes, "[e]ven before the death of Aristotle in 322 BC, the classical concept of the 'political' was being undermined by a new set of conditions. The emergence of the Macedonian Empire in the fourth century inaugurated an era of large-scale organization which later reached its fullest expression in the Roman world-state."[19] There had been empires before this point (Egypt, Babylon, Persia, etc.), but these empires had created vassals of other countries while, for the most part, leaving those countries with their own cultures in play. The Macedonian Empire aimed to change other countries' cultures (an approach that reveals that they believed that persons in other cultures had the capacity to be converted), expressing the idea that Greek thought could reside elsewhere (an approach that therein insists that "elsewhere" be preserved and maintained in order to be livable). Not to take away from Aristotle's genius, but he could write what he did because he had learned from those who preceded him and he was writing in a context in which that learning was undergoing explosive geographic growth and transformation.

Aristotle wrote during a period of relative climatic stability during which ancient Greece experienced mild and rainy winters and hot, dry summers (though the fourth century BCE was marked by a comparatively colder and wetter period than the previous few centuries had been). That he wrote about volcanic eruptions (which occurred repeatedly, including on Mt. Etna, during his lifetime) and weather patterns (including developing reflections on evaporation, cooling, and cloud formation) signals his interest in climatic activity; that he did not write about sudden changes to climate signals its stability.

More importantly, though, he also wrote during a time when human perceptions of the nonhuman natural world were changing; indeed,

[19] Sheldon S. Wolin, *Politics and Vision*, expanded ed. (Princeton, NJ: Princeton University Press, 2004), 64.

some of those changes are the products of his own reflections. In the century prior to Aristotle, Greek scholars such as Parmenides and Hippocrates were reshaping understandings of the relationship between human beings and the natural world. Aristotle continued to develop a trajectory that had already begun and initiated the study of zoology, listing more than 550 species of animals. While he distinguished between types of animals (e.g., those with blood and those without blood), he also strongly distinguished between human beings and animals, arguing that animals lack the ability to reason and, as such, could be thought of as resources to be exploited. Add to this his attention to the work of forest wardens in preventing deforestation and erosion in his *Politics* and we get a more complete sense of the way that Aristotle thought about human relations to the nonhuman natural world (or, perhaps more properly, human separation from the nonhuman natural world) and, therein, why he thought about wars as he did. Wars were less a component of human beings' connections to the nonhuman natural world and more a product of interhuman activities. They were, in a word, political. And because they were political, they could be shaped by the human capacity to reason – which is to say, they could be governed by the rules that human beings used to make sense of their own lives apart from the forces of nature. Aristotle's early use of just war language coincides with and is contingent upon the way he understood the relationship between human beings and the nonhuman natural world.[20]

CICERO AND THE ROMAN CLIMATE OPTIMUM

Over 200 years after Aristotle, Cicero played an important role in acquainting Rome with Greek philosophy, including the idea that a war could be limited by criteria obtained from natural law. More importantly, though, Cicero watched Rome expand from a small city-state with republican ideals to an empire stretching across most of the known world. Where Alexander's vision had been to export Greek culture to the "barbarian world," Rome's vision was to export its politics: to rule the world by extending both its military might and its systems of laws and roads throughout. Implicit in such an approach is the idea that

[20] For more on how Aristotle (and, more generally, Greeks just prior to and during Aristotle's life) interacted with and understood their relation to the nonhuman natural world, see Lukas Thommen, *An Environmental History of Ancient Greece and Rome*, trans. Philip Hill (Cambridge, UK: Cambridge University Press, 2009).

those who are distant from Rome – both geographically and culturally – nevertheless have the capacity to understand and obey the *lex Romanus*: "It is no mean manifestation of Nature and Reason that man is the only animal that has a feeling for order, for propriety, for moderation in word and deed ... It is from these elements that is forged and fashioned that moral goodness which is the subject of this inquiry."[21]

Cicero, then, wrote during the axial point at which Rome's expansion functioned not only to unite much of Europe and the Mediterranean politically, but also to unite it geographically and metaphysically: even within the fiercely hierarchical structures of Roman civilization, everyone (within some bounds of gender and class) had a place and nobody's place prevented them from being recognized as a human being capable of reason. Indeed, the capacity to reason was a prerequisite for recognizing the place one had within the context of Roman rule. Such a vision shaped not only prudential concerns (there is no point in destroying something you wish to be in charge of later) but also moral ones (there is no need to treat ones' enemies as subhuman in order to justify conquering them, given that they'll end up playing a subservient role in the empire later). An empire could be pluralistic, so long as everyone in it recognized the legitimate authority of Rome. Cicero was responding to Roman expansion and a new political vision as much as shaping it. Indeed, given his preference for republican systems, his tendency to sway with the political winds, and his eventual death at the hands of the empire, it can be hard to think of him as much of a driver of political culture at all.

Cicero – not to mention the Roman Empire – did have the good fortune to come of age in a climatically advantageous time. The Roman Climate Optimum (RCO), which began in the latter half of the third century BCE, was marked by milder weather as the Mediterranean ecotone shifted northward. The RCO favored Roman agricultural practices and, with them, the possibilities for the kind of increased caloric intake necessary to initiate and run an empire: The metabolic costs not only of wars of expansion but the maintenance of such an empire are heavy, and "[a]longside trade and technology, the climate regime was a silent, cooperative force in the seemingly virtuous circle of empire and prosperity."[22] The legions could be fed and cities could grow (Cicero was, himself,

[21] Cicero, *De Officiis* I, trans. W. Miller, in Arthur F. Holmes, ed., *War and Christian Ethics: Classic and Contemporary Reading on the Morality of War*, 2nd ed. (Grand Rapids, MI: Baker Academic, 2005), 26.

[22] Kyle Harper, *The Fate of Rome: Climate, Disease, and the End of an Empire* (Princeton, NJ: Princeton University Press, 2017), 15.

a proponent of urban growth – partly because he had made money through housing speculation). Armies conquered and peoples gathered. *Pax Romanus* may have been a precondition for the promotion of *lex Romanus*, but it was, itself, made possible by *clima Romanus* – a climate favorable to Roman agricultural practices and political ambitions.

AUGUSTINE AND NATURE'S WAR ON ROME

The Roman Climate Optimum would come to an end in the fourth century CE. The Mediterranean ecotone shifted back southward; as it did so, caloric capacity in the empire diminished and the metabolic costs of maintaining the empire became increasingly unpayable. Drought and low cereal yields, combined with political challenges and failures, reshaped the empire in the middle of the third century and then, again, at the end of the fourth century, eventually bringing an end to the Western portions of the empire. Overlapping with climate changes, diseases ravaged the empire, exacting their greatest costs where people were most closely gathered. In the middle of the second and of the third centuries and, finally, during the Justinian plagues of the sixth century, rapidly spreading diseases brought the remaining (eastern) part of the Roman empire to its pitiful end.[23] This was the world in which Augustine lived and toward which he wrote.

Critics sometimes read Augustine as one of the great apologists for the Constantinian church – as someone who spent considerable energy justifying the unjustifiable by being willing to unite Christian faith to political power, albeit with a peculiar turn toward the interior life.[24] Such a reading of Augustine, though, paints in too broad a brush. It is true that he lived less than a century after the Edict of Milan (313) made Christianity a legal religion and during the time when the Roman emperor Gratian made it the official religion of the empire. And it is true that one of his greatest works, *The City of God*, develops a particular understanding of the relation between church and state that reverberates to the present day.[25] Less recognized but

[23] Harper's book *The Fate of Rome* is stronger on these issues than Thommen's. Both, though, are helpful in telling the environmental history of the Roman Empire.

[24] See, e.g., Craig A. Carter, *The Politics of the Cross: The Theology and Social Ethics of John Howard Yoder* (Grand Rapids, MI: Brazos Press, 2001), 158. On reconciling Augustine's turn toward the interior self with his attention to the world around him, see Charles Mathewes, "Augustinian Anthropology: *Interior intimo meo*," *Journal of Religious Ethics* 27.2 (Summer 1999): 195–221.

[25] One speculates that where cities are particularly vulnerable to antagonistic military, political, and natural forces, the attractions of the city of God would grow.

equally important to note is that even as he was writing during a period of Christianity's political ascent, he was also writing during a period when the empire was collapsing. He was still a Manichean when the Battle of Adrianople (378) initiated the decline of the empire. He wrote the *Confessions* only a few years after Theodosius I, the last emperor of a united empire, died (395), and he wrote *The City of God* concurrent with the sack of Rome (410). He died in Hippo as that city was being besieged by Vandals (430).

Seemingly stable ideas about not only the martial superiority of Rome but also the supremacy of a Roman vision of the world were being thrown into question at profoundly existential levels. What could it mean not only that "the barbarians" were winning but also that they were doing so by rejecting the ideas and vision that shaped the Roman social imaginary – or that their ways of engaging the natural world seemed more successful than Roman ways of doing so? Not only from the perspective of the bishop's seat in Hippo, but also from that of seats throughout the empire, the world was coming undone. Indeed, *The City of God* is, in part, a response to those who were blaming the recent Roman attachment to Christianity for the ills of the empire. Small wonder that Augustine gave so much attention to the heresies of Pelagius, Arius, the Donatists, and others: From his perspective, the church undoubtedly seemed to be in danger of falling apart as well. Such a perspective may help explain why he was willing to call on the civil authorities to persecute the Donatists: Not only were times desperate on all fronts but nobody's motives were beyond reproach because everyone seemed to favor their own faltering perspectives over the stability of a coherent and unifying – if rapidly receding – Roman social imaginary. And small wonder, therein, that he gave so much attention to motives in his work on just war thought. The failings of the Roman social imaginary brought on, in part, by the southward shift of the Mediterranean ecotone made such attention almost inevitable.

GRATIAN AND TWELFTH-CENTURY ENVIRONMENTAL STABILITY

Little is known of Gratian beyond the facts that he lived in the first half of the twelfth century in Bologna, Italy, and advanced canon law through his *Concordia discordantium canonum*. Yet knowing even this means knowing a great deal. The investiture controversy, resolved via the Concordat of Worms in 1122 CE, reveals new tensions between church and state. The rise of the Cistercian order in the first decades of the century reveals deep disquiet – and the possibility to do something about it – with the church's established monastic traditions. Conflicts with other religions, expressed in the first Crusade (which had just concluded as the century began) and the second (which would begin midway through the century), reveal complicated changes in the way the church and empire thought about politics, war, religion, and their neighbors. Conflicts within the church would, by the end of the century, lead to inquisitions that cast Waldensians and Cathars as heretics.

And, at the same time, the first half of the twelfth century was a kind of little renaissance: Anselm of Canterbury, Peter Lombard, Joachim of Fiore, Bernard of Clairvaux, Peter Abelard, and Hildegard of Bingen – among others – all shared the intellectual stage with Gratian. Some of them began teaching at a new kind of institution that arose at the time: the university (the universities of Bologna [1088], Oxford [1096], and Paris [1150] were all founded near this time). The sheer volume of churn in Europe was greater in this half-century than any between the fall of Rome and the rise of the Reformation. Had Gratian not attempted to bring order to one part of the intellectual component of this churn through the creation of canon law, someone else may well have done so.

During this time of relative environmental stability (i.e., no great epidemics, volcanic eruptions, earthquakes, or climate changes), Europeans also sought to bring greater order to the nonhuman natural world. Woodlands were cleared for plowing even as the heavy moldboard plow gained far wider use and horses began to be used to pull those plows. Wetlands along the Mediterranean and, especially, northern coastal Europe were drained and fields created. Increased cereal production (as well as increased erosion and its attendant effects) followed. Nonnative species such as rabbits, fallow deer, and carp spread through the continent. Artificial fishponds proliferated throughout Western Europe. Coal mining in the French-speaking Low Countries began. Mines were dug deeper and brass-making expanded. Laws regarding property rights were laid out and complicated, including in Bologna. Gratian himself noted that while God "gave humankind all things in common, 'by the law of custom or statute, this belongs to me, that belongs to someone else'."[26] It is almost as if Europeans – especially those in the south and west – were compensating for political upheavals

[26] Richard Hoffman, *An Environmental History of Medieval Europe* (Cambridge, UK: Cambridge University Press, 2014), 241.

by reinvesting in efforts to subdue the nonhuman natural world and bring greater order and value to it.

THOMAS AND THE REORDERING OF POLITICS AND NATURE

On the face of it, Thomas lived during comparatively stable political times: The papacy was strong and the states were comparatively stable. Yet things roiled under the surface. The coherence of centuries-long political theory was coming undone as a result of the political machinations of a church that could hardly be understood as standing above (and therein pure from) the affairs of the states. The Crusades, begun 125 years earlier, were not only still ongoing, but also their conceptual effects – most notably, in this case, the recovery of Aristotelian thought in the West via its reintroduction by Muslim scholars such as Averroes (Ibn Rushd) and Avicenna (Ibn Sina) – had begun to ripple through Western society, and the Mongol invasion of Eastern Europe was near its peak. Thomas was particularly swayed by Aristotle's work, using it to shape his vision of a political order – an order that included attention to the rules of war – that stabilized both church and state, therein helping to end a period of confusion and dissolution about the political shape of the medieval social imaginary. As Wolin points out, the church during the time just prior to Thomas hadn't so much maintained a divide between the religious and the political – as it claimed to do – as blurred the two even while it maintained a vision of natural ordering that emphasized their distinctiveness. With Aristotle as his guide, Thomas "not only restored the political order to repute; he ... [gave] it a sharpness of identity, a clarity of character, that had been lacking for several centuries."[27] And this order not only included just war thinking per se but also offered an orderly structure to just war thought. Thomas lived during far more interesting times with regard to the movement from one medieval social imaginary into another than is sometimes recognized, and his development of the Augustinian/Aristotelian synthesis through which that change reveals itself bears expression in his thoughts about war and peace.

Thomas did live during comparatively stable climatic times. The Medieval Warm Period may have been coming to an end but it had not yet reached its conclusion. The Black Death would not arrive until the better part of a century had passed after his death. He also, though,

[27] Wolin, *Politics and Vision*, 126.

lived at a time during which human understandings of the natural world were changing. The Aristotelian vision of the nonhuman natural world as a world of resources had grown in prominence and Thomas reinforced this vision through rationale he developed in his own writings. Perhaps the European nonhuman natural world didn't change much. European perceptions of that world, though, changed dramatically – and as Hoffman has highlighted, it is as important to understand how people perceive the nonhuman natural world as it is to understand that world in the telling of environmental histories. That Thomas lived during a fairly stable period in Europe's environmental history doesn't undermine my claim that we need to pay attention to that history when thinking about the movement of the Christian just war tradition through time so much as it reinforces the complexities involved in telling that history. After all, for understandings of causation to be complex enough to address the many physical and conceptual forces at work in the world, they must be composed of multiple and varyingly significant forces, some of which are driven by political forces, some of which are driven by environmental changes, and some of which are driven by changing interactions with and perceptions of the nonhuman natural world.

RAMSEY, NEW WORLD ORDERS, AND THE EXPLOITATIONS OF NATURE

Having already discussed Vitoria and Grotius and their contexts in much greater detail, I now jump forward to Paul Ramsey (with the blessing of the conventional narrative of the history of the Christian just war tradition). In Ramsey, we find a writer caught up in the midst of a dramatic change in the social imaginary. Indeed, Ramsey is so recent and his thought so analytic that we might lose sight of the momentousness of the context in which he wrote. Clearly, he wrote after (and was shaped by) two world wars, themselves eponymous and dramatic examples of the magnitude of war and internationalized military systems by the middle of the twentieth century. Moreover, those wars – and especially the second – are also marked by events that immediately follow their conclusion; namely, new international orders and pursuits of peace, the beginning of the Cold War, the establishment of modern Israel, and, simultaneously, the rebirth of new projects in establishing international peace on the basis of human rights and dignity, of which the United Nations is only the most prominent. Attention to these developments is important because they are symptomatic of larger global changes in which wars have increasingly been fought by coalitions of forces over deeply held ideological differences and often within contexts in which the nominal cause was the right to self-rule. The war that most preoccupied Ramsey, after all, was Vietnam, itself a sterling example of coalition, ideological difference, and the demands for self-rule in a postcolonial world.[28]

The environmental history of the twentieth century is both more familiar and, perhaps because more familiar, less reflected upon. Just war theorists write of the use of nuclear weapons without giving much attention to the fact that in harnessing the atom, human beings, for the first time in human history, developed the capacity to bring about the end of most life on earth. Not only has the exponentially increasing use of fossil fuels during the twentieth century shaped one of the greatest crises in human history, but it has also reshaped farming, housing, travel, labor, and whole economic and political systems. Along with splitting the atom, scientific advances in chemistry and biology led to new kinds of deadly and indiscriminate weapons. The nonhuman natural world we live and fight in expanded into a third dimension as we took to the skies, including for military purposes. And between 1900 and Ramsey's death in 1988, the human population of the planet almost tripled, from 1.6 billion people to 5.1 billion, a fact that has far-reaching implications for shelter, pollution, and food, water, and energy supplies, all of which reshape the causes, types, and implications of conflicts. Using Ramsey, I make the point one last time: The great scholars of just war thought ought to be understood primarily as responding to dramatic social changes, many of which are shaped by their interactions with the nonhuman natural world, and these changes have brought just war concerns to the surface in the face of changing understandings of self, state, and other.[29]

ENVIRONMENTAL HISTORY ENRICHING THE CHRISTIAN JUST WAR NARRATIVE

My redescription of just war traditions does more than clarify why the conventional narrative about the history of just war thought has been so

[28] Indeed, as John Mueller has argued, the very shape and understanding of war underwent dramatic changes in the twentieth century. See, among his other books and articles, John Mueller, *The Remnants of War* (Ithaca, NY: Cornell University Press, 2004).

[29] It strikes me that this approach to situating the great just war thinkers within the tradition does more than help explain why they arose when they did and what questions and concerns drove their thought. It helps explain why just war thinkers have also been so likely to write against heresy. Heresy, after all, is another expression of what happens when social imaginaries change and settled convictions about the world around us are disrupted.

compelling in spite of its weaknesses. It also suggests places we might go to look for Christian just war thinking that have been disregarded, especially by that narrative. We could look, for instance, at the revolutions at the end of the eighteenth century and their impact. Although the conventional narrative skips from Grotius to Ramsey because it cannot identify many great men writing on just war between the two, we ought to be able to find just war thinking near the surface of thought when the social imaginary changed during the American and French revolutions. And, upon closer reflection, we do, provided we are willing to set that conventional narrative aside. In this book, I take as a prototypical example the Constitution of the United States of America. That enormously influential document – a document that has had an impact far greater than its author, James Madison, has himself had – manifests Christian just war thinking during a period of dramatic social transition. But it is a document, not a person.

Or, to take another example, turn to a scholar who wrote at that time but who has been largely ignored in the conventional telling of the Christian just war tradition: Immanuel Kant. Kant wrote about war and its limits yet he is seldom included in the conventional telling of the just war tradition because he writes neither to reaffirm traditions of earlier thought nor to express historic understandings of the Christian faith. Yet he perhaps did more to shape modern thought in the West than any other North Atlantic philosopher. Kant developed the idea that reason alone was sufficient to realize that war is a choice, not a natural necessity, and therefore that political systems can and should be created that establish a "kingdom of ends" and thereby constrain political and martial behavior.[30] That Kant is not included by most just warriors says more about either their cramped reading of his work or an implicit division between "religious" and "secular" thought that ought not go unchallenged by scholars trying to think after the collapse of secularism in the West. It is hard to escape the conclusion that he's excluded from the list of just war thinkers because he doesn't rely on the inherited language for thinking about the relation between justice and war. Such, though, is rather paltry criticism of a scholar whose entire project was

[30] Though this is not the place to develop the matter, Kant's arguments toward perpetual peace do not actually constitute an idealist's expression of pacifism but, instead, a system of restraints. Kant was sufficiently cynical about the human capacity for self-deception and sufficiently emphatic about the responsibility to punish misconduct that he leaves space for some forms of and reasons for war. See Immanuel Kant, "Toward Perpetual Peace" in Holmes, *War and Christian Ethics*, 274–283.

motivated by the attempt to have the courage to "think for oneself."[31] Though the core ideas of his thought fit within neither the just war tradition nor the Christianity of his time, those ideas have driven changes to both the tradition and the faith since his time.

Moreover, once we expand the range of significant forces reshaping the Christian just war tradition beyond the brains of its best-known thinkers, we are better equipped to investigate the way the tradition has actually changed. We can attend to other texts (I think I could build a convincing argument that Augustine's *City of God,* in its enormously consequential reframing of understandings of the state and how God works in it, has more profoundly shaped the tradition than his explicit writings on war have). We can attend to the impact of ideas (peace, positive law, political theology) that are not obviously linked to particular thinkers in shaping the tradition.[32] We can more closely attend to the impact of technologies (e.g., the printing press, the cotton gin, aircraft, the Internet) in shaping the tradition. We can attend to the impact of practices (including those related not only to whether and how to fight, such as the drilling of mass armies, but also to the harvesting of food,[33] patterns for educating youth,[34] etc.) on the tradition. And, as I have argued in these latter chapters, we can attend more closely to the way changes in the nonhuman natural world and human interpretations of the nonhuman natural world have shaped the tradition.

Expanding attention to the range of forces that have reshaped the Christian just war tradition also bears other fruits in that it offers a transportable method for research. Giving attention to when and how social imaginaries change could be useful in work being done by comparative just war theorists: How, for instance, have the Islamic or Jewish just war traditions changed over time and what drove those changes? Giving attention to the way social and environmental change have the

[31] See Immanuel Kant, "An Answer to the Question, 'What is Enlightenment?'" in *Political Writings* from the Cambridge Texts in the History of Political Thought, ed. H. S. Reiss (Cambridge, UK: Cambridge University Press, 1991), 54–60.

[32] See, e.g., John Mueller's arguments about the growing political power of the idea of pacifism at the end of the nineteenth and beginning of the twentieth centuries in Mueller, *The Remnants of War*, ch. 2.

[33] The Peace of God and the Truce of God were both aimed, partly, at protecting farmers and livestock.

[34] Questions about whether and how military training is allowed and/or promoted in schools reveal not only how a society thinks about war (including the rules of war) but also how it thinks about its youth (including when they are permitted and/or obligated to fight). One is reminded of the Duke of Wellington's apocryphal comment that, "[t]he battle of Waterloo was won on the playing fields of Eton."

potential to disrupt stable polities could be useful in discerning just war traditions that have not, to this point, been seen as such: How, for instance, did the dynasties of ancient China maintain order while advancing a moral vision and waging wars during change?[35] Giving attention to major climatic events in human history as they impinged on human societies would help us focus in on particular periods of time where the precursors to just war thinking might become visible: What criteria did the Akkadian or Egyptian empires use in making prudential and moral decisions about going to war when they faced massive volcanic eruptions (Akkad in 2200 BCE) or devastating drought (Egypt in 1200 BCE)?[36] Along the way, such research would give those who study just war traditions the opportunity to interact with scholars in a wide range of disciplines: paleo- and historical climatologists, environmental historians, archaeologists and anthropologists, and others.

As a final example of its constructive power, my interpretation of just war thought helps to explain both why there have been so many recent books and articles attempting to rethink the tradition and why so few of those books and articles seem especially innovative. It is clear from these texts that their authors are alert to the difficulty of attempting to apply just war thinking as they have received it (most problematically, as a theory) to a post–Cold War world in which questions about nonstate players, humanitarian interventions, and new kinds of weapons predominate. What is less clear in these texts is whether their writers are sufficiently cognizant of the fact that the very changes in the global society that are shaping such questions will also shape how they understand (or fail to understand) just war thinking itself. This book hints that entrance into the Anthropocene signals the most significant shift in social imaginaries we have seen in millennia; that the shifts we are seeing/will see over the next several decades are driven by our changing relationship with the nonhuman natural world, including our perceptions of it; and that, therefore, to imagine how just war thinking will apply in this new social imaginary is to step back from the standard application of a theory and its established criteria to explore the way the moral and prudential considerations that constitute just war thought may express themselves in light of this new imaginary.

[35] See Ping-cheung Lo and Sumner Twiss, *Chinese Just War Ethics: Origin, Development, and Dissent* (Abingdon-on-Thames: Routledge, 2015).

[36] See Brian Fagan, *The Long Summer: How Climate Changed Civilization* (New York, NY: Basic Books, 2004).

On my telling of the history of the Christian just war tradition, that tradition sits within a larger history in which interpretations of ideas and actions change over time due to natural and social forces. Or, perhaps more accurately, on my telling, the Christian just war tradition sits within a larger narrative about the possibilities and limits of traditions. I want more history, not less history, and certainly not something other than history. I want to argue about how traditions change over time and how these changes are shaped by significant shifts in the way those within them understand the world around them; about the way they understand themselves, the state, the other, and the relationships between self, state, and other; and about the way such arguments intersect with human engagements with the nonhuman natural world.[37]

Said differently, I want to argue that our understandings of the traditions change as social imaginaries change, and that these changes (a) help us resist telling narratives of progress or decline (contra the conventional telling); (b) prevent us from saying that the understanding they had then is better than the one we have now because it is closer to the original meaning of the ideas that constitute the tradition; (c) better account for the ironies, indirections, and discursive jumps that actually constitute a theological reading of the history of just war traditions; and (d) actually help us see places where just war thought is at work in ways that we have not yet recognized as being part of the just war tradition. Or, to say that differently still (and far more briefly, if more abstrusely), I want to situate the just war tradition within a tradition of explicating traditions. As we move further into the Anthropocene – into not only a new age but also a new geological epoch – understanding how and why traditions change will be vital to the projects of interpreting that new epoch. And as we face climate-shaped conflicts, understanding just war thinking will be vital to addressing those conflicts.

Throughout this book, I have used attention to the Christian just war tradition's transition from its late medieval to its early modern form in order to argue that making sense of that transition helps us make sense not only of modernity but also of modernity's blind spots in understanding and responding to climate-shaped violence. In this chapter, I have suggested that the arguments I raised in those chapters can apply

[37] Or, more specifically, those that they cannot identify as fundamentally like themselves, whether through common commitments to a shared state or some other theological or philosophical idea that connects them to the other such as "also created in the image of God" or "always to be treated as an end in him- or herself and never only as a means."

to making sense of why and how the Christian just war tradition has changed throughout its history in order to lay groundwork for thinking about how the Christian just war tradition may (and even should) transition into the Anthropocene. What I have not yet done is shift from the historical/analytic to the constructive mode: Given the move out of modernity and in light of this understanding of how the Christian just war tradition changes over time, what should Christian just warriors be thinking about as they face the wars of the Anthropocene in a changing social imaginary? Combining the insights of this book with those of *Christian Pacifism for an Environmental Age*, that work now begins.

Afterword

In the spring of 2011, unrest in Syria reached the point of protests and uprisings in the country's urban centers. These were soon followed by armed rebellions and violent responses by the Assad regime to put these rebellions down. A decade later, the war in Syria continues; between its start and the time of writing, a wide range of state and nonstate players from the international stage have been involved for good and ill purposes, the Assad regime has used internationally banned weapons on its own civilians, and refugees from Syria have played a role in reshaping politics in Europe and, to a lesser extent, North America.

Narratives about the causes and processes of the war in Syria and its impacts are too numerous to name here. One, though, warrants attention in the context of a book about the way climate and conflict interact. In the March 17, 2015, edition of the *Proceedings of the National Academy of Sciences of the United States of America*, five scientists argued that three successive years of the worst droughts in instrumental history in Syria, between 2007 and 2010, played a contributing factor in the unrest that would become a war and that these droughts were caused, at least in part, by anthropogenic climate change.[1] The droughts, they argue, exacerbated agricultural – and, therein, economic and political – insecurity and drove young men from rural areas to urban ones in search of work and incomes; in the cities, these men joined other desperate people

[1] Colin P. Kelley, Shahrzad Mohtadi, Mark A. Cane, Richard Seager, and Yochanan Kushnir, "Climate Change in the Fertile Crescent and Implications of the Recent Syrian Drought," *PNAS* 112.11 (March 17, 2015): 3241–3246, https://doi.org/10.1073/pnas.1421533112 (accessed November 15, 2019).

(refugees from the war in Iraq and other impoverished citizens) such that by the end of 2010, 20 percent of Syria's urban populations were made up of internally displaced persons (IDPs) and refugees. These desperate people, experiencing great need and seeing the Assad regime's failure to address (or even give much attention to) those needs and observing uprisings in Tunisia, Egypt, Libya, and other countries in the region, came together to protest. Conflagration followed. At the conclusion of their article, the scientists wrote:

> An abundance of history books on the subject tell us that civil unrest can never be said to have a simple or unique cause. The Syrian conflict, now civil war, is no exception. Still, in a recent interview, a displaced Syrian farmer was asked if this was about the drought, and she replied, "Of course. The drought and unemployment were important in pushing people toward revolution. When the drought happened, we could handle it for two years, and then we said, 'It's enough.'" This recent drought was likely made worse by human-induced climate change, and such persistent, deep droughts are projected to become more commonplace in a warming world.[2]

The article, with its ominous conclusion, was greeted with widespread attention and discussed in prominent science and popular journals as well as newspapers and magazines. Given the complexities of causation – which complexities the authors noted with regard to both civil unrest and climate change – the attention received by the article was sometimes laudatory and sometimes critical.[3] Over time, though, the narrative that climate change played a role in the Syrian conflict has grown just as the conflict and the need to attend to anthropogenic climate change have as well.

Part of what made the article important was the connection it made between climate change and conflict. Part of what made the article complicated was that this connection relies on understandings of causation that are subtle, difficult to disaggregate, and subject to dispute. And part of what made those understandings of causation complicated was

[2] Kelley et al., "Climate Change in the Fertile Crescent," 3245.

[3] See, e.g., Lina Eklund and Darcy Thompson, "Differences in Resource Management Affects Drought Vulnerability across the Borders between Iraq, Syria, and Turkey," *Ecology and Society* 22.4 (2017): 9, https://doi.org/10.5751/ES-09179-220409 (accessed November 15, 2019); and Jan Selby and Mike Hulme, "Is Climate Change Really to Blame for Syria's Civil War?" *The Guardian* (November 29, 2015), https://www.theguardian.com/commentisfree/2015/nov/29/climate-change-syria-civil-war-prince-charles (accessed November 15, 2019). It should be noted that the Kelly et al. article was not the first to make the link between Syrian unrest and climate change. See, e.g., Francesco Femia and Caitlin Werrell, "Syria: Climate Change, Drought, and Social Unrest," Center for Climate and Security (February 29, 2012), https://climateandsecurity.org/2012/02/syria-climate-change-drought-and-social-unrest/ (accessed November 15, 2019).

that the social imaginary through which such understandings are given expression draws a line between the natural and political sciences and locates concepts such as "cause" within the former science even though causation turns out to be a far murkier subject in the natural sciences than we would like and a far more significant subject in the latter ones that we often recognize. Entering the Anthropocene likely means complicating "cause" even more.

The project of which this book forms a part also began in 2011 during a fall sabbatical. Having recently published a book on religious language in the public sphere, I had hoped to circle around to an article on just war thinking I'd written during the build-up to the war in Iraq and try to turn it into something more significant. I was having trouble, though, finding an entry point that felt right. One morning that summer, I heard an NPR interview with a US Navy officer discussing the problems that climate changed posed for the military. Having a long-term interest in environmental issues and environmental ethics, I found in that interview the stimulus to begin research on a book about how climate change and environmental degradation in the twenty-first century would reshape the causes, types, and understandings of war we face moving forward.

Almost immediately, I fell into a research hole far deeper and wider than I had first presumed. On the one hand, I discovered a range of scholarship that, embarrassingly, was almost entirely unknown to me. The scholarly worlds of environmental security studies, paleo- and historical climatology, and environmental history, among other realms, needed exploration. Likewise, I needed to refamiliarize myself with more familiar territories, including newer research in environmental ethics, political theology, and just war, pacifism, and just peacemaking programs. I am still exploring.[4]

[4] One of the risks of being an ethicist who dabbles in a wide range of scholarly research is that I am constantly playing catch-up with important scholarship in other fields. That has certainly proven to be the case with this book as well. As fitting as it is for this project, for instance, I did not discover Geoffrey Parker's *Global Crisis: War, Climate Change and Catastrophe in the Seventeenth Century* (New Haven, CT: Yale University Press, 2013) until after the manuscript for this book was complete. What makes this late discovery especially embarrassing is that Parker's book was busy racking up awards – some of which I pay a casual kind of attention to – after its publication and during my own research on and writing of this book. Other pertinent books I have discovered post-writing include Philip Blom's *Nature's Mutiny: How the Little Ice Age of the Long Seventeenth Century Transformed the West and Shaped the Present* (New York: Liveright, 2019) and Simon Winchester's *Land: How the Hunger for Ownership Shaped the Modern World* (San Francisco, CA: Harper, 2021).

The hardest discovery was, for me, realizing that almost nobody had written anything substantively linking religious ethics to climate change and violence. Obviously, I could find significant resources that linked any two of those three, but not all three. What began as a more typical scholarly endeavor (enter a preexisting conversation and contribute to it while shaping a bit of one's own space) became, instead, an exercise in cartography: Mapping out largely unexplored terrain in order to provide people – including myself – with the resources, I thought they would need to make sense of the strange world in which they were walking. And what began as a single book about climate and conflict in the twenty-first century became, instead, three books. The first, *Christian Pacifism for an Environmental Age*, mapped out a novel path by which to make sense of the history of Christian pacifism such that its wisdom would prove meaningful as we exit modernity and enter the environmental age and its conflicts. This book, the second, tries to do a similar thing for the Christian just war tradition.

Now, at the end of two books, I am finally in a location to write the book I had intended to write all along: *War in a Warming World: Religion, Resources, and Refugees.* The three words in the subtitle of the next book are more than just an alliterative conceit. They name three focal points that not only warrant close attention as wars shaped by life in the Anthropocene become more common but also will require a different kind of attention than modern constructions of war allow. Having been framed by the modern social imaginary of a secular age, the roles religion plays in shaping understandings of and interactions with both the nonhuman natural world and conflict have been diminished and distorted by that framing. Having been framed by the modern social imaginary of free markets, the idea that all that is not human can be reduced to "resource" has concealed the complex interdependencies upon which the human and nonhuman natural world rely. Having been framed by the modern social imaginary of the nation-state, the stability/migration dynamic has been freighted with politico-legal implications that make refugees different from IDPs and make both different from other human beings even as violence and environmental changes make refugees and IDPs more likely. One purpose of this book has been to explore the origins of such distortion, concealment, and freighting.

The three words in the subtitle of the next book also return us to the conflict in Syria, where religion (including multiple world religions and multiple expressions of each of those world religions), resources (including water, land, grains, and oil), and refugees (including those that fled to neighboring countries as well as those who made their way to more distant lands) have interacted in complicated ways and thus contributed toward making the war there more incoherent and horrific and less isolated or amenable to resolution. If politics involves practicing the art of the possible, wars such as the one in Syria reveal the degree to which the politics of the modern social imaginary make wars increasingly impossible. And Syria is more coherent within that social imaginary than some of the conflicts we face.

Where conflicts don't rise to modern definitions of war (e.g., low-intensity conflicts with fewer than 1,000 deaths per year),[5] aren't fought by recognizable armed forces (e.g., conflicts other than those waged between nation-states or civil wars), ignore the rules of war (e.g., those violating international law or commitments to just war traditions), are fought sporadically rather than continuously (e.g., recurrent conflicts over land claimed by two or more parties that seldom have the resources to commit to the use of coercive force), or are fought using unusual weapons (e.g., makeshift weapons or technologically sophisticated means for controlling space or opposing groups), the modern social imaginary struggles. And such conflicts will, I suspect, increasingly be the rule rather than the exception in the environmental age.

All of which makes recognizing the origins of that modern social imaginary and the ways it understands war more helpful. After all, not only is war a social phenomenon, but it is also a social construction, and gaining a better understanding of how the wars of modernity were constructed may help clarify how the wars of the environmental age are being constructed. And all of which makes attending to the shapes and sizes of the wars of the environmental age more pressing. We need all the philosophical, religious, legal, political, economic, and scientific tools that we can use to prevent, inhibit, ameliorate, mitigate, and otherwise respond to the wars in a warming world.

With the previous volume, this book offers both resources and orientations toward those ends. Among those resources are not only the revised narratives about the Christian pacifist and just war traditions but also a particular way of describing the movement of traditions through time; not only an exploration of the way environmental change and political change have interacted in the past but also a particular way

[5] The 1,000 deaths per year threshold, associated with Singer and Small's "Correlates of War" project, is not part of a universally agreed-upon framework for defining war. It is, though, a useful heuristic for describing modern wars. See Melvin Small and J. David Singer, *Resort to Arms: International and Civil War, 1816–1980* (Beverly Hills, CA: SAGE, 1982).

of embedding the motivating forces that drive political change within a larger framework of environmental change; not only attentiveness to the role religious thought has played in making sense of wars and the nonhuman natural world but also a particular way of theologically construing wars and the nonhuman natural world within a larger framework in which divine action and, seemingly, divine absence play a role. And among the orienting features of the two books are the increasingly pressing obligations to integrate the wisdom of multiple disciplines, to recognize the interdependencies within the human and nonhuman natural world, and to pursue methods of research that emphasize relevance in naming problems and exploring responses to them. With the next volume, I hope that these resources and orientations bear fruit. The wars of the warming world require nothing less.

Select Bibliography

Aberth, John, *An Environmental History of the Middle Ages: The Crucible of Nature*. London: Routledge, 2013.

Agamben, Giorgio, *State of Exception*. Translated by Kevin Attell. Chicago, IL: University of Chicago Press, 2005.

"Americans Are Poorly Informed about Basic Constitutional Provisions | The Annenberg Public Policy Center of the University of Pennsylvania," The Annenberg Public Policy Center of the University of Pennsylvania, 2017. www.annenbergpublicpolicycenter.org/americans-are-poorly-informed-about-basic-constitutional-provisions/.

Anghie, Antony, *Imperialism, Sovereignty and the Making of International Law*. Cambridge, UK: Cambridge University Press, 2005.

Antonio, Edward E., "Ecology as Experience in African Indigenous Religions" in Linda E. Thomas (ed.), *Living Stones in the Household of God: The Legacy and Future of Black Theology*, 155. Minneapolis, MN: Fortress Press, 2004.

Aquinas, Thomas, *The Summa Contra Gentiles III.ii, 112*. Translated by the English Dominican Fathers from the Latest Leonine Edition. London: Burns, Oates & Washbourne, Ltd., 1923.

Aquinas, Thomas, *Summa Theologica*. Translated by Fathers of the English Dominican Province. New York: Benziger Brothers, 1911–1925.

Arendt, Hannah, *On Revolution*. New York, NY: Penguin Books, 1990.

Aristotle, *Poetics*. Translated by Gerald F. Else. Ann Arbor, MI: University of Michigan Press, 1970.

Austin, Jay and Carl E. Bruch (eds.), *The Environmental Consequences of War: Legal, Economic, and Scientific Perspectives*. Cambridge, UK: Cambridge University Press, 2000.

Barker, Juliet, *Agincourt: Henry V and the Battle that Made England*. London: Little Brown Book Group, 2005.

Barth, Karl, Geoffrey W. Bromiley, and Thomas F. Torrance (eds.), *Church Dogmatics II.2*. Edinburgh: T&T Clark, 1957.

Bartholomeusz, Tessa, *In Defense of Dharma: Just War Ideology in Buddhist Sri Lanka*. London: Routledge, 2002.

Beauchamp, Tom and James Childress, *Principles of Biomedical Ethics*, 4th ed. New York, NY: Oxford University Press, 1994.

Behringer, Wolfgang, "Climatic Change and Witch-Hunting: The Impact of the Little Ice Age on Mentalities." *Climate Change* 43 (1999): 335–351.

Bell, Daniel, *Just War as Christian Discipleship: Recentering the Tradition in the Church Rather than the State*. Grand Rapids, MI: Brazos Press, 2009.

Bellamy, Alex, *Just Wars: From Cicero to Iraq*. Malden, MA: Polity Press, 2006.

Bellamy, Alex, "When Is It Right to Fight? International Law and Jus Ad Bellum." *Journal of Military Ethics* 8, no. 3 (2009): 231–245.

Bennett, Jane, *Vibrant Matter: A Political Ecology of Things*. Durham, NC: Duke University Press, 2010.

Bennett, Judith, *Natives and Exotics: World War II and the Environment in the South Pacific*. Honolulu, HI: University of Hawaii Press, 2009.

Bergmann, Sigurd, *Creation Set Free: The Spirit as Liberator of Nature*. Grand Rapids, MI: Eerdmans, 2005.

Berkebile, Richard, "New Generation Warfare and the Just War Tradition." *Interagency Journal* 9, no. 3 (2018): 17–32.

Berlin, Isaiah, *Four Essays on Liberty*. Oxford: Oxford University Press, 1969.

Bethke-Elshtain, Jean, *Women and War*. Chicago, IL: University of Chicago Press, 1995.

Biggar, Nigel, *In Defence of War*. Oxford: Oxford University Press, 2013.

Blom, Philip, *Nature's Mutiny: How the Little Ice Age of the Long Seventeenth Century Transformed the West and Shaped the Present*. New York, NY: Liveright Publications, 2019.

Blumenberg, Hans, *The Legitimacy of the Modern Age*. Translated by Robert M. Wallace. Cambridge, MA: MIT Press, 1983.

Boff, Leonardo, *Cry of the Earth, Cry of the Poor*. Maryknoll, NY: Orbis, 1997.

Bok, Sissela, *Lying: Moral Choice in Public and Private Life*. New York, NY: Pantheon Books, 1978.

Bourdin, Bernard, *The Theological-Political Origins of the Modern State: The Controversy Between James I of England and Cardinal Bellarmine*. Translated by Susan Pickford. Washington, DC: Catholic University of America Press, 2010.

Brady, Lisa, *War upon the Land: Military Strategy and the Transformation of Southern Landscapes during the American Civil War*. Athens, GA: University of Georgia Press, 2012.

Brandom, Robert, *Making It Explicit: Reasoning, Representing, and Discursive Commitment*. Cambridge, MA: Harvard University Press, 1994.

Brandom, Robert, *Tales of the Mighty Dead: Historical Essays in the Metaphysics of Intentionality*. Cambridge, MA: Harvard University Press, 2002.

Braudel, Fernand, *The Mediterranean and the Mediterranean World in the Age of Philip II*. Translated by Siân Reynolds. New York, NY: Harper & Row, 1996.

Brown, Wendy, *Undoing the Demos: Neoliberalism's Stealth Revolution*. Brooklyn, NY: Zone Books, 2015.

Butler, Charles, *The Life of Hugo Grotius with Brief Minutes of the Civil, Ecclesiastical, and Literary History of the Netherlands*. London: John Murray, 1826.

Carlyle, Thomas, *On Heroes, Hero-Worship, and the Heroic in History*. Whitefish, MT: Kessinger Publishing, 2006.

Carter, Craig A., *The Politics of the Cross: The Theology and Social Ethics of John Howard Yoder*. Grand Rapids, MI: Brazos Press, 2001.

Carter, J. Kameron, *Race: A Theological Account*. New York, NY: Oxford University Press, 2008.

Cavallar, G., "Vitoria, Grotius, Pufendorf, Wolff and Vattel: Accomplices of European Colonialism and Exploitation or True Cosmopolitans?" *Journal of the History of International Law* 10 (2008): 181–209.

Cavanaugh, William, "'A Fire Strong Enough to Consume the House': The Wars of Religion and the Rise of the State." *Modern Theology* 11, no. 4 (1995): 397–420.

Cavanaugh, William, Jeffrey W. Bailey, and Craig Hovey (eds.), *An Eerdmans Reader in Contemporary Political Theology*. Grand Rapids, MI: Eerdmans, 2012.

Chalecki, Elizabeth and Lisa Ferrari, "Engineering the Climate – or Deploying Disaster? Applying Just War Theory to Geoengineering." *New Security Beat: The Blog of the Environmental Change and Security Program* (2018). www.newsecuritybeat.org/2018/06/engineering-climate-or-deploying-disaster-applying-war-theory-geoengineering/ (accessed April 22, 2020).

Chaturvedi, Sanjay and Timothy Doyle, *Climate Terror: A Critical Geopolitics of Climate Change*. New York, NY: Palgrave Macmillan, 2015.

Churchill, Winston and Richard Langwort (eds.), *Churchill by Himself: The Definitive Collection of Quotations*. New York, NY: Public Affairs, 2008.

Churn, Natalie and Cecile Lawrence (eds.), *Movements in Time: Revolution, Social Justice, and Times of Change*. Newcastle upon Tyne: Cambridge Scholars Publishing, 2012.

Clough, David and Brian Stiltner, *Faith and Force: A Christian Debate About War*. Washington, DC: Georgetown University Press, 2007.

Cohen, Elizabeth, *The Political Value of Time: Citizenship, Duration, and Democratic Justice*. Cambridge, UK: Cambridge University Press, 2018.

Cohen, Felix, "The Spanish Origin of Indian Rights in the Law of the United States." *Georgetown Law Journal* 31 (1942): 1–21.

Cone, James, "Whose Earth Is It Anyway?" *Cross Currents* 50, no. 2 (2000): 36–46.

Coopland, George William, *The Tree of Battles of Honoré Bonet: An English Version with Introduction*. Cambridge, MA: Harvard University Press, 1949.

Corey, David and J. Daryl Charles, *The Just War Tradition: An Introduction*. Wilmington, DE: ISI Books, 2012.

Crosby, Alfred, *Ecological Imperialism: The Biological Expansion of Europe, 900–1900*. Cambridge, UK: Cambridge University Press, 1986.

Curran, Charles and Richard A. McCormick, *Readings in Moral Theology No. 7: Natural Law and Theology*. New York, NY: Paulist Press, 1991.

Dalby, Simon, *Security and Environmental Change*. Malden, MA: Polity Press, 2009.

Davidovic, Jovana, "What Do We Owe Refugees? Jus Ad Bellum, Duties to Refugees from Armed Conflict Areas and the Right to Asylum." *Journal of Global Ethics* 12, no. 3 (2016): 327–364.

Davis, James Calvin, *In Defense of Civility: How Religion Can Unite America on Seven Moral Issues that Divide Us*. Louisville, KY: Westminster John Knox Press, 2010.

Davis, Natalie, "The Sacred and the Body Social in Sixteenth-Century Lyon." *Past and Present* 90, no. 1 (1981): 40–70.

Deane-Drummond, Celia, "Public Theology as Contested Ground: Arguments for Climate Justice" in Celia Deane-Drummond and Heinrich Bedford-Strohm (eds.), *Religion and Ecology in the Public Sphere*, 203. London: T&T Clark, 2011.

Deane-Drummond, Celia and Heinrich Bedford-Strohm, *Religion and Ecology in the Public Sphere*. London: T&T Clark, 2011.de Pizan, Christine, *The Book of Deeds of Arms and of Chivalry*. Translated by Sumner Willard. University Park, PA: Penn State University Press, 1999.

Descartes, Rene and Ralph M. Eaton (eds.), *Descartes: Selections*. New York, NY: Charles Scribner's Sons, 1927.

de Vitoria, Francisco, *De Indis Et De Iure Belli Relectiones*. Edited by Ernest Nys. Buffalo, NY: William S. Hein & Co, 1995.

Dippel, Horst (ed.), *Constitutions of the World from the Late Eighteenth Century to the Middle of the Nineteenth Century: Sources on the Rise of Modern Constitutionalism*, 5 vol. Berlin: De Gruyter, 2008.

Douglas, Mark, *Believing Aloud: Reflections on Being Religious in the Public Sphere*. Eugene, OR: Cascade, 2010.

Douglas, Mark, "Changing the Rules: Just War Theory in the 21st Century." *Theology Today* 54, no. 1 (2003): 529–545.

Douglas, Mark, *Christian Pacifism for an Environmental Age*. Cambridge, UK: Cambridge University Press, 2019.

Douglas, Mark, *Confessing Christ in the Twenty-First Century*. Lanham, MD: Rowman and Littlefield, 2005.

Douglas, Mark, "Theodicies for Slow-Moving Tragedies." *Sci-Tech†: The Journal of the Presbyterian Association on Science, Technology, and the Christian Faith* 28, no. 3 (2019): 1–4.

Drake, Brian Allen, *The Blue, the Gray, and the Green: Toward and Environmental History of the Civil War*. Athens, GA: University of Georgia Press, 2014.

Duggan, Lisa, *The Twilight of Equality? Neoliberalism, Cultural Politics, and the Attack on Democracy*. Boston, MA: Beacon Press, 2003.

Dworkin, Gerald, *The Theory and Practice of Autonomy*. New York, NY: Cambridge University Press, 1991.

Eklund, Lina and Darcy Thompson, "Differences in Resource Management Affects Drought Vulnerability Across the Borders between Iraq, Syria, and Turkey." *Ecology and Society* 22, no. 4 (2017): 9. www.ecologyandsociety.org/vol22/iss4/art9/.

Ela, Jean-Marc, *My Faith as an African*. Translated by John Pairman Brown and Susan Perry. Maryknoll, NY: Orbis, 1988.

Epstein, Steven, *The Medieval Discovery of Nature*. Cambridge, UK: Cambridge University Press, 2012.

Escott, Paul D., *Rethinking the Civil War Era: Directions for Research*. Lexington, KY: University of Kentucky Press, 2018.

Fabre, Cécile, *Cosmopolitan War*. Oxford: Oxford University Press, 2012.

Fagan, Brian, *The Long Summer: How Climate Changed Civilization*. New York, NY: Basic Books, 2004.

Farrell, Michael, *Modern Just War Theory: A Guide to Research*. Lanham, MD: Scarecrow Press, 2013.

Fassbender, Bardo and Anne Peters (eds.), *The Oxford Handbook of the History of International Law*. New York, NY: Oxford University Press, 2012.

Fehler, Timothy, Greta Grace Kroeker, Charles H. Parker, and Jonathan Ray (eds.), *Religious Diaspora in Early Modern Europe: Strategies of Exile*. New York, NY: Routledge, 2015.

Femia, Francesco and Caitlin Werrell, "Syria: Climate Change, Drought, and Social Unrest." The Center for Climate and Security, 2012, https://climateandsecurity.org/2012/02/syria-climate-change-drought-and-social-unrest/.

Fisher, David, *Morality and War: Can War Be Just in the Twenty-first Century?* Oxford: Oxford University Press, 2011.

Fisher, Louis, *Presidential War Power*, 3rd ed. Lawrence, KS: University of Kansas Press, 2013.

Fonseca, Jiménez, "The Colonization of American Nature and the Early Development of International Law." *Journal of the History of International Law* 12 (2010): 189–225.

Forge, John, "Proportionality, Just War Theory and Weapons Innovation." *Science and Engineering Ethics* 15, no. 1 (2008): 25–38.

Foucault, Michel, *The History of Sexuality*, Vol. 1. Translated by Robert Hurley. New York, NY: Vintage, 1990.

Foucault, Michel, *Power/Knowledge: Selected Interviews and Other Writings 1972–1977*. Edited by Colin Gordon. Brighton, Sussex: The Harvester Press, 1980.

Freist, Dagmar and Susanne Lachenicht (eds.), *Connecting Worlds and People: Early Modern Diasporas*. New York, NY: Routledge, 2016.

Frequently Asked Questions, *Interculturalstudies.Org*. www.interculturalstudies.org/faq.html (accessed January 12, 2018).

Fry, Douglas M (ed.), *War, Peace, and Human Nature: The Convergence of Evolutionary and Cultural Views*. New York, NY: Oxford University Press, 2013.

Gat, Azar, *War in Human Civilization*. Oxford: Oxford University Press, 2006.

Gerth, Hans H. and C. Wright Mills (eds.), *From Max Weber: Essays in Sociology*. New York, NY: Oxford University Press, 1946.

Gleditsch, Nils Petter, "Armed Conflict and The Environment: A Critique of the Literature." *Journal of Peace Research* 35, no. 3 (1998): 381–400.

Goodwin-Gill, Guy and Jane McAdam, *The Refugee in International Law*, 3rd ed. New York, NY: Oxford University Press, 2007.

Green, Rochelle, "Nostalgia, Past and Future Hopes" in Natalie Churn and Cecile Lawrence (eds.), *Movements in Time: Revolution, Social Justice, and Times of Change*, 89. Newcastle upon Tyne: Cambridge Scholars Publishing, 2012.

Green, Steven K., *Inventing A Christian America: The Myth of the Religious Founding*. New York, NY: Oxford University Press, 2015.

Greenwood, Ryan, "War and Sovereignty in Medieval Roman Law." *Law and History Review* 32, no. 1 (2014): 31–63.

Grim, John and Mary Evelyn Tucker, *Ecology and Religion*. Washington, DC: Island Press, 2014.

Grimi, Elisa (ed.), *Tradition as the Future of Innovation*. Newcastle upon Tyne: Cambridge Scholars Publishing, 2015.

Grober, Ulrich, *Sustainability: A Cultural History*. Translated by Ray Cunningham. Cambridge, UK: UIT Cambridge Limited, 2012.

Grotius, Hugo, "The Law of War and Peace" in Larry May, Eric Rovie, and Steve Viner (eds.), *The Morality of War: Classical and Contemporary Readings*, 66–79. Upper Saddle River, NJ: Pearson Prentice Hall, 2006.

Grotius, Hugo, "Natural Laws and the Laws of War" in Arthur F. Holmes (ed.), *War and Christian Ethics: Classic and Contemporary Readings on the Morality of War*, 2nd ed., 237. Grand Rapids, MI: Baker Academic, 2005.

Grotius, Hugo, *The Rights of War and Peace*. Translated by Jean Barbeyrac. Edited by Richard Tuck. Indianapolis, IN: Liberty Fund, 2005.

Gula, Richard, *Reason Informed by Faith: Foundations of Catholic Morality*. New York, NY: Paulist Press, 1989.

Gustafson, James, *Ethics from a Theocentric Perspective, Vol. 2: Ethics and Theology*. Chicago, IL: University of Chicago Press, 1984.

Gustafson, James, *An Examined Faith: The Grace of Self-Doubt*. Minneapolis, MN: Fortress Press, 2004.

Gustafson, James, *Intersections: Science, Theology, and Ethics*. Cleveland, OH: Pilgrim Press, 1996.

Hamilton, Clive, Christophe Bonneuil, and François Gemenne, *The Anthropocene and the Global Environmental Crisis: Rethinking Modernity in a New Epoch*. London: Routledge, 2015.

Hammill, Graham, *The Mosaic Constitution: Political Theology and Imagination from Machiavelli to Milton*. Chicago, IL: University of Chicago Press, 2012.

Hammill, Graham and Julia Reinhard Lupton (eds.), *Political Theology and Early Modernity*. Chicago, IL: University of Chicago Press, 2012.

Hanawalt, Barbara and Lisa J. Kiser (eds.), *Engaging with Nature: Essays on the Natural World in Medieval and Early Modern Europe*. Notre Dame, IN: University of Notre Dame Press, 2008.

Hare, Richard M., *Freedom and Reason*. Oxford: Oxford University Press, 1977.

Harper, Kyle, *The Fate of Rome: Climate, Disease, and the End of an Empire*. Princeton, NJ: Princeton University Press, 2017.

Hartley, Leslie P., *The Go-Between*. London: Hamish Hamilton, 1953.

Hastings, Adrian (ed.), *The Oxford Companion to Christian Thought*. Oxford: Oxford University Press, 2000.

Haymond, John A., "Laws of War: The Origins of Restraint." *MHQ: The Quarterly Journal of Military History* 31, no. 4 (2019): 18.

Heng, Geraldine, *The Invention of Race in the European Middle Ages*. Cambridge, UK: Cambridge University Press, 2018.

Hessel, Daniel J., "Founding Era Jus Ad Bellum and the Domestic Law of Treaty Withdrawal." *Yale Law Journal* 125, no. 8 (2016): 2394–2444.

Hodgson, Peter and Robert H. King (eds.), *Christian Theology: An Introduction to Its Traditions and Tasks*, 2nd ed. Philadelphia, PA: Fortress Press, 1985.

Hoffmann, Richard, *An Environmental History of Medieval Europe*. Cambridge, UK: Cambridge University Press, 2014.

Holmes, Arthur (eds.), *War and Christian Ethics: Classic and Contemporary Reading on the Morality of War*, 2nd ed. Grand Rapids, MI: Baker Academic, 2005.

Honig, Bonnie, "Dead Rights, Live Futures: A Reply to Habermas's 'Constitutional Democracy.'" *Political Theory* 29, no. 6 (2001): 792–805.

Honig, Bonnie, *Emergency Politics: Paradox, Law, Democracy*. Princeton, NJ: Princeton University Press, 2009.

Honig, Bonnie, *Public Things: Democracy in Disrepair*. New York: Fordham University Press, 2017.

Hsiang, Solomon, Kyle C. Meng, and Mark A. Cane, "Civil Conflicts Are Associated with the Global Climate." *Nature* 476 (2011): 438–441.

Hughes, J. Donald, *What Is Environmental History?* New York, NY: Polity Press, 2006.

Hunt, Lynn, *Inventing Human Rights: A History*. New York, NY: W. W. Norton and Co., 2007.

Jenkins, Willis, "After Lynn White: Religious Ethics and Environmental Problems." *Journal of Religious Ethics* 37, no. 2 (2009): 283–309.

Jenkins, Willis, "The Environment" in Michael Kessler and Shaun Casey (eds.), *The Oxford Handbook of Political Theology*. New York: Oxford University Press, forthcoming.

Jenkins, Willis, *The Future of Ethics: Sustainability, Social Justice, and Religious Creativity*. Washington, DC: Georgetown University Press, 2013.

Jennings, Willie, *The Christian Imagination: Theology and the Origins of Race*. New Haven, CT: Yale University Press, 2011.

Joas, Hans, *War and Modernity*. Translated by Rodney Livingstone. Cambridge, UK: Polity Press, 2003.

Johnson, Barbara, *Moses and Multiculturalism*. Berkeley, CA: University of California Press, 2010.

Johnson, James T., *Can Modern War Be Just?* New Haven, CT: Yale University Press, 1984.

Johnson, James T., *Ethics and the Use of Force: Just War in Historical Perspective*. Burlington, VT: Ashgate, 2011.

Johnson, James T., *Ideology, Reason, and the Limitation of War: Religious and Secular Concepts, 1200–1740*. Princeton, NJ: Princeton University Press, 1975.

Johnson, James T., "Just War Theory: What's the Use?" *Worldview* 19, no. 7–8 (1976): 43.

Johnson, James T., *Just War Tradition and the Restraint of War: A Moral and Historical Inquiry*. Princeton, NJ: Princeton University Press, 1981.

Johnson, James T., *Morality and Contemporary Warfare*. New Haven, CT: Yale University Press, 1999.

Johnson, James T., *The Quest for Peace: Three Moral Traditions in Western Cultural History*. Princeton, NJ: Princeton University Press, 1987.

Johnson, James T., *Sovereignty: Moral and Historical Perspective*. Washington, DC: Georgetown University Press, 2014.

Johnson, James T. and John Kelsey (eds.), *Cross, Crescent, and Sword: The Justification and Limitation of War in Western and Islamic Tradition*. Westport, CT: Greenwood Press, 1990.

Kahl, Colin, *States, Scarcity, and Civil Strife in the Developing World*. Princeton, NJ: Princeton University Press, 2009.

Kant, Immanuel, *Groundwork for the Metaphysics of Morals*. Translated by Allen W. Wood. New Haven, CT: Yale University Press, 2002.

Kant, Immanuel, *Perpetual Peace*. New York, NY: Macmillan, 1903.

Kantorowicz, Ernst, *The King's Two Bodies: A Study in Medieval Political Philosophy*. Princeton, NJ: Princeton University Press, 1957.

Kaveny, Cathleen, *A Culture of Engagement: Law, Religion, and Morality*. Washington, DC: Georgetown University Press, 2016.

Kaveny, Cathleen, *Law's Virtues: Fostering Autonomy and Solidarity in American Society*. Washington, DC: Georgetown University Press, 2012.

Kaveny, Cathleen, *Prophecy without Contempt: Religious Discourse in the Public Square*. Cambridge, MA: Harvard University Press, 2016.

Kelley, Colin P., Shahrzad Mohtadi, Mark A. Cane, Richard Seager, and Yochanan Kushnir, "Climate Change in the Fertile Crescent and Implications of the Recent Syrian Drought." *Proceedings of the National Academy of Sciences of the United States of America* 112, no. 11 (2015): 3241–3246.

Keyser, Paul T. and John Scarborough (eds.), *The Oxford Handbook of Science and Medicine in the Classical World*. New York, NY: Oxford University Press, 2018.

Klare, Michael, *Resource Wars: The New Landscape of Global Conflict*. New York: Owl Books, 2001.

Klepp, Silja, "Climate Change and Migration" *Oxford Research Encyclopedia of Climate Science*. New York, NY: Oxford University Press, 2016. https://oxfordre.com/climatescience/view/10.1093/acrefore/9780190228620.001.0001/acrefore-9780190228620-e-42?rskey=M62RCF&result=1 (accessed March 18, 2020).

Kling, Jennifer, "Who Owes What to Refugees?" *Journal of Global Ethics* 12, no. 3 (2016): 327–346.

Koskenniemi, Martti, "Empire and International Law: The Real Spanish Contribution." *University of Toronto Law Journal* 61, no. 1 (2011): 1–36.

Koskenniemi, Martii, "Imagining the Rule of Law: Rereading the Grotian 'Tradition.'" *European Journal of International Law* 30, no. 1 (2019): 17–52.

Kruse, Kevin M., *One Nation under God: How Corporate America Invented Christian America*. New York, NY: Basic Books, 2015.

Kumler, Aden, *Translating Truth: Ambitious Images and Religious Knowledge in Late Medieval France and England.* New Haven, CT: Yale University Press, 2011.

Kunstler, Howard, *The Long Emergency: Surviving the End of Oil, Climate Change, and Other Converging Catastrophes of the Twenty-First Century.* New York, NY: Grove Press, 2006.

Lachenicht, Susanne (ed.), *Religious Refugees in Europe, Asia, and North America (6th–21st Century).* Hamburg: Atlantic Cultural Studies, 2007.

Lachenicht, Susanne and Kirsten Heinsohn, *Diaspora Identities: Exile, Nationalism, and Cosmopolitanism in Past and Present.* Frankfurt: Campus Verlag, 2015.

Landes, David, *Revolution in Time: Clocks and the Making of the Modern World*, revised and enlarged ed. Boston, MA: Harvard University Press, 2000.

Lang, Anthony, Cian O'Driscoll, and John Williams (eds.), *Just War: Authority, Tradition, and Practice.* Washington, DC: Georgetown University Press, 2013.

Laudan, Larry, *Progress and Its Problems: Towards a Theory of Scientific Growth.* Berkeley, CA: University of California Press, 1977.

LeGoff, Jacques (ed.), *Time, Work and Culture in the Middle Ages.* Chicago, IL: University of Chicago Press, 1982.

Levine, Peter, "Who First Said 'We Are The Ones We Have Been Waiting For'?" *Peter Levine: A Blog For Civic Renewal*, 2011. http://peterlevine.ws/?p=6105.

Lofgen, Charles, "War-Making Under the Constitution: The Original Understanding." *Yale Law Journal* 81, no. 4 (1972): 672–702.

Losonczi, Péter, Mika Luoma-Aho, and Aakash Singh (eds.), *The Future of Political Theology: Religious and Theological Perspectives.* Surrey, England: Ashgate, 2011.

Lugo, Luis (ed.), *Sovereignty at the Crossroads? Morality and International Politics in the Post-Cold War Era.* Lanham, MD: Rowman and Littlefield Publishers, 1996.

Lynch, Amanda and Siri Veland, *Urgency in the Anthropocene.* Cambridge, MA: MIT Press, 2018.

MacIntyre, Alasdair, *After Virtue*, 2nd ed. Notre Dame, IN: Notre Dame University Press, 1984.

Mancilla, Alejandra, "What We Owe before Property: Hugo Grotius and the Suum." *Grotiania* 36, no. 1 (2015): 63–77.

Martin, Terence, *Truth and Irony.* Washington, DC: Catholic University of America Press, 2015.

Masuzawa, Tomoko, *The Invention of World Religions: Or How European Universalism Was Preserved in the Language of Pluralism.* Chicago, IL: University of Chicago Press, 2005.

Mathewes, Charles, "Augustinian Anthropology: Interior intimo meo." *Journal of Religious Ethics* 27, no. 2 (1999): 195–221.

Mathewes, Charles, *Evil and the Augustinian Tradition.* New York, NY: Cambridge University Press, 2001.

Mathewes, Charles, *A Theology of Public Life*. Cambridge, UK: Cambridge University Press, 2007.

May, James (ed.), *Principles of Constitutional Environmental Law*. NP: American Bar Association, 2013.

Maza, Sarah, *Thinking about History*. Chicago, IL: University of Chicago Press, 2017.

McCarty, Brett, "Medicine as Just War? The Legacy of James Childress in Christian Ethic." *Journal of the Society of Christian Ethics* 38, no. 2 (2018): 57–74.

Metz, Johann Baptist, *Faith in History and Society: Toward a Practical Fundamental Theology*. New York, NY: Crossroad, 2007.

Miéville, China, *Between Equal Rights: A Marxist Theory of International Law*. Leiden: Brill, 2005.

Miller, William Lee, *The First Liberty: America's Foundation in Religious Liberty*. Washington, DC: Georgetown University Press, 2003.

Minteer, Ben and Robert E. Manning, "An Appraisal of the Critique of Anthropocentrism and Three Lesser Known Themes in Lynn White's 'The Historical Roots of Our Ecologic Crisis.'" *Organization and Environment* 18, no. 2 (2005): 163–176.

Moltmann, Jürgen, *The Crucified God*. Minneapolis, MN: Fortress, 1993.

Mueller, John, *The Remnants of War*. Ithaca, NY: Cornell University Press, 2004.

Muldoon, James, "The Contributions of the Medieval Canon Lawyers to the Formation of International Law." *Traditio* 28 (1972): 483–493.

Murphy, James, *War Ends: Human Rights, International Order, and the Ethics of Peace*. Washington, DC: Georgetown University Press, 2014.

Murray, John Courtney, *Religious Liberty: Catholic Struggles with Pluralism*. Louisville, KY: Westminster John Knox Press, 1993.

Nash, Roderick Frazier, *Wilderness and the American Mind*, 4th ed. New Haven, CT: Yale University Press, 2001.

Nellen, Henk, *Hugo Grotius: A Lifelong Struggle for Peace in Church and State, 1583–1645*. Translated by J. C. Grayson. Leiden: Brill, 2015.

Niebuhr, Reinhold, *Moral Man and Immoral Society: A Study in Ethics and Politics*. Louisville, KY: Westminster John Knox Press, 2013.

Niemann, Susan, *Evil in Modern Thought: An Alternative History*. Princeton, NJ: Princeton University Press, 2002.

Nixon, Rob, *Slow Violence and the Environmentalism of the Poor*. Boston, MA: Harvard University Press, 2011.

Null, Schulyer and Lauren Risi, "Navigating Complexity: Climate, Migration, and Conflict in a Changing World." The Wilson Center, 2016. www.wilsoncenter.org/publication/navigating-complexity-climate-migration-and-conflict-changing-world.

Nussbaum, Arthur, *A Concise History of the Law of Nations*. New York, NY: Macmillan, 1954.

O'Brien, Kevin, *The Violence of Climate Change: Lessons of Resistance from Nonviolent Activists*. Washington, DC: Georgetown University Press, 2017.

O'Connell, Mary, *The Power and Purpose of International Law: Insights from the Theory and Practice of Enforcement*. New York, NY: Oxford University Press, 2011.

O'Donovan, Oliver, *The Just War Revisited*. Cambridge, UK: Cambridge University Press, 2003.

O'Donovan, Oliver, *The Ways of Judgment*. Grand Rapids, MI: Eerdmans, 2005.

Orend, Brian, *The Morality of War*. Toronto: Broadview Press, 2006.

Orend, Brian, *War and International Justice: A Kantian Perspective*. Waterloo, ON: Wilfrid Laurier University Press, 2000.

Osgood, Robert, *Limited War Revisited*. Boulder, CO: Westview Press, 1979.

Oster, Emily, "Witchcraft, Weather and Economic Growth in Renaissance Europe." *Journal of Economic Perspectives* 18, no. 1 (2004): 215–228.

Parenti, Christian, *Tropic of Chaos: Climate Change and the New Geography of Violence*. New York, NY: Nation Books, 2011.

Parker, Geoffrey, *Global Crisis: War, Climate Change and Catastrophe in the Seventeenth Century*. New Haven, CT: Yale University Press, 2013.

Pasnau, Robert, *Thomas Aquinas on Human Nature*. Cambridge, UK: Cambridge University Press, 2002.

Patterson, Eric and Nathan Gill, "The Declaration of the United Colonies: America's First Just War Statement." *Journal of Military Ethics* 14, no. 1 (2015): 7–34.

Pearson, Chris, *Mobilizing Nature: The Environmental History of War and Militarization in Modern France*. Manchester: Manchester University Press, 2016.

Pickstock, Catherine, *After Writing: On the Liturgical Consummation of Philosophy*. Hoboken, NJ: Wiley-Blackwell, 1998.

Pieper, Josef, *Tradition: Concept and Claim*. Translated by E. Christian Kopff. Wilmington, DE: ISI Books, 2008.

Piketty, Thomas, *Capital in the Twenty-First Century*. Boston, MA: Belknap Press, 2017.

Ping-cheung Lo and Sumner Twiss, *Chinese Just War Ethics: Origin, Development, and Dissent*. Abingdon-on-Thames: Routledge, 2015.

Porras, Ileana, "Appropriating Nature: Commerce, Property, and the Commodification of Nature in the Law of Nations." *Leiden Journal of International Law* 27, no. 3 (2014): 641–660.

Pulido, Laura, "Rethinking Environmental Racism: White Privilege and Urban Development in Southern California." *Annals of the Association of American Geographers* 90, no. 1 (2000): 12–40.

Ramsey, Paul, *The Just War: Force and Political Responsibility*. Lanham, MD: Rowman & Littlefield Publishers, 2002.

Reiss, Hans Siegbert (ed.), *Kant's Political Writings*. Cambridge, UK: Cambridge University Press, 1991.

Rengger, Nicholas, "On the Just War Tradition in the Twenty-First Century." *International Affairs* 78, no. 2 (2002): 353–363. https://doi.org/10.1111/1468-2346.00255.

Richards, John, *The Unending Frontier: An Environmental History of the Early Modern World*. Berkeley, CA: University of California Press, 2003.

Robbins, Jeffery, *Radical Democracy and Political Theology*. New York, NY: Columbia University Press, 2011.

Russell, Fredrick H., *The Just War in the Middle Ages*. Cambridge, UK: Cambridge University Press, 1975.

Salter, John, "Hugo Grotius: Property and Consent." *Political Theory* 29, no. 4 (2001): 537–555.
Sandel, Michael, *What Money Can't Buy: The Moral Limits to Markets*. New York, NY: Farrar, Straus and Giroux, 2013.
Santner, Eric, *The Royal Remains: The People's Two Bodies and the Endgames of Sovereignty*. Chicago, IL: University of Chicago Press, 2011.
Scheuerman, William, *Liberal Democracy and the Social Acceleration of Time*. Baltimore, MD: Johns Hopkins University Press, 2004.
Schmitt, Carl, *Die Lage der europäischen Rechtswissenschaft*. Tübingen: Universitäts-Verlag, 1950.
Schmitt, Carl, *The Nomos of the Earth in the International Law of the Jus Publicum Europaeum*. Translated and introduced by G. L. Ulmen. New York, NY: Telos Press, 2003.
Schmitt, Carl, *Political Theology: Four Chapters on the Concept of Sovereignty*. Translated by George Schwab. Chicago, IL: University of Chicago Press, 1985.
Schmitt, Carl, *Political Theology II: The Myth of Closure of any Political Theology*. Translated and introduced by Michael Hoelzl and Graham Ward. Malden, MA: Polity Press, 2008.
Schneewind, Jerome, *The Invention of Autonomy: A History of Modern Moral Philosophy*. Cambridge, UK: Cambridge University Press, 1997.
Schumpter, Joseph, *Capitalism, Socialism and Democracy*. New York, NY: Harper and Row, 1950.
Scranton, Roy, *Learning to Die in the Anthropocene: Reflections on the End of a Civilization*. San Francisco, CA: City Light Books, 2015.
Selby, Jan and Mike Hulme, "Is Climate Change Really to Blame for Syria's Civil War?" *The Guardian*, 2015. www.theguardian.com/commentisfree/2015/nov/29/climate-change-syria-civil-war-prince-charles.
Serra, Antonio T., *The Principles of Political and International Law in the Work of Francisco de Vitoria*. Madrid: Ediciones Cultura Hispanica, 1946.
Sherwood, Yvonne, "Francisco de Vitoria's More Excellent Way: How the Bible of Empire Discovered the Tricks of [the Argument from] Trade." *Biblical Interpretation* 21, no. 2 (2013): 215–275.
Shils, Edward, *Tradition*. Chicago, IL: University of Chicago Press, 1981.
Shirer, William, *The Rise and Fall of the Third Reich: A History of Nazi German*. Reissued ed. New York, NY: Simon and Schuster, 2011.
Shotwell, Alexis, *Against Purity: Living Ethically in Compromised Times*. Minneapolis, MN: University of Minnesota Press, 2016.
Slomp, Gabriella, *Thomas Hobbes and the Political Philosophy of Glory*. New York, NY: St. Martin's Press, 2000.
Smail, Raymond Charles, *Crusading Warfare, 1097–1193*, 2nd ed. Cambridge, UK: Cambridge University Press, 1995.
Small, Melvin and J. David Singer, *Resort to Arms: International and Civil War, 1816–1980*. Beverly Hills, CA: SAGE, 1982.
Smith, Ted, *Weird John Brown: Divine Violence and the Limits of Ethics*. Stanford, CA: Stanford University Press, 2015.
Sobrino, Jon, *Christology at the Crossroads: A Latin American Approach*. Translated by John Drury. Maryknoll, NY: Orbis, 2002.

Sölle, Dorothee, *Political Theology*. Minneapolis, MN: Fortress, 1974.

Stacy, Michelle, "Clash of the Time Lords." *Harper's Magazine*, 2006.

Stout, Jeffrey, *Democracy and Tradition*. Princeton, NJ: Princeton University Press, 2004.

Stout, Jeffrey, *Ethics after Babel: The Languages of Morals and their Discontents*. Princeton, NJ: Princeton University Press, 2001.

Stout, Jeffrey, "Justice and Resort to War: A Sampling of Christian Ethical Thinking" in James Turner Johnson and John Kelsey (eds.), *Cross, Crescent, and Sword: The Justification and Limitation of War in Western and Islamic Tradition*, 29. Westport, CT: Greenwood Press, 1990.

Stump, Eleonore, "Non-Cartesian Substance Dualism and Materialism without Reductionism." *Faith and Philosophy* 12, no. 4 (1995): 514.

Sutter, Paul S., "The World with Us: The State of American Environmental History." *Journal of American History* 100, no. 1 (2013): 94–119.

Taylor, Charles, *Modern Social Imaginaries*. Durham, NC: Duke University Press, 2004.

Taylor, Charles, *A Secular Age*. Cambridge, MA: Harvard University Press, 2007.

Taylor, Charles, *Sources of the Self: The Making of Modern Identity*. Cambridge, MA: Harvard University Press, 1989.

Terpstra, Nicholas, *Religious Refugees in Early Modern Europe: An Alternative History of the Reformation*. New York, NY: Cambridge University Press, 2015.

Thomas, Keith, *Religion and the Decline of Magic: Studies in Popular Beliefs in Sixteenth- and Seventeenth-Century London*. New York, NY: Penguin Books, 2003.

Thomas, Linda (ed.), *Living Stones in the Household of God: The Legacy and Future of Black Theology*. Minneapolis, MN: Fortress Press, 2004.

Thommen, Lukas, *An Environmental History of Ancient Greece and Rome*. Translated by Philip Hill. Cambridge, UK: Cambridge University Press, 2009.

Tierney, Brian, *The Idea of Natural Rights: Studies of Natural Rights, Natural Law, and Church Law 1150–1625*. Grand Rapids, MI: Eerdmans, 1997.

Tilly, Charles, *Coercion, Capital and European States*. Cambridge, MA: Blackwell, 1990.

Todorov, Tzvetan, *The Conquest of America: The Question of the Other*. New York, NY: Harper Collins, 1982.

Townes, Emilie, *In a Blaze of Glory: Womanist Spirituality as Social Witness*. Nashville, TN: Abingdon, 1995.

Tucker, Richard P. and Edmond Russell (eds.), *Natural Enemy, Natural Ally: Toward an Environmental History of Warfare*. Corvallis, OR: Oregon State University Press, 2004.

von Clausewitz, Carl, *On War*. Translated by Michael Howard and Peter Paret. Princeton, NJ: Princeton University Press, 1989.

von Clausewitz, Carl, *On War*. Translated by J. J. Graham. NP: Value Classic Reprints, 2017.

von Ungern-Sternberg, Antje, "Religion and Religious Intervention" in Bardo Fassbender and Anne Peters (eds.), *The Oxford Handbook of the History of International Law*. New York: Oxford University Press, 2012, 308.

Walsh, Edmund and James Mackintosh, *The Spanish Origin of International Law: Lectures on Francisco De Vitoria (1480–1546) and Francisco Suarez (1548–1617)*. Edited by James Brown Scott. Washington, DC: School of Foreign Service, Georgetown University, 1928.

Walzer, Michael, *Arguing about War*. New Haven, CT: Yale University Press, 2004.

Walzer, Michael, *Just and Unjust Wars: A Moral Argument with Historical Illustrations*. New York, NY: Basic Books, 1977.

Walzer, Michael, *The Revolution of the Saints: A Study in the Origins of Radical Politics*. Cambridge, MA: Harvard University Press, 1982.

Walzer, Michael, *Spheres of Justice: A Defense of Pluralism and Equality*. New York, NY: Basic Books, 1984.

Walzer, Michael, "The Triumph of Just War Theory (and the Dangers of Success)." *Social Research* 69, no. 4 (2002): 925–944.

Watson, Stephen, *Tradition(s): Refiguring Community and Virtue in Classical German Thought*. Bloomington, IN: Indiana University Press, 1997.

Watson, Stephen, *Tradition(s) II: Hermeneutics, Ethics, and Dispensation of the Good*. Bloomington, IN: Indiana University Press, 2001.

Weber, Max, *The Protestant Ethic and the Spirit of Capitalism*. New York, NY: Oxford University Press, 2010.

Welzer, Harald, *Climate Wars: Why People Will Be Killed in the 21st Century*. Translated by Patrick Camiller. Malden, MA: Polity Press, 2017.

Whetham, David, *Just Wars and Moral Victories: Surprise, Deception and the Normative Framework of European War in the Later Middle Ages*. Leiden: Brill, 2009.

White, James Boyd, *When Words Lose Their Meaning: Constitutions and Reconstitutions of Language, Character, and Community*. Chicago, IL: University of Chicago Press, 1984.

White, Lynn, "The Historical Roots of Our Ecological Crisis." *Science* 155, no. 3767 (1967): 1203–1207.

White, Sam, *A Cold Welcome: The Little Ice Age and Europe's Encounter with North America*. Cambridge, MA: Harvard University Press, 2017.

Williams, Robert, *The American Indian in Western Legal Thought: The Discourses of Conquest*. New York, NY: Oxford University Press, 1990.

Wilson, Eric, *The Savage Republic: De Indis of Hugo Grotius, Republicanism, and Dutch Hegemony within the Early Modern World-System (c. 1600–1619)*. Leiden: Martinus Nijhoff Publishers, 2008.

Winchester, Simon, *Land: How the Hunger for Ownership Shaped the Modern World*. San Francisco, CA: HarperCollins Publishers, 2021.

Wisniewski, Benjamin, "Soldier, Officer, Citizen: Applying Just War Theory to Police Use of Force." Ph.D. thesis, University of Missouri, 2017. https://irl.umsl.edu/thesis/300/.

Wolin, Sheldon, *Politics and Vision: Continuity and Innovation in Western Political Thought*. Princeton, NJ: Princeton University Press, 2004.

Index

1648's Peace of Westphalia, 99
Abolitionism, 177
Abrahamic traditions, 209
"Acts of God" and the Natural World, 267
Advances in Weaponry, 100
Age of Conquest, 53
agency, 57, 58, 97, 100, 103, 125, 127–128, 172, 182, 196, 213–214, 216, 224, 231–243, 244, 250, 274, 294
agricultural revolution, 129, 298
agriculture, 63, 68, 129, 265, 275
Alan Gewirth, 277
Alasdair MacIntyre, 10, 11, 13, 295, 300, 303
Alberico Gentili, 91, 82, 100, 111, 184, 235
Albert Einstein, 126
Alexander the Great, 307
Alfred W. Crosby, 278
alienation, 69, 70, 77, 120
An Eerdmans Reader in Contemporary Political Theology, 179
Andreas Vesalius, 30, 69
Anglican, 108, 187
Annabel Brett, 46
Anselm of Canterbury, 312
anthropocentrism, 247, 285
anthropogenic climate change, 321–322
Anti-Federalist, 136
Antonio Cassese, 51
Antony Anghie, 105
Antony of Montesino, 33
Archbishop William Courtenay, 263
Aristotle, 29, 66, 111, 270, 307–308, 313
  and Early Just War Language, 2, 308
  and Greek Political Thought, 306
  and Human Relations with the Non-Human Natural World, 76, 308
  and the Expansion of Greece, 307
  and the Greek Climate, 307
  and War, 6
  and Zoology, 308
Arius, 311
Arminianism, 87
Augustine, 1, 2, 6, 40, 66, 99, 125, 133, 208, 272, 305, 310–311
  and the Church and State, 241, 310
  and the Constantinian Church, 202, 310
  and the End of the Roman Empire, 310
  The City of God, 133, 228, 310, 317
Augustinian, 2, 3, 91, 271, 295, 303, 305, 313
authority and agency, 182
autonomy, 24, 25, 102, 105, 140–141, 153, 160, 218, 225–226, 228–229, 231, 233, 234
  and feudalism, 231
  and Immanuel Kant, 225
  and Jean-Jacques Rousseau, 225
  and John Locke, 225
  and John Stewart Mill, 225
  and Modern Thought, 24, 237, 240, 243, 289
  and modernity, 70, 139, 225
  and Rene Descartes, 225
  and the Separation of Peoples and Peoples and Things, 226, 232

Averroes, 313
Avicenna, 313
axial periods, 19, 20, 21, 299
  and Just War Tradition, 20, 26
  and Traditions, 20, 22, 303
Azar Gat, 67, 223

Barbara A. Hanawalt, 262
Barbara McClintock, 126
Bartolomé de las Casas, 3, 32–33, 65, 235
Baruch Spinoza, 193, 227
bellum iustum, 74
Benjamin Franklin, 176
Bernard of Clairvaux, 312
biblical inerrancy, 176
Blackstone, 145
Bonnie Honig, 132
Brian Fagan, 129
Buddhism, 209

Calvinism, 87, 136, 141, 177, 188–189
capitalism, 21, 70, 205, 219–220, 237
Carl Schmitt, 109, 161, 179–193, 201, 207, 209
  and Constitutional States, 185
  and Constitutions, 182, 184–186, 200
  and Erik Peterson, 191
  and European jurisprudence, 183
  and his Wicked Politics, 184
  and International Law, 109, 182
  and Language, 187
  and Law, 182, 184–185
  and liberal constitutionalism, 185
  and phenomenology, 181, 189
  and Political Theology, 109, 179–181, 183, 191, 193
  and Political Theology (book), 182, 185
  and Political Theory, 181
  and Political Thought, 182
  and Secularity, 179, 190
  and Sovereignty, 161, 181, 182, 185–191, 194
  and Space, 183
  and Temporal Deferral, 200–201
  and the Exception, 181–186, 190, 191
  and the Great Man Myth of History, 186
  and the motorization of lawmaking, 200
  and the Secular, 191, 209
  and the Sovereign, 161, 182–185, 189, 194
  and the term Secular, 180–181
  and the Trinity, 191
  and Theology, 179, 181–182, 184–191, 209
  and Thomas Hobbes, 181, 207
  The Nomos of the Earth in the International Law of the Jus Publicum Europaeum, 182
Cartesian dualism, 73
casus belli, 15, 117, 263
Cathleen Kaveny, 139
Catholicism, 56, 81, 177, 235
  and Time, 208
Catholicism vs. Protestantism, 87, 187
causation, 124, 125–126, 130, 283, 314, 322
  and Human Agency, 127, 244, 250
  and religion, 103
  and the Sciences, 126, 323
Celia Deane-Drummond, 289
centripetal questions and Centrifugal Answers, 296
changes in the global society and just war thinking, 318
Charles Butler, 86
Charles Mathewes, 304
Charles Taylor, 18, 55, 80, 93, 99, 180, 193, 298, 295
  and the Secular Age, 55, 180
Charles Tilly, 229–230
checks and balances, 139, 146–147, 186, 197
Christendom and the Natural World, 276, 284
Christian just war thinking, 23, 26, 69–70, 77–79, 83–84, 87, 109, 121, 127, 132–133, 142, 171, 217, 222–223, 235, 237, 240, 242, 273, 277, 283
  and Colonization, 275, 283, 284
  and Complicity, and the Malformation of Thought, 222
  and Human Interactions with the Natural World, 239, 248, 249, 256, 260, 264, 274, 283
  and Modern Thinking, 222
  and Modern Thought, 24, 73, 106, 124, 233–234, 238, 240–241, 244, 249, 273, 284, 290
  and Other, Self and State, 259
  and Property Ownership, 42
  and Technological Changes, 231
  and the Environment, 25, 78, 240–241, 252, 275, 284

and the Natural World, 239, 241, 249
and the Non-Human Natural World, 42
and War, 42
Christian just war tradition, 6, 7, 14, 18, 19, 26, 31, 40, 79, 80, 85, 91, 98–99, 106, 108, 124–125, 206, 216, 234, 286, 290, 292, 297, 301–302, 305–306, 316–317, 319–320, 324
and Climate Shaped Violence, 319
and Colonization, 277, 283–285, 277
and Confession, 242
and Documents, 142, 149
and Environmental Destruction, 241
and Environmental History, 25, 250–252, 259, 185
and Historical Context, 5, 6, 7, 8, 17, 19, 66, 316
and History, 18, 24, 242, 297–298, 305, 314
and International Law, 80, 91, 104, 106, 127–128, 130, 153, 206, 234, 241
and its Failings, 72, 108, 149, 217, 241
and Justice, 296
and Modernity, 15, 19, 24, 25, 67, 72, 93, 222, 225, 234, 237, 239, 241, 245, 286, 289, 290, 292, 298, 319
and Positive Law, 153
and Sects of Christianity, 12, 108
and Silence, 238
and the "Other," 22, 79, 106, 127, 317
and the Anthropocene, 244–245
and the Climate, 217, 283
and the Constitution, 133, 142, 148, 151, 153, 156
and the Environmental Age, 287
and the Environmental Crisis, 290
and the Myth of Progress, 216–217
and the Non-Human Natural World, 25, 42, 78, 134, 156, 249, 250, 252, 275, 280, 287, 292
and Theological Anthropology, 79, 131, 223
through Time, 7, 23–25, 27, 42, 59, 66, 77, 80, 151, 206, 221, 229, 242, 259, 277, 283–284, 286–287, 289–290, 292, 297, 314, 319
Christian Just War Tradition, 6, 14
Christian just warriors, 25, 241, 297, 320
and "Larger Spheres of Interaction," 253, 292
and the Natural World, 259, 264, 266, 268, 280, 292
Christian pacifism, 324
Christian Pacifism for an Environmental Age, 12, 21–22, 295, 320, 324
Christian pacifist tradition
and Fidelity, 22, 296
Christian thinking
and Resort to War, 223
Christianity, 2, 56, 74, 81, 107, 203, 209, 242, 246–247, 249, 261, 295, 301, 310–311, 317
and Indigenous Persons, 38, 39
and Sovereignty, 188, 190
and the Natural World, 71, 238, 243, 247
and Transformation, 202, 301
and War, 2
as an Expression of Religion, 175
as the Dominant Religious Force in America, 177
Christine de Pizan, 261, 269
and Ambush, 261
Christology, 179, 188
Christopher Columbus, 32, 49, 69, 76
Christopher Marlowe, 193
church
and Empire, 33, 102, 312
Cicero, 2, 86, 270, 305, 309
and Roman Expansion, 308, 309
War and Natural Law, 2, 270
citizens vs consumers, 220
civil law, 46
Civil Rights, 177
claims to human rights, 138
clima Romanus, 310
climate catastrophes and Social and Economic Inequality, 169
climate change, 21, 119, 158, 163–168, 169, 170–172, 201, 212–215, 217, 221, 258, 289, 291, 295, 321–322, 323–324
and Colonization, 282
and Europe, 258, 282, 310, 312
and Inequality, 168–169, 170, 172, 168
and its Scale, 163
and Scarcity of Resources, 118
and the Military, 217, 323
and the Natural Sciences, 289
and War, 129, 130, 170, 291, 322–324
as Catalyst, 163
Climate Terror and Militarization, 169
Climatic stress, 129

clocks and monasteries, 203
coalition, 315
codes of chivalry, 99, 272
  and the Natural World, 266
coercive authority, 192
coercive force, 36, 47, 296, 325
colonization, 3, 20, 24, 30, 73, 77, 82, 102, 106, 108, 114, 127, 128, 146, 159, 168, 175, 177, 225, 232, 252, 267, 275–278, 283–284, 285
  and Contrasting National Behaviors, 44, 256
  and Economics, 75, 104, 128
  and Land, 253, 278, 284
  and the Catholic Church, 48
  and the Deficiencies of Europe, 128
  and the Religious Wars of Europe, 107
communication, 47, 76, 121, 193, 195, 263
community, 10, 14, 35, 45–46, 93, 97, 102–103, 122, 133, 139–141, 161, 163–165, 172, 174, 193, 195–197, 204
conceptions of the Self, 225
confession, 242
conflict in Syria, 217, 321–322, 324
  and drought, 322
  and urbanization, 172
  and Climate Change, 322
  and Drought, 322
  resources religion and refugees, 324
conscientious objection, 107, 143
conservative American Christian Beliefs, 176
constituting power, 174, 186
Constitution
  Belgium, 138
  Brazil, 138
  Corsica, 137
  France, 137
  Great Britain, 137
  Norway, 137
  Polish-Lithuanian, 137
  Spain, 138
  Sweden, 137
  the Netherlands, 137
constitutional law, 150, 185–186
constitutional law and state sovereignty, 185
constructivism, 292
contemporary neoliberalism and the environmental crisis, 220
Contract theory, 93
conversion, 39
corporate action, 137
covenantal theology, 194
cuius regio, eius religio, 103
cultural imperialism, 65, 38
culture and nature, 246, 251

Darwinian, 124
David Gray Adler, 145
de-fetishizing freedom, 221
deism, 185, 227
Desiderius Erasmus, 28–30, 55, 89
deterritorialization, 170
Different Perspectives on the Natural World, 239, 259, 284
Discontinuity and Tradition, 10, 67, 300, 304, 305
discrimination, 2, 3, 17, 100, 122, 127, 146–147, 149, 152, 167, 170, 254, 257, 300
Diversity of Moral Traditions, 9, 223
divine action, 302–303
  and the Non-Human Natural World, 302–303, 326
  and war, 326
divine command, 57, 189
divine law, 36, 38, 112, 181, 228, 270
Divine right, 69, 155, 180, 227
  Human Rights, 161
divine sovereignty, 187–189, 190
  and politics, 188
  Divine Law, the Attributes of God and the Powers and Purposes of the State, 181
divine violence, 302
Doctrine of Double Effect, 3
domestic law, 137, 149, 283
dominium, 74–75, 258
Dorothee Sölle, 178
dualistic narratives, 84
due process, 139, 150

Eastern Orthodoxy, 247
economic and political systems, 220, 315
  and the Environmental Crisis, 129
Economics, the Environment and Resources, 22
Edmond Burke, 135
Edward E. Antonio, 246
Edward Shils, 293, 296
  and Flaws, 294

Tradition, 293
Tradition and Natural Forces, 293–294, 296
Elisa Grimi, 295
embodied persons, 61
Emmerich Vattel, 143, 148, 184, 261
encomienda, 33, 49
Enlightenment, 3, 88, 188, 237, 276
Enticement and Obedience, 195
environmental change, 317, 325
environmental collapse, 289
environmental degradation, 71, 211, 221, 274, 323
environmental ethics, 246, 323
environmental history, 246, 250–252, 276, 284–286, 323
and Hybridity, 250
and the History of the Christian Just War Tradition, 25, 250, 314
and War, 250, 290, 315
environmental issues, 163, 323
environmental issues and the law, 158
environmental law, 275
environmental legislation, 246
environmental security studies, 78, 119, 323
eschatology, 206, 209, 241
essentialism, 292
ethical and cultural creation, 223
ethnocentrism, 87, 104
European colonizing projects, 276
European Colonizing Projects and Environmental History
Access to Natural Resources, 276
and Changes in Warfare, 280
and Changes to Warfare, 282
and Climate Change, 282–283
and Invasive Species, 278
Favorable Environmental Conditions, 276
European financial markets, 75
European imagination, 27
European property law, 77
evangelical, 39, 176, 211, 307
evangelization, 34, 39, 57
evolutionary biology, 299
explicating traditions, 319

faith and national identity, 103–104
false witness, 176
fear of the impact of religion, 176
Federalist, 136
Fernando Vazques de Menchaca, 30
forms of government, 232
fossil fuels, 157, 254, 315
Francis Bacon, 30, 71, 273
The Proficience and Advancement of Learning, 126
Francis I of France, 101
Francis Lieber, 3
"Code for the Government of Armies in the Field," 148, 154, 283
Francisco de Vitoria, 3–4, 6, 30–67, 70, 75–77, 79–84, 89, 91–92, 100, 105, 113–114, 131, 154, 224, 235, 237–238, 256, 272, 277, 281, 297, 305–306, 314
and Christianity in War, 74
and cultural imperialism, 65
and Indigenous Persons, 28, 31, 33–34, 36 39, 41, 43, 48–50, 52–55, 58, 60–66, 69–70, 73, 75, 224
and International law, 30, 31–32, 44–45, 51–52, 54, 65–66, 82–83, 92
and irreconcilable just cause, 54
and Just War Thinking, 28, 31, 34, 40, 41, 43–44, 52, 54, 57–58, 73, 83, 84
and Justification for War, 54, 63, 256
and New Laws, 42, 82, 256
and Property Law, 37, 75–77, 255
and Religious Persons, 61
and Rules of War, 41, 51
and Slavery, 70
and Social Changes Due to the Age of Exploration, 66
and Spanish Conquest, 46, 49
and the modern state, 70
and the Non-Human Natural World, 42, 76, 224
and the Other, 27, 34, 44, 50–51, 54, 63–66, 72, 77, 79
and the Salamanca School, 69, 74, 75
and Universal Law, 36, 46–47
commodification, colonization and racism, 224
de Indis, 29, 32–34, 37–42, 44, 48, 49–50, 52, 55–59, 63, 76
De Iure Belli, 27, 32–37, 40–42, 76
Indigenous Peoples in the Americas and West Indies, 37
on Indigenous Peoples in the Americas and West Indies, 38, 61, 224

Francisco de Vitoria (cont.)
  Relectio de Indis, 46
  Relectio de potestate civili, 46
  Relectiones, 27, 29–31, 59–60, 91, 256
  the Non-Human Natural World, 75
Francisco Suárez, 271
Fredrick the Wise, 101
free will, 188, 219, 227
freedom, 24, 25, 75, 80, 102, 112, 142, 177, 182–184, 188–189, 195, 213, 218–219, 221, 224–225, 269, 298, 306
  liberty and goods, 218
  political systems and capitalism, 219

Gabriella Slomp, 207
Gaius, 270
Galileo, 31, 227
Georg Agricola
  Control over Nature, 257
  De re metallica, 257
Georg Wilhelm Friedrich Hegel, 3, 6–7, 13
  The Hegelian Theory of History, 7
George Washington, 176
Giorgio Agamben, 179
globalization, 21, 169
  and religion, 120
God, 2, 7, 22, 36, 37, 47, 60, 61, 76, 82, 87, 89, 94, 125, 136, 153, 161, 175–177, 179–182, 186–189, 191, 202, 204–210, 219, 227–228, 233, 236, 238, 241–242, 246, 248, 254, 259–260, 268–270, 280, 293, 296, 301–302, 304–305, 310–312, 317–319
  and Christian Just War Tradition, 7
  and Stability, 259, 304
  and the Christian Just War Tradition, 234, 301
  and Time, 204–205, 301, 303
  and War, 37
Governance in Crisis vs. Order, 137
Graham Hammill, 174, 179–180, 186, 192–195
  and Political Theology, 179
  and the Mosaic Constitution, 192
Gratian, 2, 37, 99, 271, 305, 310–312
  and Canon Law, 2, 311–312
  Concordia discordantium canonum, 311
Greece, 20, 307
Greta Grace Kroeker, 120

Hans Joas, 237, 239
Hans Kelsen, 184
Harald Welzer, 289
Henry V, 263, 264
heresy and social imaginaries, 315
Herman Witsius, 208
Hermann Conring, 102
hierarchy, 17, 161, 171, 272
Hildegard of Bingen, 312
Hinduism, 209
Hippocrates, 308
Honoré Bonet, 125, 261, 264, 272
  and the Hierarchy of Resources, 261
  The Tree of Battles, 261, 264, 266
hope, 174, 206, 208, 215, 241, 301
  and optimism, 213, 215–217, 221
  and Optimism, 213
  and politics, 167, 207, 214
  and the Anthropocene, 213, 214
  and theology, 214
  optimism and politics, 215
  optimism and time, 215
  and history, 214
Hugo Grotius, 1, 3–6, 23, 30–31, 42, 47, 51, 58–59, 79–118, 120–122, 131–132, 143–144, 151–152, 154, 224, 227, 235, 237–238, 264, 272, 274, 277, 297, 305–306, 309, 314, 316
  and Christian Just War Tradition, 80, 85, 91, 93, 98–99, 103
  and Colonization, 3, 104, 108
  and Developing Technology, 104
  and Humanism, 87, 100
  and Indigenous Persons, 104
  and International law, 31, 51, 80, 82–85, 87, 91–92, 94–95, 104–105, 110, 112, 122
  and Intra-European Strife, 81, 83, 274
  and Just and Unjust Processes in Conflict, 91
  and Just War Thinking, 83–85, 87, 93, 98, 106, 109, 112, 115, 118, 120
  and Justification for War, 101, 113, 115–117
  and Natural Law, 82, 83, 85, 89–90, 92, 94, 98, 104, 110–111, 117
  and Natural Rights, 91, 111–112

and Ownership, 80, 114
and Problems in his Work, 117
and Problems with his Work, 92, 120
and Property, 90, 91, 93, 101, 105, 107, 109, 112–116
and Reason, 88, 90, 107, 109–112
and Rebellion, 109
and Refugees, 113, 120–121, 122
and Rules of War, 112
and Secularization, 103
and Slavery, 117
and Social Contract Theory, 94, 97
and States, 81, 90, 91, 95, 99, 102–104, 109, 115, 121
and the Movement from Just War Tradition to Theory, 84, 108
and the Movement of Peoples, 81, 120
and the Non-Human Natural World, 77, 116–118, 120
and the Other, 106
and the Self, 84–85, 87, 93–94, 96, 100, 104–105, 118, 120, 131
and the State, 83, 88, 90, 92, 101
and Trade and Ownership, 114–115
and Transitioning Social Imaginaries, 83, 87
and Westphalia, 51, 82, 85, 101, 105, 235
De Jure Praedae, 113, 117
De Mare Liberum, 113
Neutrality and Bias, 92
On the Rights of War and Peace (De Juri Belli ac Pacis), 3, 81, 277
Religion and Morality, 92, 99
Self, State and Other, 83, 95, 105
Hugo Krabbe, 184
Huldrych Zwingli, 204
human behavior and nature, 157, 158
human beings and the natural world, 76, 114, 228, 239, 243, 247, 258, 267, 273, 308
human relations to the natural world, 156, 257, 308
human rights, 4, 6, 13, 49, 90, 105, 112, 122, 132, 156, 160, 161, 186, 226, 227, 237, 314
and global ethics, 156–157
and just war thinking, 156
human sociality, 35, 38, 42, 47, 75, 115, 289
and combat, 289
Human Societies and Individual Ideology, 229
human-and-nature-in-interaction paradigm, 239
humanism, 29, 57, 87, 100, 281
humanitarian intervention, 43, 170, 287
Humanity, Defense and Ownership, 75
Humans as Agents in the World, 243
hybridity, 250

identity, 68, 102–103, 105, 122, 158, 162, 176, 181, 191, 196, 199, 313
Ileana Porras, 77, 116
immanence, 24, 102, 139, 190, 225, 228, 229, 231–234, 237, 239, 240, 243, 275
and Augustine, 228
and Heidegger, 228
and Humans, 228
and Kant, 228
and Plato, 228
and teleology, 228
and the power to bring about change, 231–232
and the US Constitution, 139
and transcendence, 231–232
Autonomy, Universalizability and Instrumentalism, 228
Immanuel Kant, 3, 152, 226
and Political Systems, 154
and Reason, 316
and Universality, 226
and War, 237, 316
impartiality, 83, 152
Imre Lakatos, 299
Individual and State Sovereignty, 155, 160, 162, 164, 167
Individual Sovereignty to Individual Morality, 157
instrumentalism, 24, 227–229, 231, 233–234, 236–237, 240
and agency, 231–232
and modernity, 139, 225, 239
and the Future, 239
nihilism, 228
intellect, 61
inter-human agreement, 46

international law, 3–5, 14, 23–24, 30, 31–32, 39, 44–45, 51–52, 54, 59, 66, 74, 77, 80–85, 87, 91–92, 94–95, 98, 102, 104–106, 109–112, 121–123, 124, 126, 128–130, 149, 152, 167, 173, 182, 206, 224, 227, 232, 235, 237, 241, 271, 283, 325
  and early international law, 51
  and human rights, 6, 23, 102, 105, 112, 121, 227
  and indigenous peoples, 32, 235
  and indigenous persons, 51–52, 65, 106
  and just war thinking, 4, 59, 84–85, 120, 123, 128–130, 173
  and just war tradition, 5, 14, 44, 52, 66, 80, 91, 104, 106, 124, 126, 130, 149, 153, 206, 234, 241
  and modern international law, 30–31, 45
  and positive law, 45, 51
  and refugees, 121–124, 224
  and secularizing the tradition, 106, 110, 127
  and western influences on international law, 65, 74
  early international law, 121, 123, 234
  modern international law, 30, 98, 241
  secularizing the tradition, 54
  Western Influence on International Law, 30, 51–52
invincible ignorance, 36, 38, 53, 54, 56, 58
Isaiah 55:8, 302
Islam, 20, 209
Islamic, 13, 83, 238, 264, 317
Islamic Just War Tradition, 13, 264, 317
Israel, 37, 47, 192, 314
ius gentium, 36–39, 42, 45–47, 50–51, 56–57, 62–65, 74–75, 115
ius natural, 42

Jacques Derrida, 179
James Boyd White, 139, 223–224, 299
James Brown Scott, 30–31
James Harrington, 193
James Kennedy, 176
James M Gustafson, 291
James Madison, 139, 316
James Turner Johnson, 2, 4–5, 17, 27, 53, 82, 91, 95, 98, 107, 112, 156, 161, 206–261
James Witherspoon, 135
Jean Bodin, 69, 161
Jean-Jacques Rousseau, 137, 154, 181, 225, 207–208, 226
Jeffery Stout, 11, 13, 48, 80, 223, 295, 296
Jerry Falwell, 175
Jewish Just War Tradition, 13, 264, 317
Joachim of Fiori, 312
Joannes de Legnano, 30
Johann Baptist Metz, 178–179, 209
Johannes Cocceius, 208
John Calvin, 101, 109, 124, 205
John Courtney Murray, 179
John Dewey, 227
John Grim, 238
John Knox, 109
John Locke, 135, 139, 145, 148, 208, 225, 235, 281
  and the US Constitution, 139, 145, 208, 235
John Mueller, 78, 230, 315, 317
John of Leyden, 203
John of Paris, 259
John Wycliffe and the Lollards, 263
Judaism, 141, 177, 209
Julia Reinhard Lupton, 179
  Political Theology and Early Modernity, 179
Julian N. Hartt, 260
Juliet Barker, 267
June Jordan, 215
Jürgen Moltmann, 178
jus ad bellum, 2, 3, 15–16, 40, 53–54, 58–59, 78, 100, 106–107, 115, 134, 144–147, 149, 153, 155, 169, 206, 234, 257, 259, 264, 283
jus ante bellum, 78, 234
jus gentium, 77, 116, 107
jus in bello, 2, 3, 15, 17, 40, 53, 59, 78, 91, 100, 106–107, 115, 122, 134, 144, 146, 148–149, 153, 157, 169, 206, 234, 254, 257, 283
jus post bellum, 78, 122, 234, 287
just cause, 2–3, 34, 36–38, 41–42, 53–59, 93, 101, 114, 122, 146, 152, 167, 170–171, 299
  and Cultural Differences, 54
  and Self Defense of Self and Property, 58, 90, 93, 101, 107, 114, 146, 264, 271

and the Self-Defense of Sovereign States, 41, 57, 101, 234
just order, 69
just war language, 42, 308
just war theory, 1, 4, 8–9, 13–17, 84, 85, 98, 106–108, 122, 125, 130, 143, 148–149, 152, 155–156, 169, 171, 173, 210, 217, 272, 286
and Colonization, 107
just war thinking, 3–4, 8–12, 14, 15, 18, 24, 26, 28, 31, 34, 37, 40–44, 52–54, 57, 58–59, 73, 78, 84–85, 93, 98, 106–107, 112, 115, 118, 120, 122–123, 128–129, 132–134, 143, 146, 148–155, 157–158, 166–167, 173, 206, 210, 224–226, 237–238, 253, 260–262, 264, 268–269, 271, 277, 279, 281, 283, 287, 296, 298–299, 313, 316, 318, 319, 323
and Artifacts, 133
and Climate-Shaped Conflict, 157, 172, 274, 319
and Colonization, 44, 127, 146, 231, 276, 280, 283
and Decision Deferral, 149
and Heresy, 315
and Justness, 210
and Last Resort, 43
and law, 15, 52, 54, 59, 84, 85, 109, 115, 123, 124, 127–130, 143, 149, 151, 153, 168
and Legitimate Authority, 210, 234
and Modernity, 44, 45, 70, 118, 130, 148, 154, 167, 222, 225, 241, 249, 273–274, 290
and Natural Law, 85, 269, 271, 273, 269
and proportionality, 43, 149
and Stable Societies, 10, 298
and the American Revolution, 316
and the Development of Modern Democracies, 168
and the Evolution of Societies, 299
and The French Revolution, 316
and the Myth of Progress, 217
and the Other, 34, 44, 54, 58, 64, 72, 73, 79, 106, 149
and the Post-Cold War World, 318
Climate-Shaped Conflict, 222
Early Modern Just War Thinking, 72–73, 133, 154, 234
just war thought, 9, 13, 16, 20, 28, 98, 134, 253, 264, 268, 271, 280–281, 284–285, 287, 300, 311, 313, 315, 318–319
and Self, State and the Other, 28
and social changes, 315
in the Non-Human Natural World, 268, 286
just war tradition, 2, 3–8, 10, 13–14, 18–20, 27, 44, 50, 52, 98, 100, 133, 149, 151–153, 155, 240, 287, 290, 299, 306
justice, 10–11, 16, 36, 41–42, 53, 56, 78, 80, 88, 90, 101, 113–115, 118–120, 125, 140, 142, 165, 167, 174, 195, 200, 215, 223, 228, 238, 242, 265, 272, 296–298, 306, 316
and war, 41, 56, 91, 101, 115
Justifications for War, 54, 63, 83, 101, 113, 116, 117, 256, 270
Justinian, 252, 310
Corpus juris civilis, 271

Kant
and War, 3
Karl Barth, 189
and Divine Command Ethic, 189
Karl Marx, 70

land ownership, 253
language, 10, 42, 47, 60, 82, 87, 95, 103, 149, 177, 187–188, 223, 237, 242, 267, 274, 296, 299
About the Natural World, 249, 286
and power, 52
and religion, 57, 92, 102–103
last resort, 3, 43, 58, 170
late medieval Christian just war thinking, 240
and environmental history, 251
Late Medieval Christian Just War Thinking to Modern Just War Thinking and the Non-Human Natural World, 239, 252, 260
late medieval Christian just war tradition, 125, 259
late medieval just war thinking, 238, 257, 272
war and morality, 238, 240

late medieval just war thought, 271
and natural law, 271
legal systems, 249
and environmental change, 276
legitimate authority, 2, 3, 16, 17, 41, 58, 83, 100, 122, 142, 145–147, 152, 156, 167, 197, 210, 254, 264, 266, 271, 300, 309
legitimate cause, 90
Leo Strauss, 179
lex Romanus, 309–310
lexicon, 109
and modern just war theory, 286
and modernity, 97
and the non-human natural world, 286
liberal jurisprudence, 185
liberation theologians, 179
libertine serfdom, 266
liberty, 135, 139, 156, 192, 218–220, 225, 281, 150
likelihood of success, 125, 127
Limited wars, 132
Lisa J. Kiser, 262
literary authorship, 192
liturgy, 242
loss of biodiversity, 22, 291
Lutheran, 103
Lynn White, 71, 246, 260, 284
and Context, 249
and Human/Nature Dualism, 246
Christianity, Human Beings and the Natural World, 76
"The Historical Roots of Our Ecologic Crisis," 284

Magna Carta, 139
maintenance of governance, 137
Malthusian, 119, 169
Margaret Mead, 215
Martii Koskenniemi, 74, 76, 112
Martin Luther, 29, 101, 109, 204
Martti Koskenniemi, 73
Mary Evelyn Tucker, 238
mass army, 279, 283
mass migration, 289
Matthew Sutcliffe, 82
Matthew Tindal, 227
Max Weber, 205, 230
Michael Walzer, 1, 4–6, 15, 46–47, 122, 138, 153, 155–156, 189, 205, 221
Just and Unjust Wars, 4
Michel Foucault, 52
microbiomes, 291
military law, 3
military might, 223, 308, 253
Military Might and Productive Capacity, 223
military training, 147, 318
and school, 317
Modern (Inert) Conceptions of the Non-Human Natural World, 285
modern democracy, 227
modern just war thinking, 72–73, 132–133, 151, 154, 156, 170, 234, 244, 279, 287
and Natural Events, 279
and politics, 72–73, 154, 234
modern just war tradition
and the natural world, 292
Modern just war tradition, 126
modern political theology
and secular theology, 193
and secularized theology, 181
modern political thought, 134, 160, 182, 186–187, 190, 192–193, 202, 109
and the natural world, 76
modern social imaginary, 18–19, 133, 163, 171, 225, 231–232, 240, 250, 324–325
and free markets, 324
and the environment, 163, 240, 250
and the state, 133, 171, 225
modern state, 24, 70, 154, 161, 164, 170, 181, 190, 192, 195, 230, 282
and climate change, 164
and the monopoly on violence, 170, 181, 230
modernity, 5, 8, 11, 15, 19, 21, 24–25, 44–45, 49, 56, 63, 65, 70–71, 73–74, 84, 93–94, 97–98, 100, 102, 104, 118, 125, 130, 138–139, 142, 148, 152, 154–155, 167, 170–171, 180, 187, 190, 207, 222, 225–226, 229–232, 234–237, 239–241, 243–245, 249, 261, 273–274, 277, 284, 288, 289, 290, 292, 295, 298, 300, 319, 324, 325
and "The Disengaged Subject," 93–94
and Capitalism, 70, 237
and Changing Economic Systems, 225, 235

and Christianity, 71, 261
and Demographic Shifts, 225
and Human Rights, 237
and Individual and National Sovereignty, 164, 167, 237
and its Beginnings, 231
and Modern conceptions of evil, 236
and Modern Democratic Polities, 237
and New Religious Identities, 225
and new Technology, 225
and political forms, 171, 230
and Secularism, 243
Montesquieu, 139
moral arguments, 10, 14, 64, 239
moral sense, 55, 268, 270
morality, 1, 3, 5, 9, 12, 15, 57, 64, 83, 92, 99–100, 106, 152, 157, 175, 212, 225, 228, 233, 237–238, 269–270, 272
and law, 3, 99, 175, 212, 270
and politics, 64
and religion, 57, 92, 175, 228

national borders, 162–163
national sovereignty, 43, 164, 170, 237
nationalism, 162
NATO, 17
natural history, 239, 250
natural law, 2–3, 36–37, 45–46, 49–54, 57, 74, 82–83, 85, 88–90, 92, 94, 98–99, 104, 107, 110–111, 117, 138, 141–142, 188, 269, 270–274, 280, 308
and canon law, 271
and Eurocentrism, 107
and God, 89
and positive law, 46, 51–52, 141–142, 250
and Roman Catholicism, 51, 54
and Roman Law, 270
natural law thinking, 54, 82, 98, 269–271, 281
natural order, 36, 43, 45, 47, 50, 272
natural resources, 37, 66, 68, 71, 104, 106, 114, 117, 159, 166, 221, 249, 257–258, 262–265, 276–277
and trade, 277
natural rights, 49, 91, 111, 271
and property rights, 91
nature, 1, 36, 38, 42–43, 45, 55, 57, 59–61, 63, 71, 77, 79, 89–90, 93, 107, 109–111, 113–114, 116–117, 134, 141–142, 182, 207, 214, 225, 227, 238–239, 241, 246–248, 250–251, 257–260, 262–263, 266, 268–270, 273, 279–280, 284, 289, 292, 308
and agency, 224
and ownership, 114
vs. culture, 292
Nazism, 179, 192
NGOs, 166
Niccolò Machiavelli, 30, 55, 64, 69, 111, 174, 186, 193, 207, 236
Nicholas Copernicus, 30
Nigel Biggar, 272
Nils Petter Gleditsch, 129
nominalism, 292
non-human natural law, 275
non-human natural world, 8, 12, 22, 25–26, 42, 75–78, 116–117, 119–120, 128–130, 134, 156–157, 224, 238–240, 243, 256, 258, 260–262, 264, 266, 268–269, 271–281, 284–287, 290–292, 297–300, 302–303, 305, 307–308, 312, 314–315, 317–319, 324, 326
non-state actors, 122, 164–167, 170
nuclear weapons
and just war theory, 4, 122

obedience, 112, 193–196
authority and community, 193, 196
objectification, 65
Oliver O'Donovan, 179, 289, 296
optimism, 213
and agency, 213, 216
and hope. *See* hope and optimism
order of nature, 57, 270
othering
and otherness, 52, 69, 71–72, 80, 196

pacifism, 51, 100, 234, 316, 317, 323, 324
paleo- and historical climatology, 323
Parmenides of Elea, 308
Pat Robertson, 176
paternalism, 39, 42, 43, 63

Paul Ramsey, 6, 66, 107, 122, 133, 153, 259, 305, 314–316
and the World Wars, 314
The Just War
Force and Political Responsibility, 4
Paulus Niavis
The Valley of Beauty, 266
Pax Romanus, 310
peace, 2, 16, 21, 51, 63, 78, 82, 85, 89, 91, 92, 95, 98, 99, 101, 108, 109, 120, 132, 148, 153, 155, 167, 202, 211, 234, 238, 242, 300, 313, 314, 316, 317
peacemaking programs, 323
Pedro Álvares Cabral, 69
Pelagius, 311
perceptions of the non-human natural world, 300, 307, 314
Personhood
The 3/5's Compromise, 137, 218
Peter Abelard, 312
Peter Lombard, 312
Peter the Chanter, 263
pietism, 208
Plato, 7, 73, 228, 306
Platonic Ideal, 67
Platonic Sensibility, 292
political atomism, 93
political authority, 16, 93, 192
political community, 133, 161, 172, 174
political geography, 170
political power, 2, 8, 9, 57, 67–69, 104, 120, 158, 164, 168, 170, 210, 225, 244, 255, 273, 310, 317
political systems, 21, 24, 45, 48, 69, 78, 101, 105, 108, 131, 154, 157, 165, 168, 169, 173, 179, 182, 183, 201, 202, 206, 207, 219–221, 230, 235, 264, 271, 281, 316
and the environment, 129, 165, 221, 271, 315
power and change, 157
Power and Individual Pushes for Change, 21
political theology, 24, 25, 109, 160, 178–181, 183, 185–187, 191, 193, 194, 199, 208–210, 217, 237, 317, 323
and modern liberalism, 183
and temporal deferral, 24
government and religion, 180, 181, 210
political order, 183
the secular and the eternal, 191, 193–195
pollution, 158, 291, 315
Pope Alexander
Inter Caetera, 32
population growth, 119, 244, 258, 282
positive law, 45, 46, 51, 112, 141, 142, 153, 154, 156, 272, 280, 317
positivism, 141, 227
post-colonial, 183, 315
power and resources, 67
pragmatism, 11
presidential immunity, 185
Problems contemporary just war thinking, 130
Problems with international law, 130
Proceedings of the National Academy of Sciences of the United States of America, 321
Progressivism, 177
proportionality, 3, 6, 38, 41, 43, 58, 115, 123, 127, 146, 149, 152, 167, 264
Protestantism, 81, 83, 177, 188, 202, 203, 208, 210
and Time, 202, 203, 208, 210
Ptolemy, 69
punitive wars, 92
puritanism, 189
and Modern Society and Economics, 205

R.M. Hare, 227
racism, 71, 104, 168, 200, 224, 250
and the environment, 71, 250
radicalization, 289
rationality, 16, 42, 51, 52, 55, 59, 61–63, 82, 88, 112
and society, 61, 62
realism, 89, 292
refugees, 65, 113, 120–124, 128, 162, 165, 168, 170, 224, 234, 253, 265, 274, 276, 321, 322, 324
and Christian Just War Tradition, 121, 124
and climate change, 78, 123, 165, 168, 170, 224, 253, 324
and conceptions of the self, 123
and Identity, 214
and international law, 121–124, 274
and just war theory, 122
and just war thought, 122, 253
and land conflict, 253
and politics, 122, 124, 165, 324

and religion, 121
and State Status, 121
and status, 234
and the state, 122, 124, 162, 170
Reinhold Niebuhr, 179
Religion, 2, 3, 33, 35, 38, 39, 57, 61, 62, 81, 82, 91, 98, 99, 102, 103, 108, 112, 127, 137, 143, 164, 171, 172, 175, 177–179, 190, 194, 219, 226, 228, 231, 235, 243, 246, 247, 295, 301, 310, 312, 324
and Culture, 98, 175, 177, 178, 205
and the Non-Human Natural World, 243, 247, 324
and the public sphere, 103, 178, 243
Politics and Culture, 177
Politics, and The Secular Age, 180, 194
religious ethics
climate change and violence, 324
religious vs. secular, 84
René Descartes, 73, 77, 84, 94, 95, 225, 226, 273
and the Dualism of Substances, 73
representation, 162, 174, 193, 195, 268
resource conflict, 289
reterritorialization, 170
revanchism, 287
revelation, 192, 301
rhetoric, 64, 138, 139, 144, 193, 195, 198, 199, 264
Richard Gula, 270
Richard Hoffman, 246
and Environmental History, 275
and Human Attitudes toward Nature, 238, 239, 275
An Environmental History of Medieval Europe, 238, 239
right intention, 2, 3, 58
right to might, 87
right to safe passage, 37
Robert Brandom, 293, 295, 306
Robert Osgood, 132
Rochelle Green, 174
Roderick Frazier Nash, 250
Wilderness and the American Mind, 258
Roger Williams, 44, 136
Roman Colonization, 258, 310
and Road Systems, 252
Roman law, 20, 46, 74, 271
ius inter homines, 46
romanticism, 239
romanticization of a nation, 65
rules of governance, 91, 135, 200
rules of war, 3, 41, 110, 112, 166, 313, 317, 325

Samuel von Pufendorf, 47, 93, 94, 96, 98, 143, 152, 184, 235
Sanjay Chaturvedi, 169
secularism, 106, 133, 243, 316
secularization, 103, 283
self, state and other, 23, 26, 83, 95, 105, 133, 138, 222, 224, 225, 229, 239–241, 244, 299, 315, 319
and Creation, 241
and Ideas, 229
and Just War Thinking, 241
self-differentiation, 219
self-rule, 315
Seneca the Younger, 2, 117
Shaka kaSenzangakhona (Shaka Zulu), 8
Sheldon Wolin, 307
siege warfare, 254, 265
slavery, 32, 70, 71, 105, 117, 154, 177, 218, 236
demora, 33
mita, 33
Slavoj Žižek, 179
social breakdown, 289
social change, 84, 120, 201, 229, 315
social constructions, 230
social contract and the state, 104, 109, 156, 183
social contract theory, 94, 97, 111, 195, 236
social flux, 300
Social Imaginaries, 19–21, 28, 45, 48, 65, 82–84, 87, 124, 125, 128, 193, 202, 214, 217, 229, 231, 238, 240, 290, 299, 311, 313, 314, 316, 318, 320, 323, 325
and just war thinking, 44, 83, 217, 225, 231, 238, 316
and the state, 45, 124, 133, 171, 225, 230, 324
medieval social imaginary, 27, 290, 313
modern social imaginaries, 289
the Anthropocene, 45, 217, 318, 320
social structures, 109, 253
and humanity, 62
sociality, 35, 37, 51, 52, 55, 59, 62, 63, 75, 88, 90, 114, 119, 289

society, 10, 12, 19, 27, 35, 43, 49–51, 63, 67, 88, 94, 95, 97, 103, 111, 112, 138, 141, 142, 150, 151, 153, 171, 180, 194, 196, 201, 212, 215, 219, 220, 228, 294, 297, 298, 300, 305, 306, 313, 317
  and change, 12, 43
  and the Climate, 212, 215
Socrates, 306
solidarity, 29, 140, 142
Solomon Hsiang, 130
Somalia, 169, 217
sovereignty, 3, 16, 17, 24, 25, 57, 59, 90–92, 95, 97, 102, 104, 105, 109, 121, 122, 127, 138, 146, 147, 152, 155–158, 160–164, 166, 167, 169–173, 180, 181, 185–191, 194, 224, 234, 274
  and citizens, 138, 146, 147, 153, 156, 160, 224
  and environmental issues, 156–158, 160, 163–165, 169, 172, 173, 274
  and the state, 3, 16, 43, 57, 90–92, 95, 102, 105, 109, 155–158, 160, 162, 167, 171, 181, 187, 224
  and violence, 170–173, 182, 224, 274
Spain, 28, 29, 31–35, 37, 39, 40, 43, 46, 48, 49, 56, 58, 63–65, 70, 81, 109, 255
  and colonization, 33, 256
  and Indigenous Persons, 31, 33–35, 37, 39, 40, 48, 49, 56, 58, 63–65, 70
Spanish conquest, 46, 49
spheres of interaction, 291, 292
  and the Christian Just War Tradition, 292
St. Ambrose, 2
St. Augustine, 37
St. Francis of Assisi, 247
St. Paul, 71
stable societies, 298
  and social imaginaries, 298
State loyalty, 74, 103
state sovereignty, 102, 122, 160, 162
  and political thought, 24, 25, 158, 171
Subsidiarity, 94
Suffrage, 177
supernatural and divine social imaginaries, 125
Susan Niemann, 236
Syria, 172, 321, 325
technologies, 7, 8, 10, 13, 14, 18, 19, 25, 26, 33, 66, 78, 83, 107, 133, 142, 159, 197, 212, 225, 229, 231, 233, 236, 240, 241, 276, 277, 280, 281, 284–287, 298–300, 303, 306, 317
  and practices, 7, 8, 10, 13, 14, 18, 19, 25, 26, 33, 231, 233, 277, 281, 285–287, 298–300, 303, 306
technology, 19, 21, 22, 67, 104, 108, 201, 240, 247, 248, 249, 282, 309
temporal deferral, 24, 196–200, 202, 206, 208, 211–213, 215, 218, 224
  and Climate Change, 212, 213
  and Early Modern Protestantism, 202, 208
  and Human Agency, 213
  and Human Rights, 206
  and Political Processes, 196–198
  and Temporal Delay, 211
  and the Anthropocene, 211–213, 224
  and the Fetishization of Freedom, 218
testimony, 242
the 30 Years War, 99
The Akkadian Empire, 318
the American Civil War, 3, 147, 150, 251
the American Revolutionary War, 134, 135
the Anthropocene, 21–23, 25, 26, 45, 73, 77, 79, 112, 118, 119, 120, 157, 162, 167, 170, 175, 211, 212, 213, 214, 217, 221, 222, 224, 240, 242, 244, 258, 297, 298, 318, 319, 320, 323, 324
  and Optimism, 213
  and Power, 213
The Apostles' Creed, 188
the Articles of Confederation, 135
the Assad regime, 321, 322
the Athanasian Creed, 188
the Augsburg Confession, 188
The Babylonian Empire, 307
the Battle of Adrianople, 311
the Bill of Rights, 136, 150, 152, 218, 275
the Black Death, 252
the Blackfriar's council, 263
The Book of Revelation, 203
the Boston Tea Party, 134
the Cathars, 238
the Cistercian order, 312
the "civilizing process," 291
the Cold War, 314

the Concordat of Worms, 255, 311
  and the investiture controversy, 311
the Constitution of the United States of America, 131–133, 142, 143, 146, 151, 154, 156, 171, 174, 195, 198, 222, 224, 235, 316
  Agency and Impotence, 196
  Amending the Constitution, 136, 137
  and a National Military, 137
  and Checks and Balances. *See* Checks and Balances
  and Christian Disquiet, 175
  and Christian Just War Tradition, 133, 142, 148, 149, 151, 153, 156, 176, 210, 217
  and Christian Political Theology, 210, 217
  and Congress, 135, 136, 144, 145, 174
  and Discrimination, 146, 147, 149
  and Domestic Law, 137, 149
  and Early Modern Just War Thinking, 132
  and Epistemic Humility, 197
  and Foreign States, 145, 149
  and Governance, 135
  and Instrumentalism, 139, 147
  and its Myth, 195, 209
  and Moral Authority, 155
  and Political Community, 133, 174, 206
  and Political Engagement, 206
  and Powerful Narratives, 195
  and Rhetoric, 138, 144, 198
  and Temporal Deferral, 196
  and the Non-Human Natural World, 156
  as a Christian Just War Document, 132
  as a Legal Document, 133, 140, 151
  Article Four, 155
  as a Living Document, 198
  a Secular State, Freedom, and Temporal Deferral, 224
  Authority and Obedience, 195, 196
  Decision Makers, 147, 196
  Diffusion of Authority, 161
  Earl Modern Christian Just War Thinking, 132
  Freedom and Goods, 218
  Freedom and Temporal Deferrals, 218
  Freedom Liberty and Goods, 218
  Governance, 137, 145, 199, 200
  Hope and Optimism, 215
  Identity and Otherness, 196
  jus ad bellum, 134, 144, 145, 146, 147, 149, 155
  Just Cause, 146
  Just War, the Military and Citizens, 146, 150
  Just War Thinking and the Myth of Progress, 209, 217
  Legitimate Authority, 145, 146, 147, 156, 197
  Natural Law, 138, 141
  Pluralistic Society, 141, 142, 171
  Proportionality, 146, 149
  Positive Law, 141, 153, 156
  Promise and Threat, 195, 196
  Secularism, 133
  Self and Community, 196
  Social Contract, 156, 196
  Social Contract Theory, 195
  Society, 138, 150, 196
  Sovereignty, 138, 146, 147, 155, 224
  State, 133, 134, 136, 138, 142, 151, 152, 155, 161, 173, 178, 181, 192, 198
  State and Citizen Sovereignty, 144, 146, 155, 156
  State and Military, 145, 146
  State and Nation, 181
  State Community, 139
  State Nation and Self, 136, 138
  Text and Interpreters, 196
  The 18th and 21st Amendments, 136, 212
  The American Dream, 195
  The Articles of the Constitution, 147, 155, 196, 197
  The Fifth Amendment, 149, 150
  The First Amendment, 143, 171, 174, 177
  The Bill of Rights, 218
  The Fourteenth Amendment, 150
  the Importance of Artifact, 133, 142, 149, 184
  the King's Two Bodies, 199
  the Limits of Wisdom, 139, 197
  the Modern Self, 141
  the People, 138, 139, 140, 141, 147, 159, 173, 175, 195, 196, 199
  The Powers of Congress, 144
  the Power of the Judicial Branch, 145
  the Power of the President, 145, 185
  The Preamble, 138, 199, 200

the Constitution of the United States of America (cont.)
The Revolutionary War, 137
The Second Amendment, 147
the Separation of Church and State, 171, 177
The Third Amendment, 148, 149
the Wisdom of Citizens, 197
Treason, 145
Universal Community, 139
Universality, 147, 225
Unalienable Rights, 175
The Copenhagen Agreement, 160
the crusades, 2, 268, 313
the Danbury Baptist Association, 177
the Declaration of Independence, 134, 141, 143, 175
the Decretists, 271
the "disengaged subject" of modernity, 93, 94
the doctrine of discovery, 232
the Donatists, 311
the Dover earthquake of 1382, 263
the Duke of Wellington, 317
the Dutch, 81, 113, 265, 282
and military innovation, 265, 282
the Dutch Republic, 85
the Earl of Derby, 263
the Early Medieval Cold Period, 252, 253
the Edict of Milan, 2, 310
The Egyptian Empire, 318
the English Bill of Rights, 139
the English Reformation, 109
the English Revolution, 20
the European Union, 17
the exception in jurisprudence and the miracle in theology, 182
The Exodus and Law on Sinai, 192, 194
the fall of Rome, 312
The Federalist Papers, 136
the Fifth Monarchists, 203
The First Continental Congress, 134
The French revolution, 161
The Geneva Convention, 6, 153
the Glorious Revolution of 1688, 137
The "Great Man" Myth of History, 7, 24, 133, 186, 305
the "Great Man" theory of history, 7
the Hague Convention, 3
the Holy Roman Empire, 43, 102, 109
the Hundred Years War, 54, 263, 265, 267
The Battle of Agincourt, 267
the Battle of Crécy, 267
the English chevauchees, 265
the International Criminal Court, 6
the Iroquois Confederacy, 139
The Justinian Plague, 252, 310
the Kellogg–Briand Pact of 1928, 3
the king's two bodies, 161
the Korean War, 16
the Kyoto Protocol, 160, 163
The Law of Nations (jus gentium), 30, 38, 56, 63, 77, 144
ius inter gentes, 46
the Little Ice Age, 127, 129, 258, 267, 274, 277, 281, 283, 285, 323
and Christian Just War Tradition, 127, 283
and Colonization, 267, 277, 281, 283, 285
the Macedonian Empire, 307
the Medieval Warm Period, 252, 313
The Military, 3, 146, 154, 279
the Moral Majority, 175
the Mosaic constitution, 192–194
the myth of progress, 6, 7, 209, 216, 217
and advancement, 216
and faith, 216
the nation-state, 3, 15, 16, 17, 21, 30, 43, 45, 47, 57, 67, 68, 81, 83, 101, 102, 107, 121, 122, 124, 127, 128, 144, 156–158, 160, 165, 167, 168, 171, 182, 203, 230, 255, 259, 276, 277, 279, 283, 324, 325
and addressing the environmental crisis, 157
and borders, 158
and the environmental crisis, 169
the Netherlands, 86, 109, 129, 258, 264, 265, 282
and Conflict in the Middle Ages, 264
and land value, 129, 264
The Nicene Creed, 188
the Paris Agreement, 160, 163
the Paris Climate Treaty, 213, 215
The Peace of God, 2, 254, 317
The Persian Empire, 307
the Protestant Reformation, 30, 171, 203, 205
The Radical Reformers, 204

the Reformed tradition, 188, 189
  and the importance of scripture, 189
the religious right, 176
The Renaissance, 29, 70, 102, 274, 281
the rights of war, 38
the Rio Declaration, 163
the Roman Catholic Counter-Reformation, 30
the Roman Climate Optimum, 252
  and the Roman Empire, 309, 310
the Roman empire, 20, 272, 310
the Roman social imaginary, 311
the Roman world-state, 307
the Salamanca School, 30, 184, 235
the Schleitheim Confession, 188
the scientific revolution, 30, 227, 285
the Second Continental Congress, 134, 135
The Secular, 80
the self, 23, 43, 69, 80, 84, 85, 87, 93–97, 99, 100, 104, 105, 118, 120, 123, 124, 128, 131, 141, 223, 224, 229, 230, 272, 310
  and community, 97, 196
  and Language, 223
  and Society, 94
  and war, 93, 124
the separation of church and state, 171
the Shoah, 184
the Spanish empire, 64, 81
the Spanish Inquisition, 48, 56, 120
the state, 16, 20, 23, 43, 63, 66, 68, 69, 90–92, 95, 101–103, 121, 131, 133, 134, 138, 144, 145, 151–153, 160, 162, 164, 167, 178, 181–184, 187, 190–192, 198, 224, 225, 236, 258, 259, 272, 317, 319
  and Just War Thinking, 69
  and Religion, 102, 103, 121, 171
  and the individual, 99
  Constitutional Law, 185
  Individual, 164
  Just War Thinking, 134, 151, 152, 158, 259
  Positive Law, 45
  Religion, 178, 317
  Religious Institutions, 178
  Rights of the State, 181
  Self-Defense, 151, 214
  the individual, 16, 92, 160, 162
the Stoics, 270
the Sudan, 217
The Supreme Court, 178
the Ten Commandments, 175
the theory of punctuated equilibrium, 299
the Thirty Years' War, 85
the Trinity, 188, 191
the Truce of God, 2, 254, 317
The United Nations, 6, 17, 102, 122, 153, 314
The United Nations Environment Programme, 159
The United Nations Security Council, 159, 168
The United Nations' Intergovernmental Panel on Climate Change, 165
the United States of America as a "Christian Nation," 175, 176, 178
the Universal Declaration of Human Rights, 6
The University of Bologna, 312
The Vietnam War, 4, 108, 315
The War in Iraq, 4, 154, 172, 322, 323
the Weimar republic, 183, 185
theodicy, 303
theological interpretation, 302
Thomas
  and Church and State, 313
Thomas Aquinas, 1, 2, 16, 28, 40, 55, 60–62, 74–76, 113, 125, 257, 259, 261, 264, 270, 271, 305, 314
  and Aristotle, 3, 29, 66, 313
  and Just War Thought, 2, 16, 264
  and Natural Law, 74, 270
  and Natural War, 270
  and the Non-Human Natural World, 76, 261, 313
  and the Papacy, 313
  and the Rules of War, 257, 261, 264
  Secunda secundae, 46
  Summa Theologica, 60, 61, 74, 261, 270
Thomas Cajetan, 28, 270
Thomas Hobbes, 114, 139, 181, 207, 208, 227
  and Glory, 207
  and Social Contract, 156
  and the Natural Man, 207
  Political Order and History, 207
Thomas Jefferson, 139, 176, 177
  Statute of Virginia for Religious Freedom, 139
Thomas Kuhn, 7
  Scientific Revolutions, 299

Thomas Müntzer, 203
time, 15, 22, 24, 88, 174, 183, 200–210, 224, 232, 233, 235, 301
  and The Constantinian revolution, 202
Timothy Doyle, 169
trade, 25, 37, 65, 75, 76, 114–119, 127, 213, 225, 254, 276, 309
tradition, 3, 5, 7, 10–13, 22, 292–294, 300, 319
  and God, 301, 304
  and Natural Events, 294
  and Social Imaginaries, 19, 20
  and stability, 12, 285, 292, 303, 304
  and the Non-Human Natural World, 292
  and transition, 11, 12
  and questions, 22, 293, 295
  Characteristics, 10, 11, 13, 234, 293, 304
  endogenous and exogenous factors, 294
  Stability, Discontinuity and the Non-Human Natural World, 300, 301, 304, 305
transcendentalism, 239
Truism, 52

Ulpian, 270
Universal Emperor (a dominus mundi), 46
universal law, 36, 46, 47, 49, 226
  and cultural bias, 49
  and indigenous cultures/sovereignty, 36, 49
  versus National/Cultural Law, 46, 49
universalism, 27, 47, 50, 52, 139, 160, 225, 233, 237
  and Indigenous Peoples, 47
  and international law, 52
  and modernity, 139
  and modern thought, 225, 237
  and the US Constitution, 139
  and time, 233
universality, 24, 106, 107, 147, 226
universality of just war tradition, 27
universality of the just war tradition, 52
University of Oxford, 312
University of Paris, 28
usury, 75, 270
utonomy, 229

Vegetius
  De Re Militari, 267, 262, 269
virtue ethics, 82
voluntarism, 97

Waldensians, 120, 312
War in a Warming World
  Religion, Resources and Refugees, 324
War in the 20th Century, 4, 314, 315
wars of the environmental age, 325
Werner Heisenberg, 126
Western Christianity and the Contemporary Environmental Crisis, 248
Western Romanticization of the Other, 65
Westphalia, 21, 47, 51, 82, 85, 101, 105, 107, 155, 235, 277
wicked paradoxes, 164, 167, 172
  Climate Change and the Wicked Paradox of Global Inequities, 167
  Climate Change and the Wicked Paradox of Non-State Actors in a Nation-State World, 164
  Climate Change and the Wicked Paradox of Sovereignty Promoting Violence, 171
William Ames, 82
William E. Scheuerman, 200, 201
  and Climate Change, 201
  the Impact of the Social Acceleration of Time on Liberal Democracies, 200, 201
William Fulbecke, 82
William of Ockham, 188
  and a Nominalist God, 188
  and Natural Law, 188
  "The Sovereign and Exception," 188
Willis Jenkins, 25, 119, 248
Winston Church, 167
World War I, 5, 183
World War II, 43, 212, 224, 251

www.ingramcontent.com/pod-product-compliance
Lightning Source LLC
Chambersburg PA
CBHW032158040425
24612CB00055B/23
* 9 7 8 1 0 0 9 0 9 8 9 3 9 *